NORTHWEST EXPOSURES

A Geologic Story of the Northwest

David Alt and
Donald W. Hyndman

Mountain Press Publishing Company
Missoula, Montana
1995

Library of Congress Cataloging-in-Publication Data

Alt, David D.
Northwest exposures : a geologic story of the Northwest / David Alt and Donald W. Hyndman.
p. cm.
Includes bibliographical references (p. -) and index.
ISBN 0-87842-323-0 (paper : alk. paper)
1. Geology—Northwest, Pacific. I. Hyndman, Donald W. II. Title
QE79.A46 1995 95-37140
557.95—dc20 CIP

Printed in the U.S.A.

Mountain Press Publishing Company
P.O. Box 2399 • Missoula, Montana 59806
(406) 728-1900 • (800) 234-5308

For Sandy and Shirley

Contents

Preface

About 700 or 800 million years ago a continent split and established a new coastline approximately along the present western border of Idaho. Everything west of that coastline was open ocean, everything east of it was the old continent. That arrangement survived for about 500 million years.

During the last 200 million years, large and small scraps of crustal rocks attached themselves to that old coastline, one after the other, assembling the geologic mosaic that is the Pacific Northwest. Meanwhile, the northern Rocky Mountains rose along the western margin of the old continent and a shallow inland sea flooded across the northern High Plains.

Seventeen million years ago immense flows of basalt spread across eastern Washington and Oregon in a brief but cataclysmic event. At the same time, giant eruptions of pale ash erupted from a new volcano in southeastern Idaho. In the years to follow new volcanoes appeared to the northeast, burning the track of the Snake River Plain and ending in the Yellowstone volcano.

Southeastern Oregon began to spread, dropping desert basins and raising the mountains that separate them. Earthquakes jolted the region with each lurching motion as a basin dropped a few more feet.

The last two million years brought an icy sheet that spread southward into our region. Ice surrounded high, jagged peaks while grinding smooth the lower mountains and straightening the valleys. The retreating glaciers left behind flat valley floors filled with outwash sand and gravel.

No one looking at the region 500 million years ago could have predicted any of this. Like all history, geologic history consists of a series of events that are related to each other in the sense that each establishes a contingency for those that follow. Very few geologic events could have happened as they did if the earlier ones had not set the stage.

Some of the events that made the great Northwest are fairly obvious and easy to understand; most are not. Most probably happened as a consequence of other events in distant places. We can see their effects within our region, but we cannot provide any profound understanding of their ultimate causes.

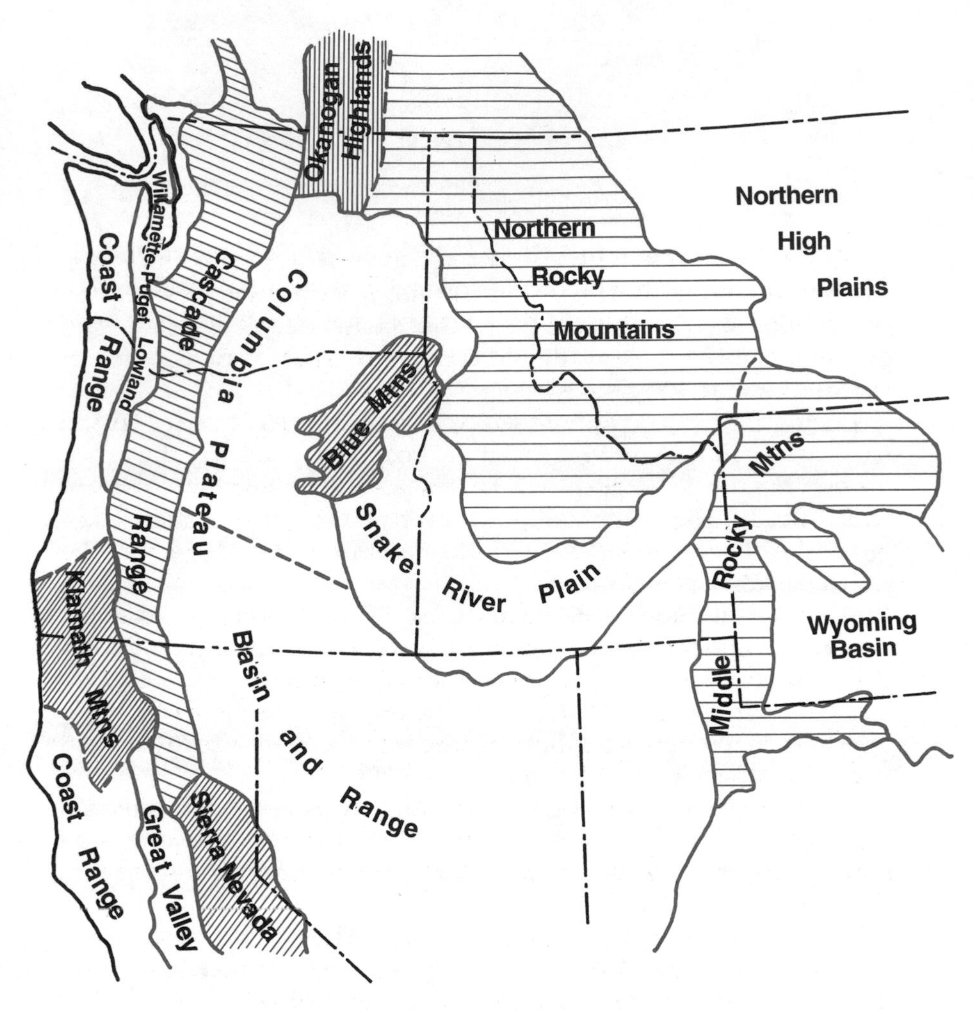

The main regions of the Pacific Northwest.

A long time ago, when we were new to geology, we thought rocks were solid and enduring. Now they seem as inconstant as figures in the fog. Look at a rock, learn a little; look at it again, and it seems completely different. It is risky to write a book about something as changeable as our knowledge of rocks. If this book is not obsolete within a decade, that will be a sad commentary on the state of geologic research.

We did our best to write this book on a strict time line, taking the rocks in the order in which they formed, one event at a time. That works fairly well for most of the rocks, but some do overlap in time, and nothing can be done about that. We tried to cover most of the major geologic events, but ours is a big region with an eventful history, so we skipped a few. And we skipped most of the minor events and local details.

This is a portrait of the underpinnings of a large region, painted with bold strokes of a broad brush. It is our own view won through many years of reading, looking, and thinking. We owe most of the raw material of this book to the hundreds of geologists who have worked in the region during the last century.

Geology is an uncertain science. In the end, you review the evidence and do your best to interpret it reasonably, knowing that many others will almost certainly disagree. We tried to fairly point out places where we know others disagree and their reasons for doing so. And we tried to proceed in an orderly way from the evidence to the interpretation, to show how geologists go from rocks in the field to their ideas about the earth.

We thank the many geologists who contributed to this book without knowing they were doing so. We especially thank our colleagues who helped us think through some difficult aspects of the geology: Jim Sears, Don Winston, Steve Sheriff, Hugh Hurlow, and Roy Hyndman. We much appreciate thoughtful comments by Darrel Cowan and Tracy Vallier on some early chapters. Careful reviews by J. Eric Schuster on later chapters on Washington and adjacent areas and by Ray Wells on western Oregon were very helpful.

Kathleen Ort ushered the manuscript through the many steps involved in editing and producing the book. Jeannie Nuckolls is mainly responsible for its elegant design. Trudi Peek drafted most of the maps and diagrams. We especially appreciate the untiring help of the highly skilled and energetic geologic editorial staff at Mountain Press.

GEOLOGIC TIME SCALE

(m.y.a.=million years ago; b.y.a.=billion years ago)

Era	Period	Epoch	Important Events in the Northwest
C E N O Z O I C		Pleistocene began 2 m.y.a.	Yellowstone hotspot Ice age ended about 10,000 years ago Lake Missoula and Bonneville floods Earlier ice age about 100,000 years ago
	–2.5 m.y.a		
	T E R T I A R Y	Pliocene began 5 m.y.a.	Second long spell of dry climate; high plains gravel
		Miocene began 24 m.y.a.	Puget Sound and Willamette Valley High Cascades begin erupting Snake River Plain hotspot track begins Basin and Range begins Flood basalts in Washington and Oregon
		Oligocene began 36 m.y.a.	Western Cascades active First long spell of dry climate; John Day and Renova beds
		Eocene began 58 m.y.a.	Blue Mountains rotate Metamorphic core complexes Small alkalic volcanoes in central Montana Clarno, Sanpoil, Challis, Lowland Creek volcanoes Straight Creek fault in North Cascades Pacific Rim, Crescent, Siletz terranes arrive
	–65 m.y.a.	Paleocene began	Fort Union sandstone and coal in Montana Extinction of dinosaurs and many other animals
M E S O	CRETACEOUS began 144 m.y.a.		Dry coastal plain in Montana; Hell Creek sandstone Overthrust Belt in Rockies Boulder batholith and Elkhorn Mountains volcanic rocks Idaho batholith Insular and Blue Mountains terranes collide North Cascades terranes collide; Skagit gneiss, Shuksan thrust Western Idaho mylonite Shallow seaway east of Rockies

Era	Period	Important Events in the Northwest
ZOIC	JURASSIC began 208 m.y.a.	Josephine ophiolite Intermountain terranes microcontinent Red shale and limestone in Montana and Idaho New coastal mountains in Idaho and eastern Washington Atlantic Ocean begins to open; collision of Pacific plate with North America Folds and granite in Kootenay Arc in northeastern Washington Warm, dry climate in Montana and south-eastern Idaho; red mudflat shale
	245 m.y.a. TRIASSIC began	First dinosaurs and birds
PALEOZOIC	PERMIAN	Phosphoria formation in southwestern Montana and eastern Idaho
	PENNSYLVANIAN	
	MISSISSIPPIAN	Madison limestone in Montana and Idaho
	DEVONIAN	Jefferson dolomite
	SILURIAN	
	ORDOVICIAN	Sandstone, then dolomite in Montana and Idaho
	CAMBRIAN	Flathead sandstone beach in Montana and Idaho; inland seaway to the east.
	570 m.y.a. PALEOZOIC began	
PRECAMBRIAN — PROTEROZOIC	800 m.y.a.	Old western part of continent rifted off in northeastern Washington and western Idaho
	1.5 b.y.a.	Formation of the Belt basin, Montana and northern Idaho Metamorphism of basement rocks in Montana and Idaho
	2.5 b.y.a. PROTEROZOIC began	
PRECAMBRIAN — ARCHEAN	2.7 b.y.a.	Formation of most basement rocks in Montana and Idaho
	3.2 b.y.a.	Stillwater Complex
	3.9 b.y.a.	Oldest rocks preserved on earth
	4.5 b.y.a ARCHEAN began	Age of the earth

▶ Part 1 ◀

BACK IN THE PRECAMBRIAN

3,500 to 570 Million Years Ago

Geology begins with the oldest rocks, which are almost 4,000 million years old. What happened on the earth before that time is mostly a question for astronomers, who do not depend on rocks for their evidence.

Formation of the earliest rocks marks the beginning of Precambrian time. The world of Precambrian time is gone, vanished into the depths of space and the abyss of time. The earth has kept a few enigmatic rocks as souvenirs of its early days, but they tell us very little, and very little of that seems familiar. Precambrian time lasted until about 570 million years ago. It ended with the abrupt appearance of the earliest ancestors of all modern animals at the beginning of the Cambrian period.

Chapter 1

THE OLD CONTINENT

3.5 to 2.7 Billion Years Ago, The Beginning

Ants scurrying on a board might reasonably suppose that it is made of paint—until they happen onto a scrape that reveals the wood beneath. We wander the land, seeing mostly the veneer of sedimentary and volcanic rocks that thinly covers most of the continental crust. Here and there that cover is missing, and we see the deep rocks of the continental crust. You can see those rocks in parts of the northern Rocky Mountains, where the forces that raised the mountains also exposed the deep rocks of the continental crust.

A continent is a raft, a thick slab of relatively light rocks that floats on the much denser rocks below, much as a board floats on a

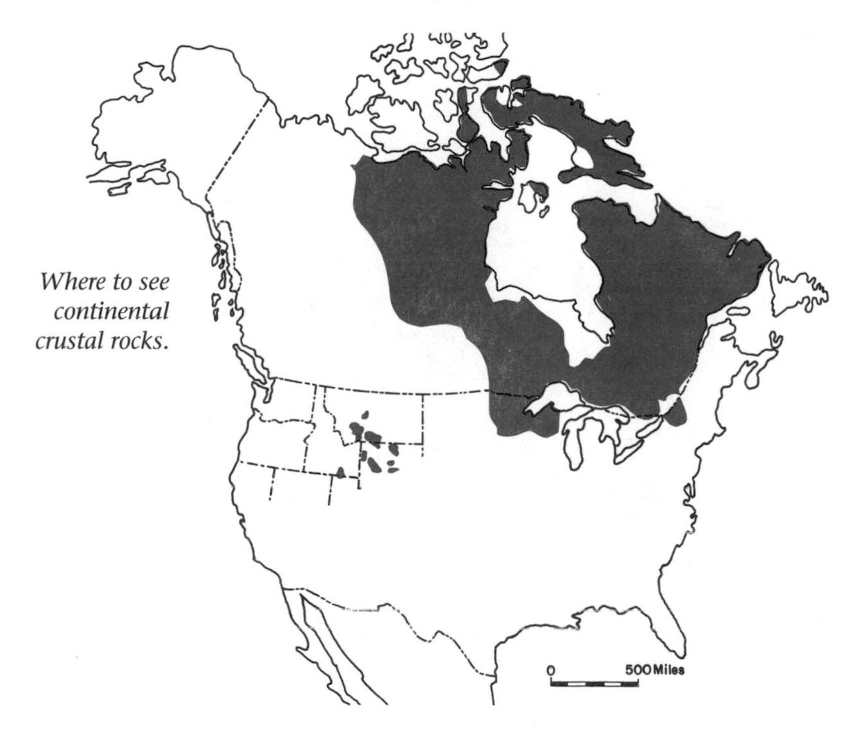

Where to see continental crustal rocks.

lake. The down and back timing of earthquake waves, which echo back to us from the base of the crust, reveals that the continental raft is typically 25 to 30 miles thick, comparatively much thinner than an apple's skin is to an apple. Oceanic crust is even thinner than continental crust. It is made of denser rocks, which do not float as high as the lighter rocks of the continents.

Most of the earth's inventory of continental crust formed before 2.5 billion years ago—rather early in Precambrian time. Although those rocks look generally similar wherever you see them, geological dating shows that they formed at different times, in great patches called age provinces. Similar rocks have formed since, and still form, but not in such great abundance and not in such broad patches. Some things evidently worked differently in early Precambrian time than thay do now, perhaps because our planet was hotter then.

Geologists call the ancient rocks of the continental crust basement rocks, because they lie beneath all the younger rocks and appear to continue down indefinitely. Nothing we see at the surface hints of anything different beneath them. Only the echoing earthquake waves reveal their lower limit.

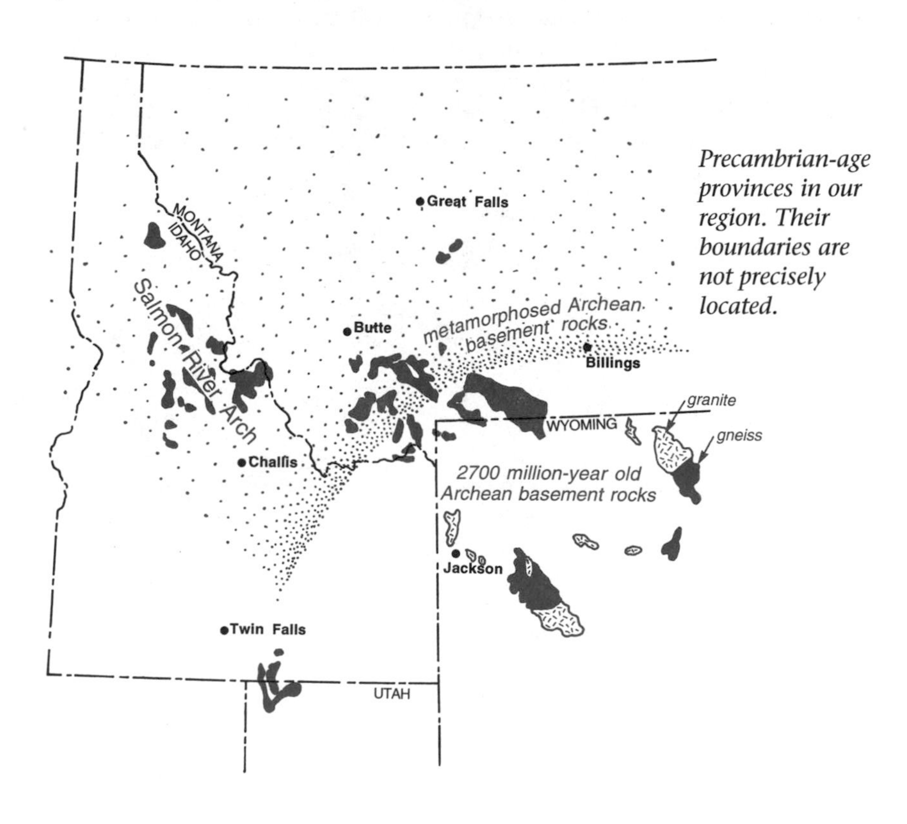

Precambrian-age provinces in our region. Their boundaries are not precisely located.

Chapter 1

THE OLD CONTINENT

3.5 to 2.7 Billion Years Ago, The Beginning

Ants scurrying on a board might reasonably suppose that it is made of paint—until they happen onto a scrape that reveals the wood beneath. We wander the land, seeing mostly the veneer of sedimentary and volcanic rocks that thinly covers most of the continental crust. Here and there that cover is missing, and we see the deep rocks of the continental crust. You can see those rocks in parts of the northern Rocky Mountains, where the forces that raised the mountains also exposed the deep rocks of the continental crust.

A continent is a raft, a thick slab of relatively light rocks that floats on the much denser rocks below, much as a board floats on a

Where to see continental crustal rocks.

lake. The down and back timing of earthquake waves, which echo back to us from the base of the crust, reveals that the continental raft is typically 25 to 30 miles thick, comparatively much thinner than an apple's skin is to an apple. Oceanic crust is even thinner than continental crust. It is made of denser rocks, which do not float as high as the lighter rocks of the continents.

Most of the earth's inventory of continental crust formed before 2.5 billion years ago—rather early in Precambrian time. Although those rocks look generally similar wherever you see them, geological dating shows that they formed at different times, in great patches called age provinces. Similar rocks have formed since, and still form, but not in such great abundance and not in such broad patches. Some things evidently worked differently in early Precambrian time than thay do now, perhaps because our planet was hotter then.

Geologists call the ancient rocks of the continental crust basement rocks, because they lie beneath all the younger rocks and appear to continue down indefinitely. Nothing we see at the surface hints of anything different beneath them. Only the echoing earthquake waves reveal their lower limit.

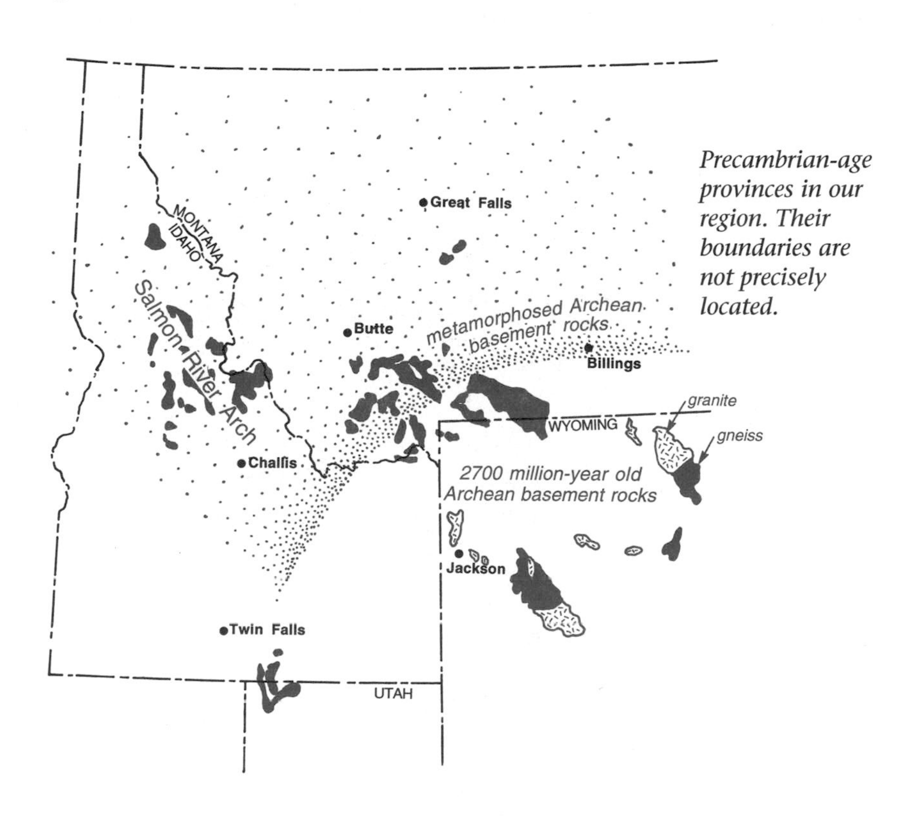

Precambrian-age provinces in our region. Their boundaries are not precisely located.

Chapter 2

BASEMENT ROCKS

2.7 to 1.5 Billion Years Ago

Many of the mountain ranges in southwestern Montana and northwestern Wyoming contain basement rocks. To see them wonderfully exposed in a splendid landscape, drive Highway 212 between Cooke City and Red Lodge, Montana, across the high Beartooth Plateau along the boundary between Montana and Wyoming. The basement rocks near the eastern end of that route are at least 3.3 billion years old; those farther west are closer to 2.7 billion years, the age of most of the basement rocks in the region.

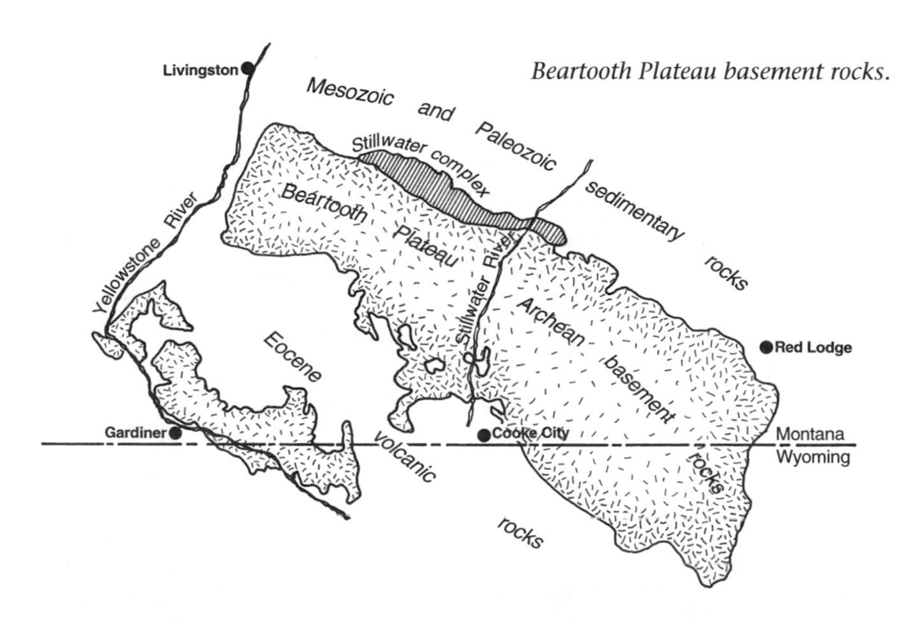

Beartooth Plateau basement rocks.

The Old Continental Basement

Rocks are free for the taking. Beside the Beartooth Highway, you can collect rocks nearly as old as any known anywhere. They may be relics of the earth's primordial crust, and taking one home is perhaps as close as you can come to owning a link to the very beginnings of the planet. But nothing in their appearance betrays their great age. For that, you need age dates based on precise analysis of the radioactive elements in the rocks.

Basement rocks in the eastern part of the Beartooth Plateau look much like those farther west in the northern Rocky Mountains. Those rocks are closer to 2.7 billion years old, and they probably belong to a different age province. Basement rocks everywhere look much alike. You can recognize them wherever you see them.

Among all the many kinds of rock, the gneisses, schists, and granites of the continental basement are perhaps the most beautiful. They are glittering, crystalline rocks composed mostly of mineral grains large enough to see and identify without the help of a magnifying glass. Most come in pale shades of pink, red, and gray; only a few are dark gray, black, or dull green. Most bear the scars of a long and complex history: swirling patterns of light and dark bands, fractures filled with dikes of igneous rock injected into them as molten magma, or mineral veins deposited from circulating hot water. Those are the badges of geologic experience.

Swirling patterns of light and dark in an outcrop of gneiss.

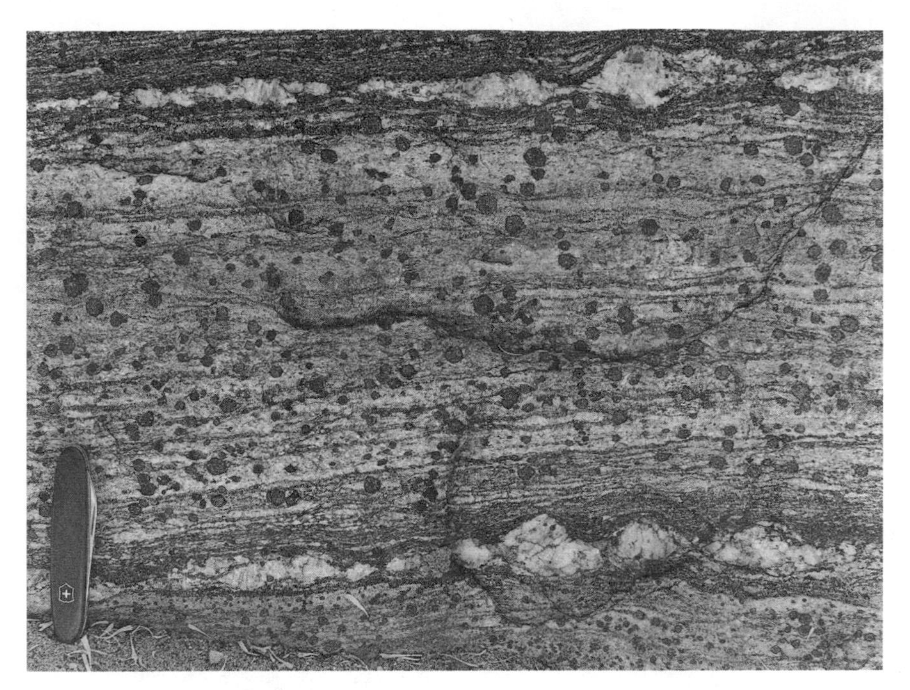

A gneiss studded with small dark garnets.

Despite their complexity and variety, most basement rocks consist mainly of the common minerals feldspar, quartz, mica, hornblende, and garnet. Those few basic ingredients nevertheless manage to assemble themselves into a wonderful variety of forms. It is all a matter of proportion and arrangement.

Feldspar generally forms blocky crystals in milky shades of pink and white. Irregular, glassy grains are probably quartz. Flaky crystals are almost certainly mica, which may be white, brown, or black; it weathers to shades of golden brown. Glossy black needles are hornblende. Everyone likes to find crystals of garnet, red as blood and with fourteen nicely geometric sides.

The mineral grains are randomly oriented in massive igneous rocks, aligned in metamorphic rocks.

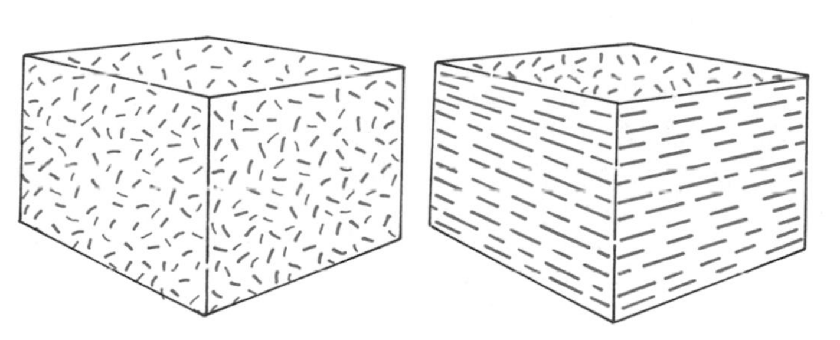

Granite and other igneous rocks are typically massive, which means that the mineral grains are randomly oriented. The rock looks the same from all directions. The minerals in metamorphic rocks are lined up, in military formation, which gives them a directional grain that somewhat resembles the grain in wood. You can easily distinguish between igneous and metamorphic rocks by looking for the massive or directional texture: if the rock has a strong directional grain, it is certainly metamorphic; if not, it is probably igneous.

Granite is the commonest igneous rock in the continental basement. It is generally very pale, pink or gray. Look closely and you will see that most of the rock consists of a patchwork of blocky crystals of milky white or pink feldspar and irregular grains of quartz. The quartz is generally transparent, like glass, but it tends to look dark because you see through the grain into the shadow behind it. Most granites are speckled with black minerals, either flakes of mica or needles of hornblende.

A chunk of granite.

Gneiss and schist are the common metamorphic rocks. Both announce their metamorphic identity in their directional grain. You can safely use the term schist to describe almost any coarsely crystalline metamorphic rock that contains enough mica to give it a generally flaky character. Gneiss is not flaky, but it is platy. Its directional grain makes it look different from different directions. Gneiss generally contains light and dark bands. Gneiss and schist grade

Gneiss, left; schist, right.

into each other; if you find it hard to decide which it is, it probably does not matter.

Gneiss and schist form as older rocks recrystallize below the surface at a dull red heat, and while changing shape under stress. Stress aligns the growing mineral grains, giving metamorphic rocks their directional grain. Schist owes its flaky layering to the tendency of mica grains to grow with their flat faces perpendicular to the direction of greatest flattening. It crystallizes as it is compressed in a direction at right angles to the layers.

Metamorphism so thoroughly transforms most basement rocks that it is nearly impossible to know what the original rocks may have been. Many rocks, including granite, pale volcanic rocks, and ordinary mudstone, contain the right mix of ingredients to recrystallize into schist or gneiss. A few basement rocks clearly betray their ancestry through their peculiar compositions: marble, for example, can form only through recrystallization of carbonate sedimentary rocks such as limestone; certain dark gneisses rich in black hornblende and red garnet have exactly the same composition as the common volcanic rock basalt.

Basement Rocks in the Northern Rocky Mountains

Age dates from the eastern end of the Beartooth Plateau leave little doubt that the rocks there are as much as 3.3 billion years old. Farther west, the pattern of basement age dates becomes a bit scattered, probably because many of the natural radioactive clocks were partially reset long after the rocks formed. Strong heating can do that.

Enough of the rocks in southwestern Montana yield dates of about 2.7 billion years to make it seem likely that they actually did form then. To put it more precisely, that was when they assumed their present form through metamorphism of even older rocks. Some of their minerals and combinations of minerals leave no doubt that they crystallized at a temperature that brought them to the verge of melting—and at considerable depth.

The 2.7 billion year old basement rocks of southwestern Montana contain scattered smaller bodies of rock that crystallized at a much lower temperature and a much shallower depth. Those bodies typically give age dates between 1.5 and 1.6 billion years. And a large area along the Salmon River in central Idaho contains basement rocks that yield the same age dates. Evidently, something happened to the older basement rocks between 1.5 and 1.6 billion years ago.

Although most basement rocks contain little that anyone would care to mine, a few of those in southwestern Montana are extremely valuable.

Southwestern Montana Talc

Talc is a pale gray or greenish magnesium silicate mineral so soft that you can scratch it with your fingernail or carve it with a pocketknife. Artists carve talc into figurines. Grind talc, and you have talcum powder, a raw material with many uses. Besides being the stuff you dust onto a baby's bottom, it is also the basis of many cosmetics. Talc helps make chocolate candies melt in the mouth instead of the hand. It makes pills big enough to handle easily. Precision ceramics made of talc support the catalyst in your car's catalytic converter. And it makes wall paint spread easily.

Unfortunately, most talc contains quite a lot of asbestos, a dangerous carcinogen. Montana talc contains no asbestos. The basement rocks of southwestern Montana support the world's largest talc-mining district, measured either in terms of tonnage or commercial value. The big open pit mines in the Ruby and Gravelly Ranges owe their prosperity to the dangers of asbestos.

The talc in southwestern Montana formed within large masses of dolomitic marble, which differs from ordinary marble in that it contains magnesium. Circulating hot water added silica, thus converting the marble into talc. That reaction happens at relatively low temperatures and at low pressure. Age dates leave little doubt that the transformation from marble to talc was part of the second metamorphic event, about 1.5 to 1.6 billion years ago.

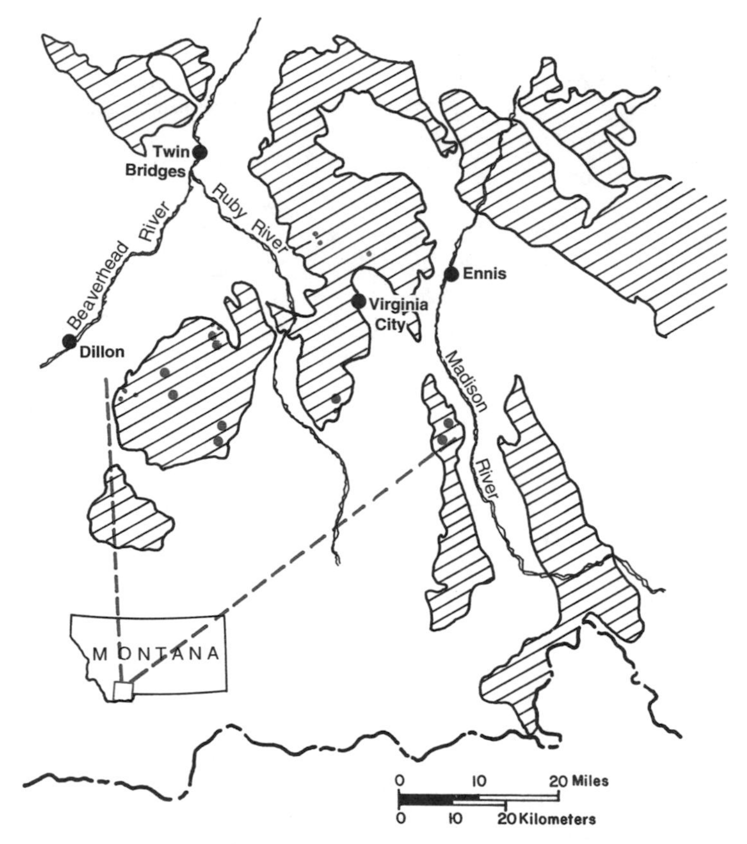

The Montana talc district.

Platinum and Chromite

The Stillwater complex is an extraordinary mass of mostly black and gray igneous rocks along the northern edge of the Beartooth Plateau. The rocks there are about 2.7 billion years old. Only a few other complexes like it exist anywhere—North America contains one in northeastern Minnesota, one in southwestern Oklahoma, and one near Sudbury, Ontario.

The top of the Stillwater complex is now gone, presumably lost to erosion. The large fault that raised the Beartooth Plateau sliced right across the complex, exposing the rocks in its deep interior along the northern edge of the plateau. From a distance, the layers visible in those rocks make them look like a thick pile of layered sedimentary rocks. A few contain valuable minerals.

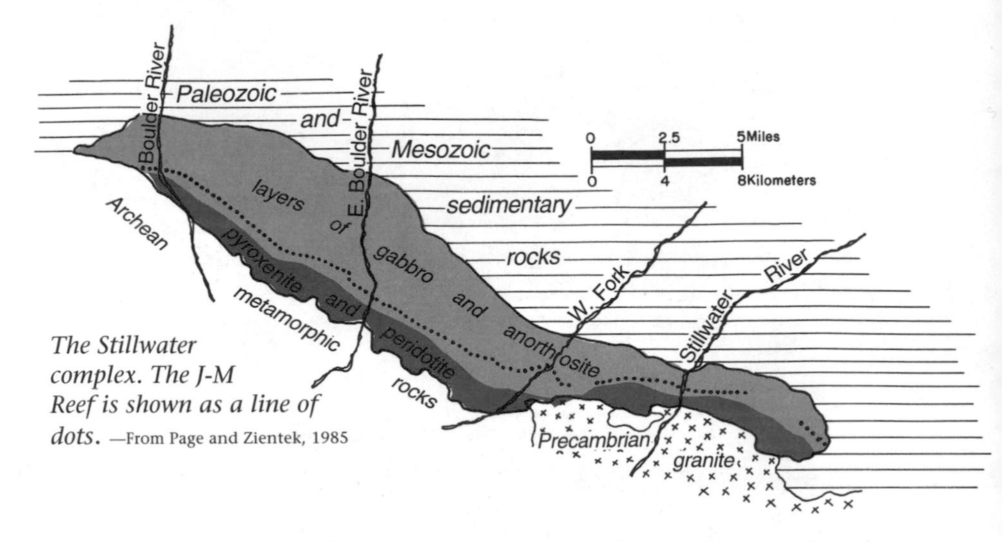

The Stillwater complex. The J-M Reef is shown as a line of dots. —From Page and Zientek, 1985

The Bushveld complex of South Africa, another of the same breed, has for decades produced much of the world's supply of platinum from a single layer called the Merensky Reef. Geologists reasoned that the closely similar Stillwater complex should contain a similar layer. They finally found it, the J-M Reef. Although it is only about 20 feet thick and not much to look at, the J-M Reef contains large reserves of palladium and similar rare elements, including iridium, platinum, and rhodium.

Some of the layered igneous rocks in the lower part of the Stillwater complex contain chromite, the only mineral source of chromium. Chromite separated from the magma to form layers and blobs of black chromite strung through rock almost as dark. The Stillwater complex contains the only large reserves of chromite known to exist in North America.

Layered rocks in the Stillwater complex. View is about six inches wide.

Chapter 3

THE BELT SEDIMENTARY BASIN

1.5 Billion to 800 Million Years Ago

Sometime between 1.5 and 1.6 billion years ago, layered sedimentary rocks began to accumulate in much of the region that would eventually become the northern Rocky Mountains. They continued to accumulate for some 800 million years to become enormously thick, more than 12 miles thick in some parts of western Montana and northern Idaho.

Geologists call that overwhelming volume of mudstone, sandstone, and limestone the Belt sedimentary rocks, because they were first noted and studied in the Belt Mountains of central Montana. It would be far more appropriate to name the Belt rocks after some locality farther west, where they are thicker and more typically developed. But venerable geologic custom dictates otherwise.

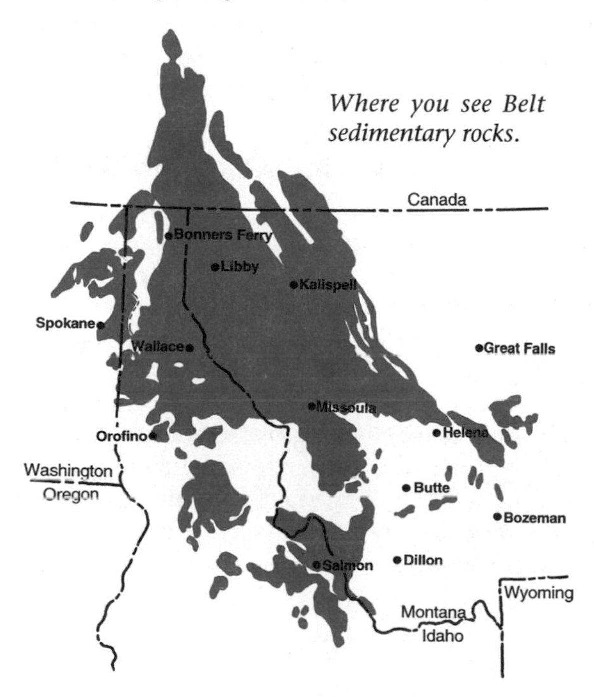

Where you see Belt sedimentary rocks.

Sedimentary Basins

What happened around 1,500 million years ago to start the accumulation of that extraordinary thickness of sediment? Despite other differences of opinion, most geologists agree that such an enormous accumulation of layered rocks could not simply pile up. Whatever the story is, it is an interesting one.

Many geologists contend that the Belt sedimentary pile was laid down along the edge of the continent and in a long arm of the sea. That seems very unlikely. A sedimentary pile laid down along a coast should follow a more nearly linear pattern, not the broad patch that the Belt formations make on the geologic map. Furthermore, few of the Belt formations look like typical coastal accumulations.

The Belt rocks more likely fill a deep and more or less circular sedimentary basin that was enclosed within the continent when they accumulated. It continued to subside as the sediments accumulated. Every continent contains such large sedimentary basins, typically a dozen or so. The Belt basin is more interesting than most because it was so torn up during the formation of the northern Rocky Mountains that we can see rocks deep within its interior, rocks that would otherwise lie far beyond the reach even of the heavy machinery used to drill deep oil wells. The Williston basin of northeastern Montana and nearby North Dakota and Saskatchewan is an example of one that escaped deformation.

That explanation, however, still leaves the question of how such basins form. Why should a roughly circular area of the continent persistently sink and accumulate sediments while nearby areas do not? There is no widely accepted answer to that question. Most geologists accept the existence of sedimentary basins without worrying too much about why they exist.

Belt Sedimentary Rocks

You can gain an impression of the tremendous volume of the Belt sedimentary rocks from almost any highway in western Montana or northern Idaho. It is possible to drive for hours past one road cut after another that exposes layered sedimentary rocks. Most are mudstones and dirty sandstones in monotonously somber shades of gray and dark gray, but some are yellow, green, red, or even purple. Most of the layers tilt, so you pass great vertical thicknesses of rock sections within relatively short horizontal distances.

Most Belt sedimentary rocks are sandstone, mudstone, or limestone, and they tend to be in stacks of thin layers. Geologists have divided the Belt sedimentary rocks into more than a dozen formations, which can be lumped into two broad groups: The older group

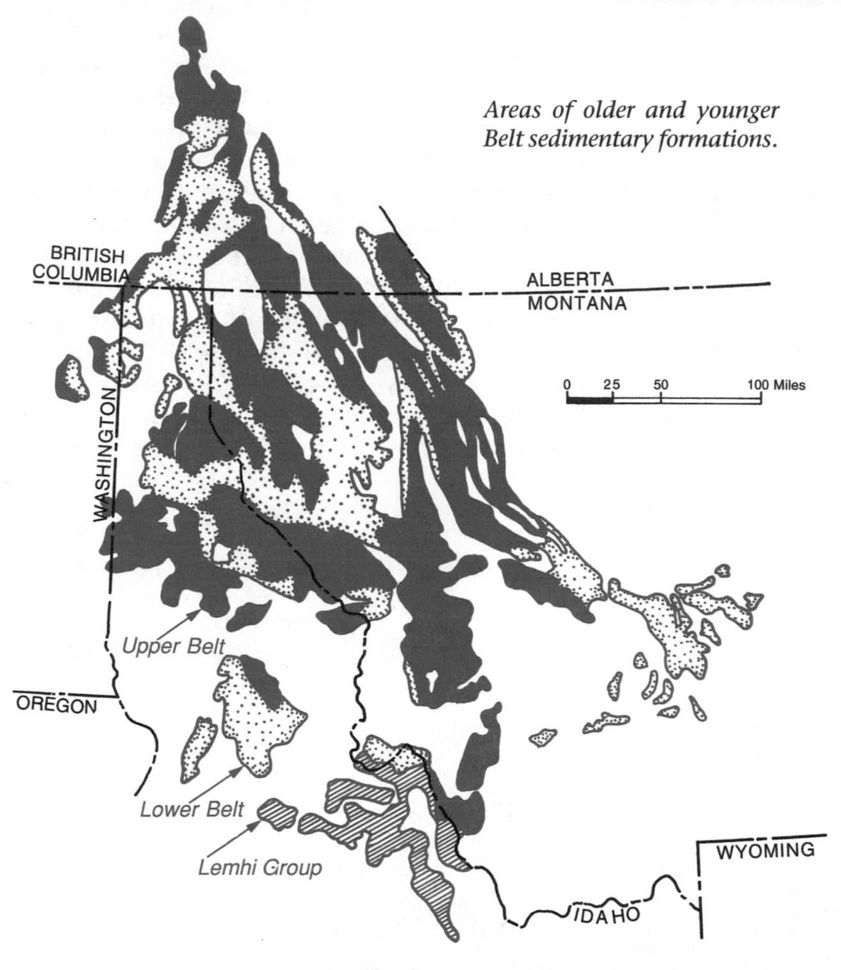

Areas of older and younger Belt sedimentary formations.

of formations consists mostly of rocks deposited under water, quite possibly under deep water. The younger group of formations includes many rocks that were laid down in very shallow water or on dry land. In general, the older formations are in the west, the younger ones in the east.

The Prichard formation, the oldest of the old formations, may be as much as 12 miles thick in northwestern Montana. It is mostly very dark gray mudstone and sandstone. Many of the layers grade from coarse sand at the base to fine mud at the top. Such graded layers are typical of deep sea muds; they form as sediment settles from a cloud of mud billowing across the ocean floor, the larger particles settling first. The Prichard formation seems to have accumulated in very deep water before sediments filled the lower part of the Belt basin. Several of the older formations above the Prichard formation also consist mainly of dark mudstones and sandstones deposited under water, but apparently not at such great depth.

The younger group of Belt formations, the Missoula group, includes many colorful mudstones and sandstones in various shades of red, green, and yellow. Intricate patterns of ripple marks and mudcracks cover many of the bedding surfaces, telling tales of sand ruffling under gentle waves or currents, and of mud cracking in the drying air. Here and there you may see delicate little prints left where sharp cubes of crystallizing salt pressed into soft mud. Those details tell of deposition under very shallow water or on dry land.

In most parts of the Belt sedimentary pile, the rocks are full of extremely thin layers, many hardly thicker than a sheet of paper.

Ripple marks on a bedding surface in Belt sandstone in Glacier National Park. —E. M. Parks, U.S. Geological Survey.

Such exquisite preservation of delicate details is common in sedimentary rocks laid down during Precambrian time, but is rather rare in younger rocks. In all probability, it tells us in an indirect way that no animals lived then to disturb the freshly deposited mud and sand. Nowhere do rocks as old as the Belt formations contain anything that even hints of animal life.

But Precambrian sedimentary rocks do contain the remains of extremely primitive plants and bacteria—in abundance. Indeed, the oldest known rocks capable of preserving such fossils do in fact contain them. They exist in formations elsewhere that are much older than the Belt rocks, as much as 3.5 billion years old. Those most ancient fossils are minute wisps as thin as hairs, which become visible only under a good microscope. They appear to be identical to the blue-green algae that still spread a green stain across damp tree trunks and scummy roadside puddles. If that is what they were, those earliest living things used sunlight to break down carbon dioxide into carbon and oxygen, just as modern plants do. They must have been green.

Mudcracks on a bedding surface west of Anaconda, Montana.

Many Belt limestones and some mudstones contain very thin sedimentary layers that appear to be the remains of a thin scum of green algae that spread across a brightly lit shallow bottom. Exactly similar algal mats still form in shallow water today and still trap thin layers of fine mud. Here and there, the algal mats develop into distinctive humpy structures called stromatolites. In some places, the stromatolites make dense clusters, or reefs.

Some Peculiarities of the Belt

Most Belt rocks are fairly ordinary mudstone, sandstone, and limestone. They look familiar. In many ways, they seem very much like younger sedimentary rocks; in other ways, they are very different.

Some of the younger Belt formations contain extraordinary numbers of bedding surfaces covered with pretty little current ripples and the jagged patterns of suncracked mud. Every blow of the hammer seems to reveal another such surface. The ripples tell of shallow water, the mudcracks of drying. But why such frequent alternations between wet and dry?

Upper Belt mudstones and limestones were deposited in a variety of environments. —Redrawn from Winston, 1989

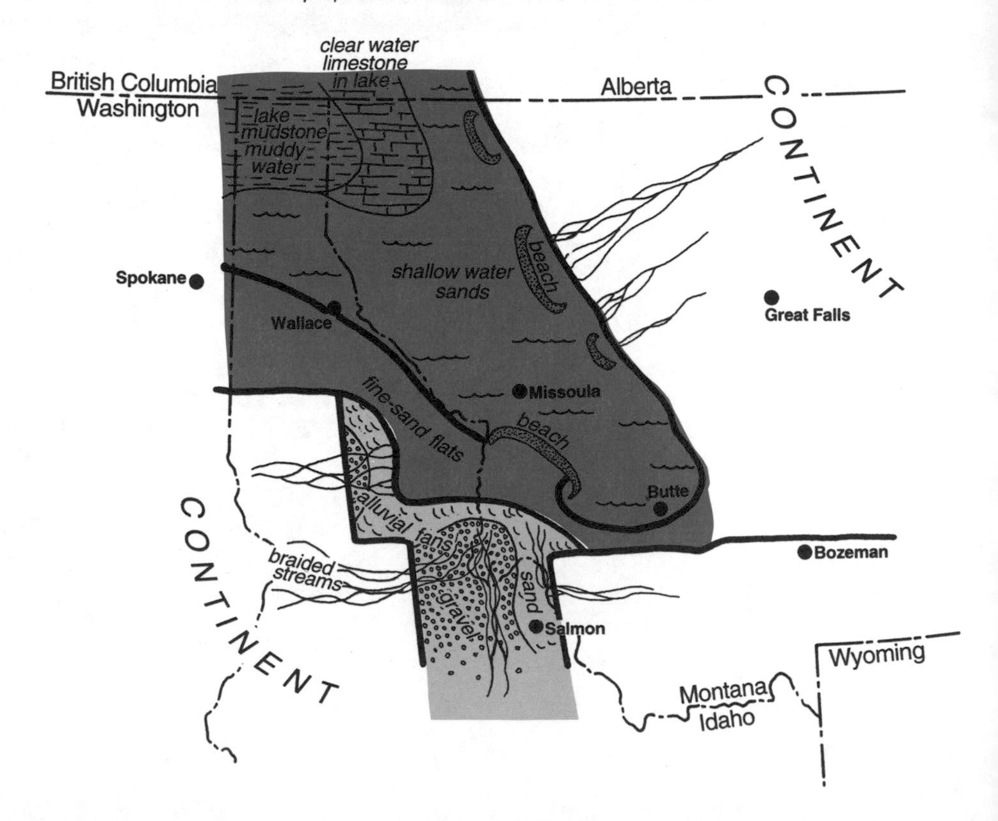

Continuous thin layers of Belt sedimentary rocks north of Helena, Montana. —C. D. Walcott, U.S. Geological Survey

It seems exceedingly strange that the Belt formations, full of sun-cracked mud, contain absolutely no hint of windblown sand, nothing that even remotely resembles an old sand dune. Obviously, the ground was dry part of the time. How then was it possible for all that Belt sandstone to accumulate on land without moving before the wind? Did the wind not blow hard enough to drive sand?

Belt formations consist mostly of sand and silt. The proportion of clay is much lower than in younger sedimentary piles. Why so little clay? Did rocks not weather into soil then, as they have for hundreds of millions of years?

You can follow individual layers of sediment in the Belt formations for long distances, even miles. The thickness and composition do not change. Yet individual layers in younger sedimentary piles generally change within a short distance. Why do the Belt formations contain such continuous layers?

What do these peculiarities mean? No one can be sure, but it is possible to conjecture.

The Vanished Precambrian World

Belt rocks open a window of sorts onto a vanished planet that existed billions of years ago, the planet that became the earth we know. Unfortunately, the view through that window is dim. The rocks tell only a few tales about the lost world of Precambrian time; they offer a flimsy basis for speculation. Some of the best evidence comes from rocks laid down before the Belt formations.

Older sedimentary rocks elsewhere contain shiny little grains of the iron sulfide mineral pyrite, which quickly weathers to rusty iron oxide when exposed to oxygen. The atmosphere evidently contained no oxygen whatsoever when those rocks formed. No shiny grains of pyrite sand exist in the Belt rocks, probably because all those blue-green algae had added some oxygen to the atmosphere by the time they accumulated. But that does not mean that the atmosphere a billion years ago contained nearly as much oxygen as the air we breathe today. Almost certainly it did not.

It is hard to find direct evidence of the atmosphere as it was when the Belt formations were laid down. Nitrogen is the most abundant gas in today's atmosphere, but it leaves no record in the rocks. Water vapor was surely an important gas in the atmosphere a billion years ago, simply because the earth has so much water on its surface. Most geologists suspect that carbon dioxide was the dominant atmospheric gas.

Sections of sedimentary rocks of all ages include large volumes of limestone, which is mostly calcium carbonate. Many sedimentary rocks also contain carbon in the form of coal, petroleum, and disseminated organic matter. Every molecule of carbon in sedimentary rocks came from atmospheric carbon dioxide. If it were somehow possible to restore all that carbon to the atmosphere as carbon dioxide, it would far exceed in quantity all the other atmospheric gases combined.

If a significant part of the carbon dioxide now in sedimentary rocks was in the atmosphere when the Belt rocks were laid down, it must have caused an intense greenhouse effect. The climate would have been very hot—but not hot enough to boil water or kill the blue-green algae. If the temperature had been that hot, we would not find ripple marks in Precambrian sedimentary rocks. Primitive algae thrive in the boiling hot springs in Yellowstone Park, so their fossils in the Belt rocks tell us little about the Precambrian climate.

If the Precambrian atmosphere was hot, it would certainly have evaporated enormous amounts of water from the oceans, winding a thick shroud of clouds around the earth. The earth of Precambrian time may have been a sort of planetary sauna bath set in a windless and suffocating atmosphere beneath skies heavy with clouds. A very hot atmosphere laden with water vapor could explain the frequent alternations between wet and dry that so abound in the younger Belt formations. Imagine a hot rain falling on a muddy surface and soon sizzling dry.

Such an atmosphere might also explain why we see no windblown sand in the Belt formations. A heavy cloud cover would rather evenly distribute incoming heat from the sun, minimizing the difference in temperature between the poles and the equator. All that water vapor might also have enabled the atmosphere to conduct heat from warm to cold places. Today, winds even out temperature differences.

Without land plants to protect it from erosion, soil might have been scarce during Precambrian time. Clay forms mainly in soils. The lack of land plants could explain the shortage of clay in Precambrian sedimentary formations.

How about those continuous layers of sediment? How do they fit in? We have no clue.

Stromatolites in Belt limestone, Glacier National Park, Montana. Primitive algae still build these curious structures.
—B. Willis, U.S. Geological Survey

The Changing Precambrian Atmosphere

Like all green plants, blue-green algae absorb carbon dioxide, then use the energy of the sun to break it down into carbon and free oxygen. They use the carbon to build their tissues, and they release the oxygen into the atmosphere as a waste product.

Normally, the carbon and oxygen recombine into carbon dioxide as the plant's tissues decompose after it dies. If accumulating sediments bury the dead plant before it can decompose, the carbon remains in the rocks and the oxygen in the atmosphere. Every molecule of free oxygen gas in the atmosphere must correspond to an atom of organic carbon in a sedimentary rock.

Early in Precambrian time, oxygen released into the atmosphere converted iron dissolved in seawater into iron oxide. The iron oxide was deposited in sedimentary iron formations, the source of most of the world's iron ore. All the iron dissolved in the oceans had been oxidized, and all the earth's iron formations deposited, long before the Belt basin began to fill with sediments. By that time, free oxygen was certainly accumulating in the atmosphere. Meanwhile, the primitive atmosphere was losing more of its carbon dioxide to new deposits of limestone.

Middle Belt sandstones and limestones may have formed in a shallow lake full of green algae.—Redrawn from Winston, 1989

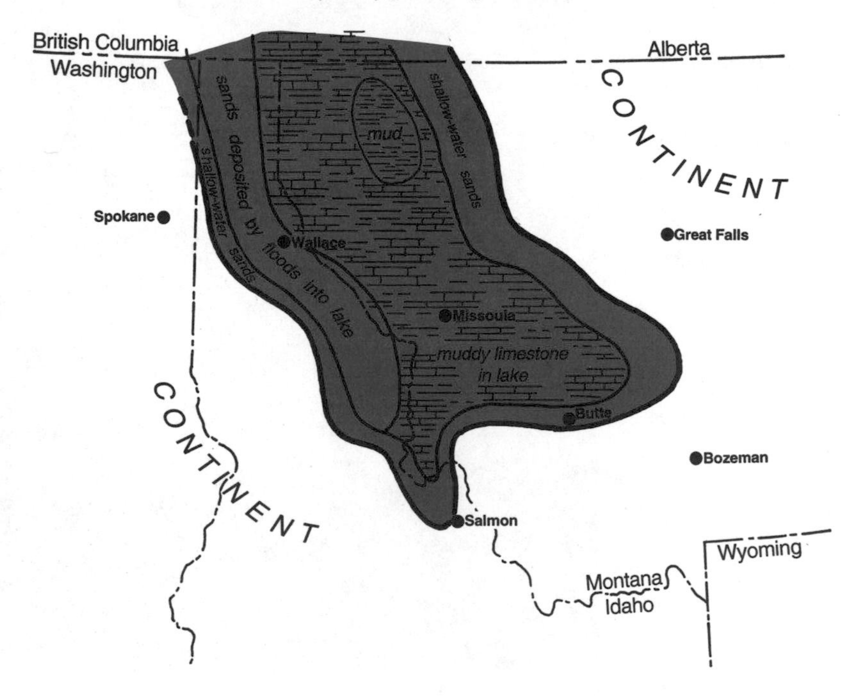

It seems likely that most of the earth's early atmosphere is now in sedimentary rocks in the form of organic carbon, iron oxide, and calcium carbonate. All of those deposits formed from carbon dioxide withdrawn from the atmosphere. The Belt rocks include enormous amounts of limestone, much of it black with disseminated organic matter. Their accumulation must have reduced the greenhouse effect, cooling the climate. As the atmosphere cooled, its water vapor content must have decreased. Eventually, the atmosphere became oxygenated enough to support animals and cool enough to permit them to thrive. None of those changes could have happened without the primitive algae. They first gave the earth a breathable atmosphere. They are the heroes of atmospheric evolution.

Diabase Sills

The Belt basin contains dozens of diabase sills neatly sandwiched between layers of sedimentary rocks. They formed as molten basalt magma rising from below the crust squirted between the sedimentary layers, then crystallized into black diabase. Many of the sills are several hundred feet thick, some more than a thousand. Most are in the Prichard formation in the lower part of the section, but quite a few stragglers injected the younger formations. Diabase is a

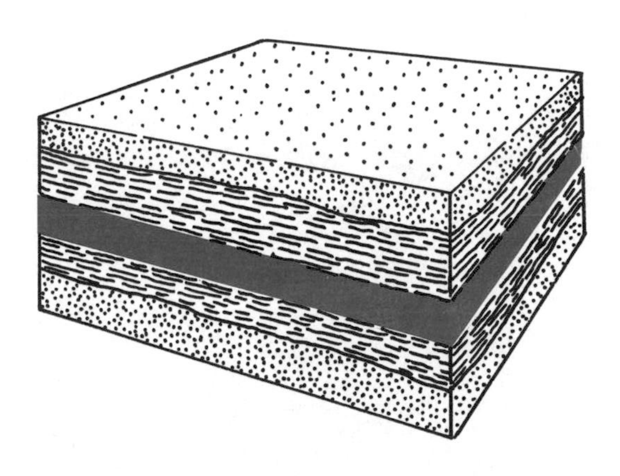

A sill is a layer of igneous rock sandwiched between layers of sedimentary rock.

rock much like basalt, but with larger mineral grains. Crystals of white plagioclase about the size of grains of rice nestle in a matrix with black pyroxene.

All sills must be younger than the sedimentary rocks they intrude, but these are probably not much younger. Some rocks at the top of the sills contain little round masses that suggest boiling of the sediments. Gas bubbles that accompany boiling can form only under low pressure near the surface, not under a great weight of overlying sediment. The sills that contain them must have intruded before most of the rock above them was deposited. Furthermore, age dates clearly show that the deeper the sills, the older they are; they evidently intruded as the sedimentary rocks accumulated. A few sills reached the surface to erupt as lava flows.

Sills are not unique to the Belt basin. All sedimentary basins that filled during Precambrian time contain numerous diabase sills, basalt lava flows, or both. It seems that they are an essential part of the basin fill. The basalt magma must have come from the earth's mantle below the continental crust, because that is where the rocks have the proper composition.

Chapter 4

THE GREAT CONTINENTAL RIFT

800 Million Years Ago

Once formed, that thick raft of basement rocks called the continental crust floats about on the earth's surface indefinitely. No geologic process can destroy a continent. Erosion may plane off a continent's upper surface but cannot remove it. The earth's internal heat can recrystallize a continent's rocks into younger generations of gneiss, schist, and granite, but the old continent will still exist. The earth's internal movements can move continents, collect pieces of continental crust into larger continents, or break continents into pieces. All those processes played a part in the development of the Pacific Northwest. First, the old continent split.

The Continent Splits

Geologists who study the Belt formations find strong evidence that much of their sediment came from the west. But no rocks old enough to supply sediment to the Belt basin now exist anywhere west of it. The old continental crust that once stretched west of the Belt basin is gone. Geologists who follow the Belt rocks west through central Idaho find them looking as though they should continue farther. But instead they come to an abrupt end near the western border of Idaho, as though they were chopped off. Farther north, they continue into northeastern Washington.

And that, it seems, is exactly what happened. It seems clear that the old continent split along a line near the present western border of Idaho, chopping the Belt basin approximately in half. Part of the continent, with the western part of the Belt basin, moved away, as a new ocean basin opened along the line of the split.

The Windermere sedimentary rocks of southeastern British Columbia and nearby northern Idaho and northeastern Washington began to accumulate about 800 million years ago. Like most forma-

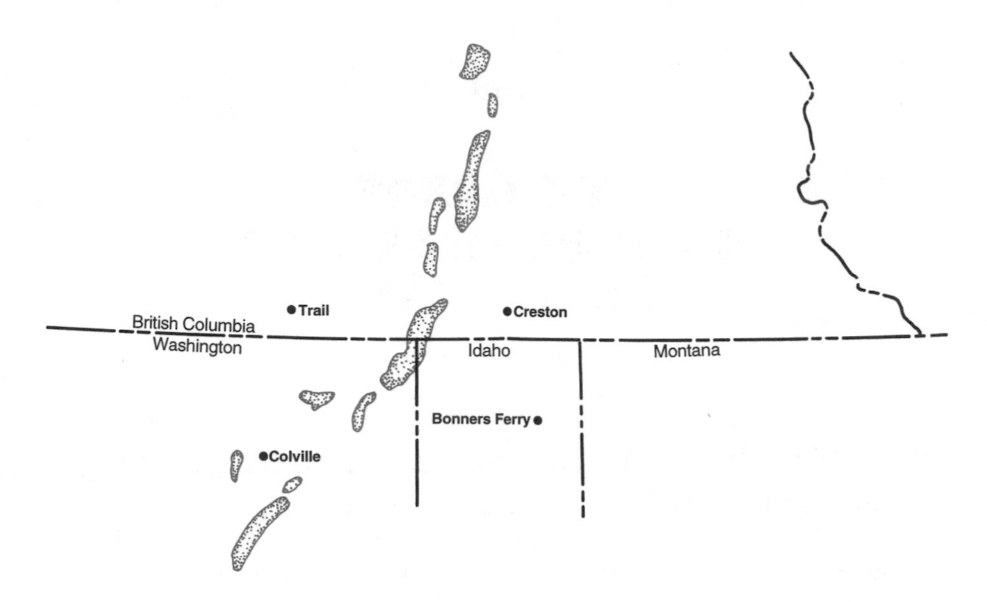

Distribution of the Windermere formations.

tions, they are named for the area where they were first described in detail—in this case, southeastern British Columbia. They look like the kind of rocks that should accumulate along an open coast. Their deposition apparently began as a rift split the continent and established a new west coast. But if the continent broke, the other piece must still exist, somewhere.

Geologists entertain variety of opinions about the missing piece of North America. Some think that it is now the northeastern part of Russia and nearby areas of easternmost Asia. The basement rocks there are the right age. They resemble those in the northern Rocky Mountains, and a series of sedimentary rocks very like those in the Belt basin lies on them. The idea seems perfectly plausible, but the point is difficult to prove.

A New West Coast

The distribution of the Windermere formations, and younger rocks that also appear to have been deposited along a coast, suggests that the new continental margin passed through southeastern British Columbia, across northeastern Washington, and then south along a line near the western border of Idaho. Later events cloud the picture near the border between Idaho and Oregon. Rocks in the Sierra Nevada of California make it seem reasonable to continue the line there.

Whatever its exact location, that new continental margin seems to have been the edge of the continent for something like 600 million years. For all of that time, the west coast of North America remained somewhere near the western border of Idaho, and an ocean stretched away to the west. Throughout most of that time, muddy sediments quietly accumulated along that coast. They finally became a broad coastal plain and continental shelf that was to play a role in the later development of the Northwest.

The new continental margin established about 800 million years ago.

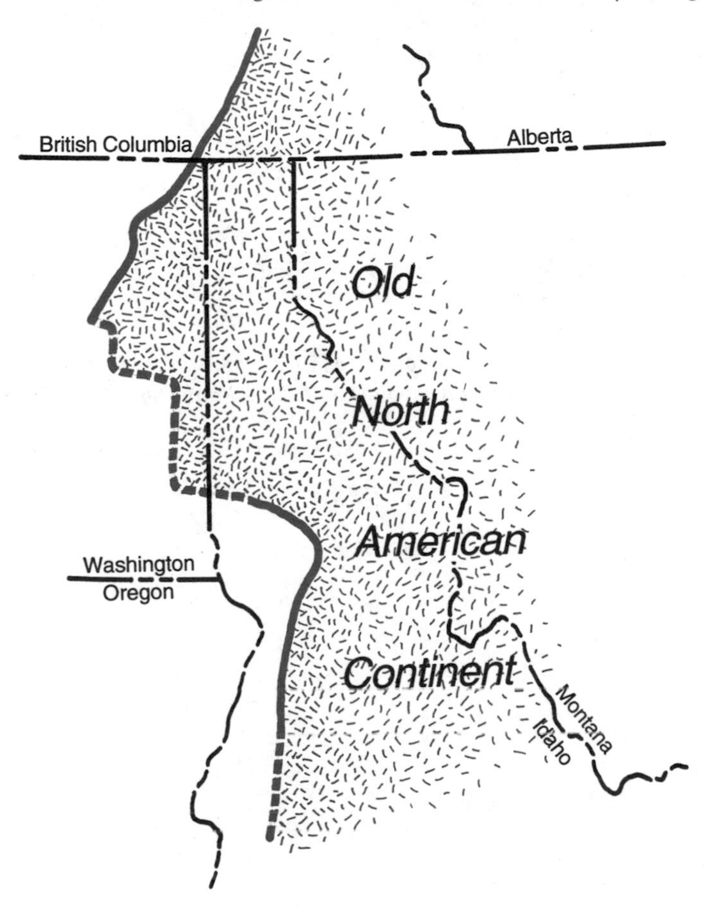

Thin laminations preserved in Precambrian Belt mudstones, Glacier National Park, Montana.

▶ Part 2 ◀

PALEOZOIC TIME

570 to 245 Million Years Ago

Precambrian time ended as Paleozoic time began, about 570 million years ago, at the beginning of the Cambrian period. Paleozoic time lasted until about 245 million years ago, ending as approximately 90 percent of all the animals then living vanished in a mass extinction late in the Permian period.

The western part of our region did not yet exist during Paleozoic time, except perhaps as a few scattered island fragments of continental crust wandering somewhere in the vastness of the western ocean. Like most parts of most continents, the eastern part of our region was above sea level during the latter part of Precambrian time. Then, during Cambrian time, the oceans overflowed onto the continents. Thick sequences of Paleozoic sedimentary rocks began to accumulate beneath shallow seawater across broad reaches of the continents, including the eastern part of our region. They contain a valuable record.

Throughout the Paleozoic era, the Williston basin of northeastern Montana, western North Dakota, and southwestern Saskatchewan tended to sink and collect sediments, even when the surrounding areas were high and dry. The region is nearly circular, and it contains more than 15,000 feet of sedimentary rocks. But the only way to see them is to drill through them.

The earth keeps its historical archives in the continental sedimentary rocks, which ride on the floating continents. Erosion may eventually remove them, but they are safely beyond the reach of most other processes that might destroy the record. The Paleozoic sedimentary formations contain a rich record of life during that time, and of some of the geologic events as well. Those formations are much disturbed in the northern Rocky Mountains, but they remain almost undisturbed east of the mountains.

In the 325 million years of Paleozoic time, three or four thousand feet of Paleozoic sediments accumulated in the eastern part of our region. They thicken westward to as much as 30,000 or more feet in parts of Idaho and western Wyoming. West of Idaho was the open ocean.

Magnetic fields, indications of which are trapped in the rocks, show that during most of Paleozoic time our region lay close to the equator. It is probably a mistake to leap from that deduction directly to inferences about the climate. Because the composition of the Paleozoic atmosphere differed from that of today, it is difficult to extend our ideas about climate that far into the past. If, for example, the atmosphere contained more carbon dioxide then than it does now, the greenhouse effect was probably still significant.

Chapter 5

EARLY PALEOZOIC TIME

570 to 250 Million Years Ago

What event in the long and disorderly history of our planet could be more deeply significant than the earliest appearance of our own ancestors, however primitive and remote? That happened at the beginning of Cambrian time. Geologists draw the boundary between Precambrian and Paleozoic sedimentary rocks where fossils of Cambrian animals appear.

The appearance of our ancestors was so abrupt and their variety so great that many geologists call it the Cambrian explosion. Those many new animals proliferated, and their descendants have ever since left their fossil remains in sedimentary formations.

The geologic periods, Cambrian through Devonian.

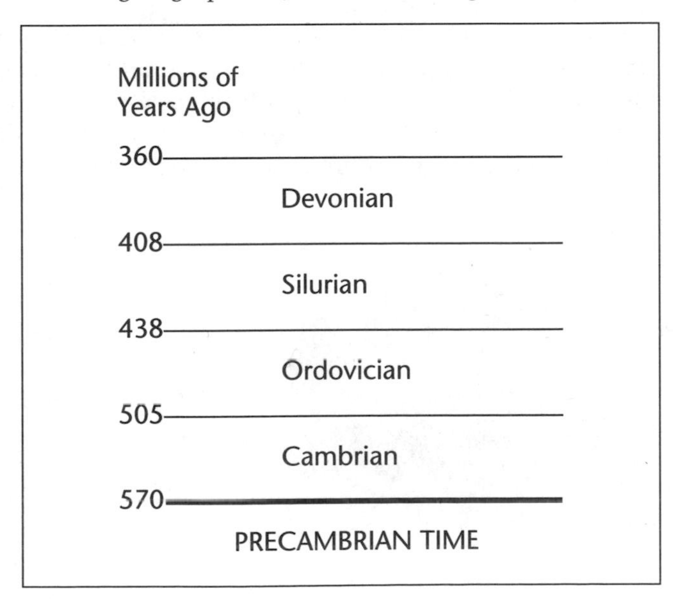

Cambrian to Devonian and Mississippian sedimentary rocks in Bighorn Canyon, southern Montana.

Animals Appear

After harboring primitive plants for several billion years, the earth produced its first animals about 600 million years ago. They were the ediacarids, strange animals like nothing now living. They lasted for 30 million years, then completely vanished, leaving no apparent descendants. Almost immediately after the ediacarids vanished, the Cambrian animals appeared.

Some geologists suggest that animals appeared very late in Precambrian time and again at the beginning of the Cambrian period because the earth had finally acquired a breathable atmosphere. After some three billion years, the scummy growths of blue-green algae had finally used enough carbon dioxide, incorporated enough reduced carbon in the crust, and released enough oxygen to create

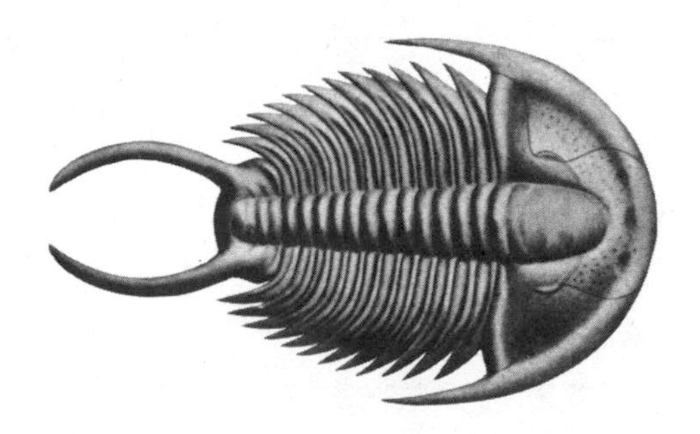

A Cambrian trilobite from southeastern Idaho. —G. R. Mansfield, U. S. Geological Survey

an atmosphere that would support animal life. And if the concentration of carbon dioxide in the atmosphere was decreasing, so was its effectiveness in maintaining a high surface temperature by means of the greenhouse effect. The atmosphere may finally have become cool enough for animals to thrive. In these matters, oxygenation and cooling probably run together.

Many of the new animals that the rising Cambrian sea brought with it were amazingly complex, considering the brevity of their ancestry. Trilobites, for example, are distant but unmistakable relatives of modern shrimps and lobsters, complicated animals with segmented shells, many pairs of jointed legs, and eyes. Other Cambrian animals included the ancestors of all the major groups of animals that live today, as well as of a number of other lineages now long extinct. It seems that the Cambrian animals found themselves on a planet that offered plenty to eat and no competition. They evolved rapidly, soon covering the earth with a wide variety of progeny.

Sea Level Rises

Early in Cambrian time, the sea began to flood inland from the old continental margin in the west, across a low and nearly flat land surface. Early Cambrian beach sandstones exist in northernmost and southeastern Idaho, and in western to central Utah. Farther west, toward the old coast, they blend into shales.

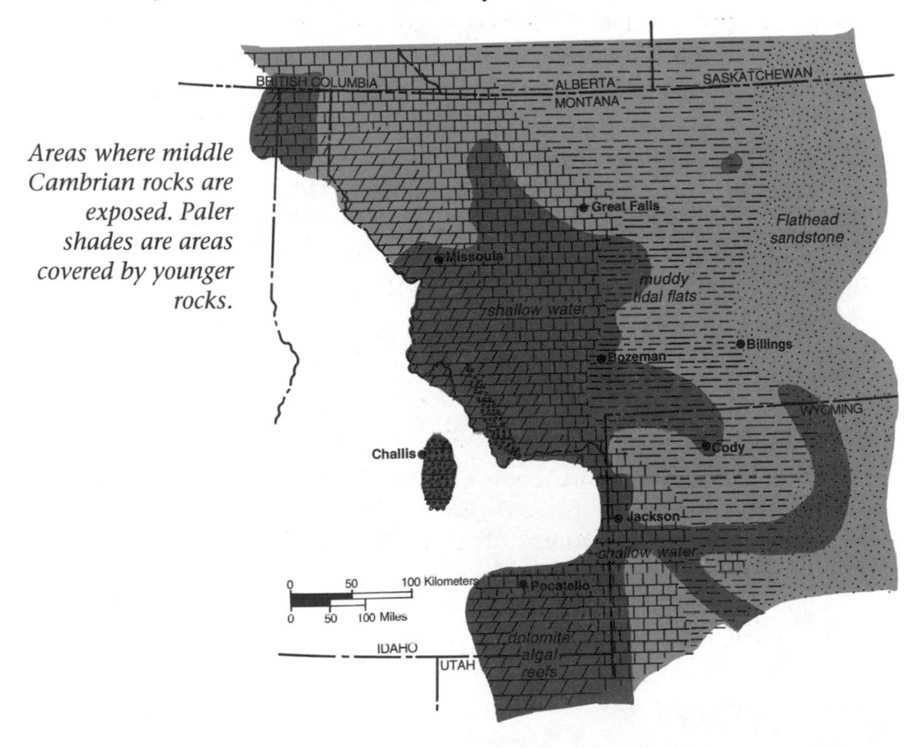

Areas where middle Cambrian rocks are exposed. Paler shades are areas covered by younger rocks.

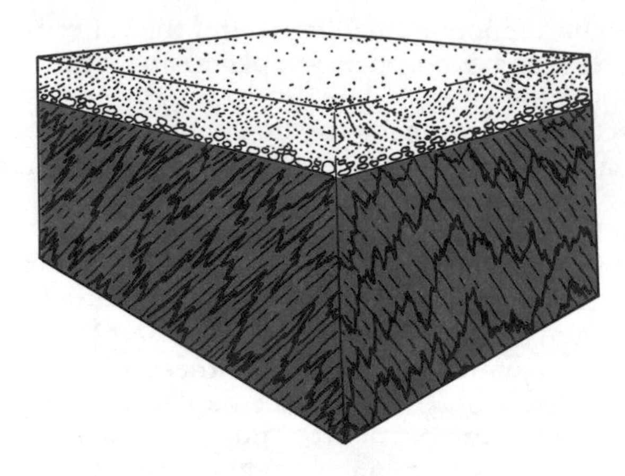

Cambrian sandstone lying on an erosion surface smoothly planed across basement rocks. The erosion surface formed while the area was above sea level.

In most areas, only the upper part of the sandstone at the base of the stack of Cambrian formations contains the fossils of animals that lived in sea water. The lower part typically contains no fossils of any kind. But it does contain the complex patterns of internal layering typical of sand deposited from running water. That makes it seem likely that the blanket of sand already covered the land before the Cambrian sea moved across it; the waves then reworked its upper part into beach sands.

By middle Cambrian time, the advancing sea was spreading beach sands across Montana. By late Cambrian time, the sea had retreated to the eastern border of Montana, still laying down that sheet of sandstone. Meanwhile, shale, limestone, and dolomite were accumulating on top of the Flathead sandstone farther west. Some of the limestone is technically dolomite, which differs from limestone in that it contains as much magnesium as calcium. The basal sandstone is commonly 100 to 200 feet thick, the shale 50 to several hundred feet thick, and the limestone several hundred feet thick. The thickness of the Cambrian sediments increases westward, to about 6,000 feet in southeastern Idaho and about 3,000 feet in northwestern Montana.

On every continent, geologists find a similar record of the sea's flooding across the land during Cambrian time. The fact that it happened everywhere at the same time leaves no doubt that sea level was rising. It is probably reasonable to assume that the amount of water on the earth's surface has remained the same since early in Precambrian time. If so, the volume of the ocean basins must have decreased during Cambrian time, then remained small for hundreds of millions of years.

Cambrian sedimentary formations, all laid down in shallow seawater, exist throughout large areas of the northern Rocky Mountains and northern High Plains. Most are buried under younger rocks, so you see them only locally in the mountains where crustal movements have brought them to the surface.

Ordovician Sedimentary Formations

After the Cambrian flood rose, the part of our region east of the new continental margin remained near sea level. At any particular time, some areas were slightly above sea level and eroding, while other areas were below sea level and receiving sediments. As you would expect, the part near the old coast, now Idaho, was below sea level more persistently than the part farther east, and it collected a deeper pile of sedimentary rocks.

Ordovician sedimentary formations are exposed in areas where the crustal movements that raised mountains brought them to the surface. These areas include parts of Idaho, southern Montana, and Wyoming. As much as several hundred feet of middle Ordovician Bighorn dolomite exists in southeastern Idaho, northwestern Wyoming, and eastern Montana. Southeastern Idaho also contains early Ordovician slates.

East of the mountains, the Ordovician formations are buried under younger sedimentary rocks. They appear in oil wells, especially in the Williston basin. The Ordovician formations there are especially important because some of them contain oil.

Areas where middle to late Ordovician formations are exposed. Paler shades are areas covered by younger rocks.

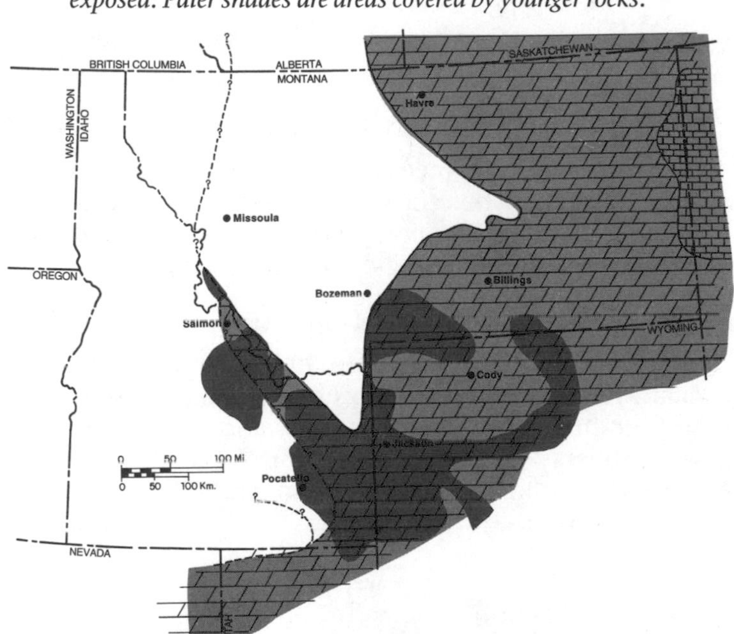

Silurian Formations

Except in the Williston basin, our region contains no Silurian sedimentary formations. It is possible that some once existed and that they were eroded off the land. But it seems more likely that none were deposited, that the region was above sea level when Silurian time began some 440 million years ago. Deep oil wells provide the only access to the Silurian rocks in the Williston basin. They include limestone mounds and reefs built by algae and corals.

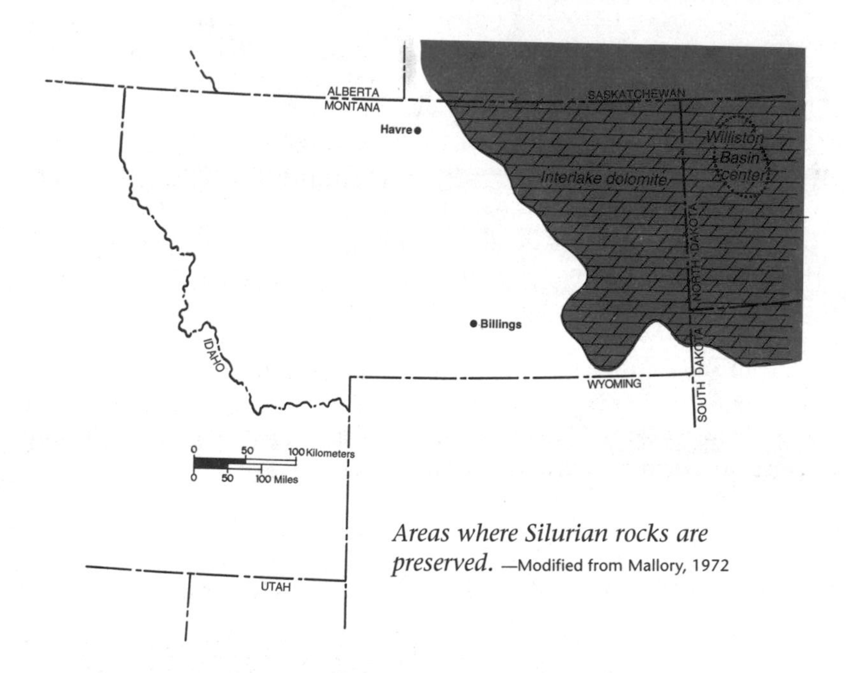

Areas where Silurian rocks are preserved. —Modified from Mallory, 1972

Devonian Formations

Most of the region remained above sea level, still eroding, through early Devonian time. Then the sea returned, either because the land sank or the sea level rose. About 400 million years ago, pale dolomite and sandy dolomite were deposited in shallow water in southeastern Idaho and nearby Utah.

Meanwhile, in Montana and northcentral Wyoming, dolomite and limestone fragments filled sinkholes, caves, and crevices in Ordovician limestone. Those rocks evidently had been above sea level and eroding during Silurian time, and it seems that the climate was probably wet. Limestone weathers mainly by dissolving in the rain. If the climate is wet, the rain erodes large holes in some limestone formations, making them as cavernous as a slab of Swiss

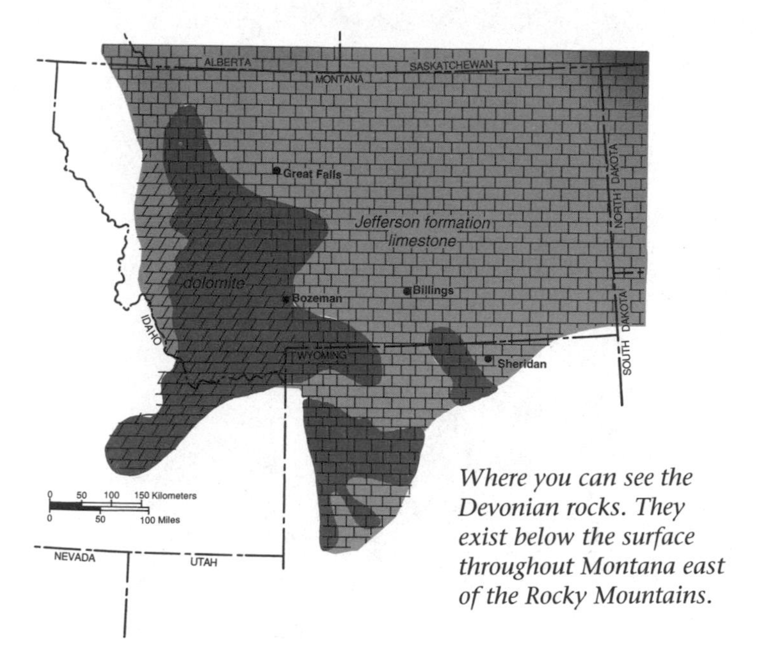

Where you can see the Devonian rocks. They exist below the surface throughout Montana east of the Rocky Mountains.

cheese. The caves eventually collapse to open sinkholes at the surface, and those collect sediments.

During middle Devonian time, most of our region accumulated several hundred feet of limestone, along with some shale and dolomite. Some of the limestone and dolomite is debris eroded from reefs and deposited in banks below them. Those are porous enough to hold large amounts of petroleum. The Williston basin was then a closed inland sea with little connection to the ocean. Evaporating water left large deposits of salt and anhydrite, a calcium sulfate mineral.

As late Devonian time began, the shallow inland sea again spread eastward across most of Montana and western Wyoming to deposit hundreds of feet of muddy limestone and dolomite, the Jefferson group of formations. In some areas, the Jefferson dolomite is so full of organic material that it is black and stinks of sewage when you break it. It is truly a putrid rock. It seems strange that a rock so loaded with organic matter should contain almost no fossils. That shallow Devonian inland sea must have been full of life but nearly devoid of animals.

Many salty modern lakes, such as the Great Salt Lake, support lush blooms of algae and other minute plants and animals, but very few larger animals. It is easy to imagine the black Jefferson dolo-

Irregular cavities in this Devonian dolomite formed when smaller magnesium atoms replaced calcium atoms in the limestone. View is about three feet high.

mite accumulating in extremely salty water blooming with minute plants and animals, a living brine soup. The formation does contain beds of salt. If the environment was poisonous to decay organisms, that might explain the preservation of the stinking organic matter.

Late Devonian formations in many other parts of the world also include black sedimentary rocks full of putrid organic matter. Indeed, such rocks are so typical of the time that geologists use the phrase "Devonian black shales" almost as though it were all one word. Why did so many black sedimentary rocks form? What peculiarity of late Devonian time favored their formation? And why did a great extinction eliminate a large proportion of the earth's animal species while the Jefferson dolomite was accumulating?

The organic matter in the Devonian sedimentary rocks is as economically important as it is smelly. The stinking Devonian sedimentary rocks of Alberta produce enormous quantities of oil and gas. For some reason, the valuable resources do not extend south into Montana, where basically similar rocks exist. Perhaps that is because the Devonian rocks in Alberta were buried more deeply than those in Montana. Their temperature may have risen high enough to cook their rich content of organic material into petroleum, while the rocks in Montana remained too cool to brew a nice batch of crude oil.

Chapter 6

THE PHANTOM ANTLER MOUNTAINS

350 Million Years Ago

Although you would never guess it by looking at the sedimentary rocks in the inland parts of our region, Devonian time brought dramatic geologic changes to the old continental margin. It seems that a mountain range rose all along the old western coast, which had by then existed for more than 400 million years.

Sometime around the middle of Paleozoic time, perhaps about 350 million years ago, something happened along the old continental margin. Geologists call it the Antler orogeny, which means the Antler mountain building event. That seems a safely equivocal name for something still so vaguely understood. As things now stand, the available knowledge does not even make a coherent geological map.

The Antler Mountains

The Antler event got its name from Antler Mountain near Battle Mountain, Nevada. Geologists working on the big gold deposits along the Carlin trend recognized that the sedimentary rocks there broke and moved along large faults during middle Paleozoic time. The patterns of breakage and of the folding that accompanied it suggest that something gave those rocks a hard push from the west.

Other sedimentary rocks similarly pushed from the west at about the same time have since been recognized in the Sierra Nevada, the Klamath Mountains, central Idaho, and British Columbia. Those widely separated places all lie close to the old western margin of North America.

Any doubt about the Antler event vanished during the 1980s, when geologists obtained age dates revealing that several large masses of granite in the Sierra Nevada and Klamath Mountains formed about 350 million years ago. It is hard to interpret the asso-

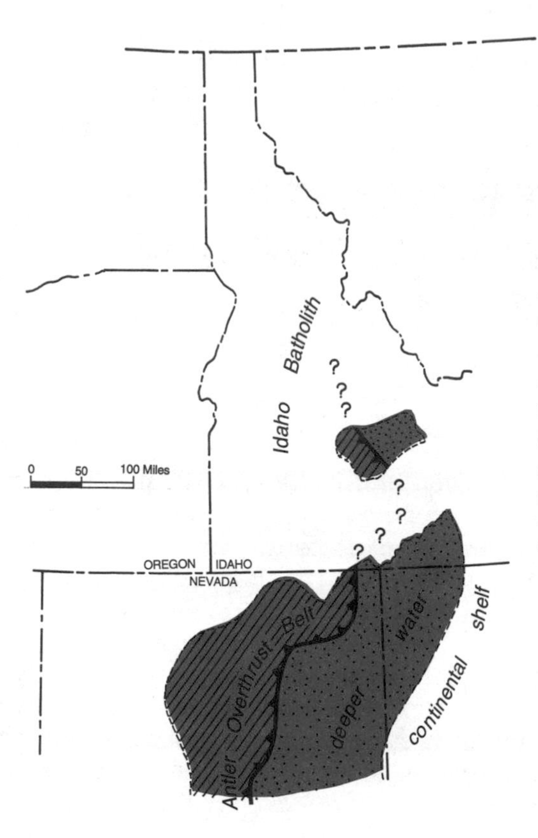

Remnants of the Antler Mountains can be traced from Nevada to central Idaho. Farther north it is lost in the Idaho batholith and western Idaho mylonite zone.
—Redrawn from Burchfiel and Royden, 1991

ciation of strongly deformed sedimentary rocks and granite as anything but the record of a long collision between the western margin of North America and the ocean floor to the west. An oceanic trench must have formed off the old continental margin as mountains rose and volcanoes erupted along the old western edge of the continent. The Antler event was an early version of happenings to come 175 and more million years later, during the latter part of Mesozoic time.

If the big picture of the Antler event now seems reasonably clear, its details are still fuzzy. Unfortunately for geologists, the mountain building of late Mesozoic time nearly obliterated the evidence of the Antler event. Separating similar rocks that formed in the same way and in the same area during different times will take many years of painstaking field work and many age dates. Many rocks that now appear to be puzzling aspects of the later story will probably turn out to be part of the earlier story. The Antler Mountains may have not so much vanished as been trampled nearly beyond recognition in the rush of later events.

Chapter 7

LATE PALEOZOIC TIME

350 to 200 Million Years Ago

During and after Antler mountain building, much of the inland part of our region remained shallowly submerged; it continued to acquire sedimentary formations. Nothing so far recognized in those rocks clearly speaks of mountains rising in the west during middle Paleozoic time, or eroding there during later Paleozoic time.

Mississippian Formations

Early in Mississippian time, about 355 million years ago, the sea again flooded across the region. The Madison limestone was laid down over most of Montana, eastern Idaho, northern Wyoming,

The geologic periods, Devonian through Permian.

Millions of Years Ago	
	MESOZOIC TIME
245	
	Permian
286	
	Pennsylvanian
320	
	Mississippian
360	
	Devonian
408	

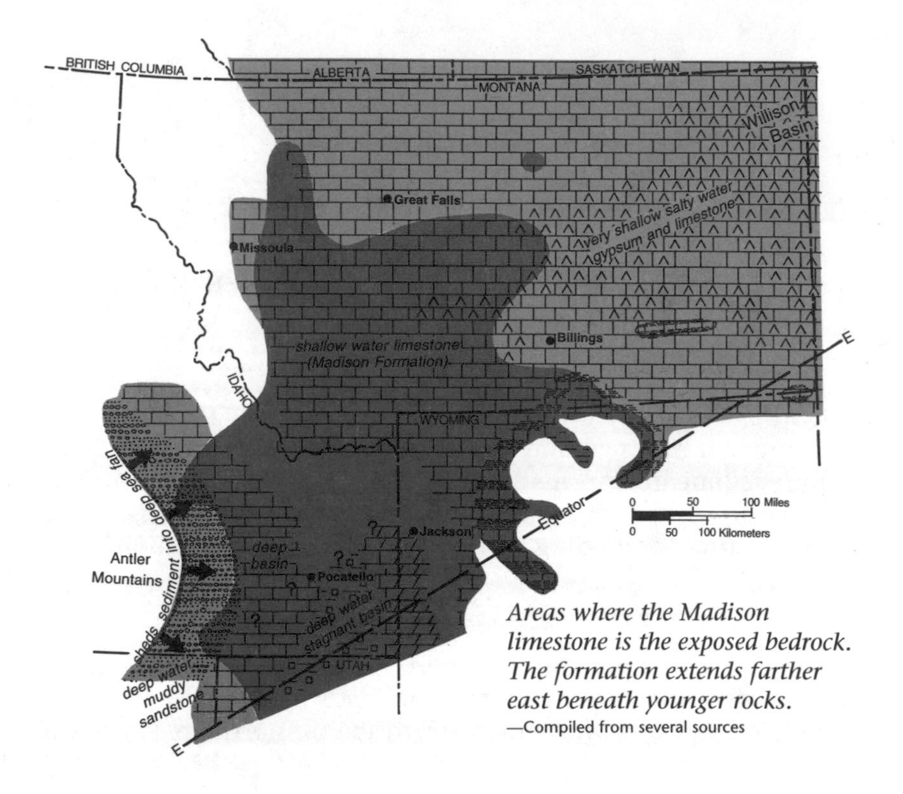

Areas where the Madison limestone is the exposed bedrock. The formation extends farther east beneath younger rocks.
—Compiled from several sources

and the Dakotas. Technically, it is a group of formations, but most geologists simply call them all Madison limestone.

Madison limestone comes in shades of gray—and in immense quantity. It is between 1,000 and 2,000 feet thick in most of Montana, but thins to about 500 feet in Wyoming—still a lot of limestone. In the Williston basin and in much of central Montana, layers of anhydrite or gypsum cover the Madison limestone. In Mississippian time, those areas were evidently isolated inland seas where evaporating water left its deposits. In much of central Montana, a series of limestone, shale, and sandstone formations—the Big Snowy group—were laid down on top of the Madison limestone.

Here and there, the Madison limestone contains oil. In the Williston basin, its lower part is locally porous enough to hold quite a lot of petroleum. In the Sweetgrass arch of northern Montana, old, long-collapsed caverns and fractures in the massive upper part of the Madison limestone provide major reservoirs of petroleum. But the petroleum migrates so easily through those large cavities that it remains trapped only in very large folds.

Cliffs of Madison limestone in central Montana. —C. D. Walcott, U.S. Geological Survey

A horn coral from the Madison limestone. —Larry French

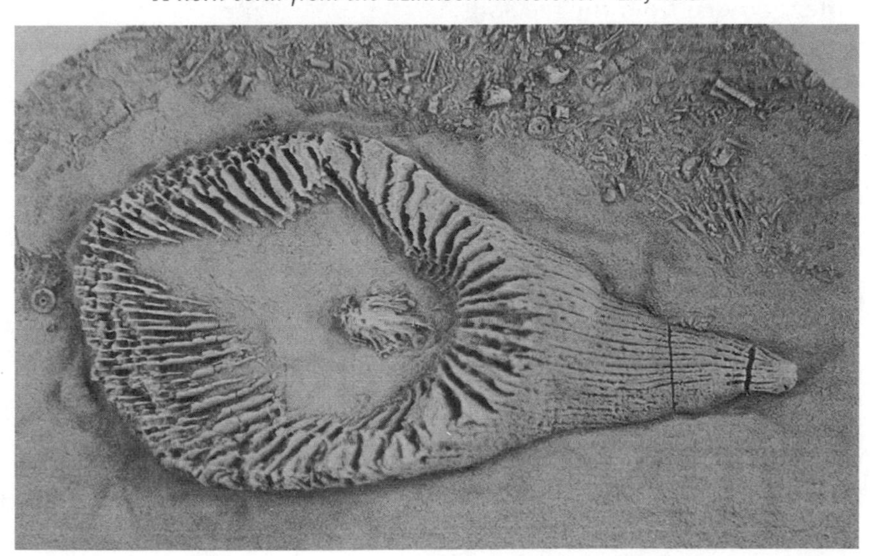

Limestone weathers by dissolving in the rain and in groundwater. But the calcareous soils typical of dry regions establish a slightly alkaline condition in which limestone becomes extremely insoluble. In the northern Rocky Mountains, the thick Madison limestone resists weathering and erosion more successfully than any other kind of rock. It stands high, making bold ridges and spectacular cliffs that contribute a great deal to the landscapes of the northern Rocky Mountains and central Montana. Watch for the pale gray rocks. The thin wash of red that splashes them here and there is a souvenir of Pennsylvanian time.

In central Idaho, the Antler Mountains shed sands and muds eastward into a deep water basin. In southeastern Idaho, phosphatic shale and limestone accumulated. The region lay close to the equator, which trended northeast from northern Utah through the southeastern part of Montana.

Pennsylvanian Formations

After the Madison limestone was laid down, large parts of our region again rose above sea level. In many places, the upper part of the Madison limestone dissolved into a network of caverns that collapsed into sinkholes. A red soil, the Amsden mudstone, developed on the limestone, filled the sinkholes, and filtered down into the caverns. You often see the Amsden mudstone in the bright splashes of red and pink on the pale gray cliffs of Mississippian limestone. In some parts of the region, the red Amsden mudstone was deposited in shallow seawater and probably consists of sediment eroded from the red soils.

Meanwhile, the Big Snowy group of formations was exposed in a wide belt that trended eastward across central Montana in a flat landscape in which streams flowed west, perhaps through steaming jungles. Deposits of sandstone were laid down in long ribbons: some filled channels eroded in the formations of the Big Snowy group, and others were beaches. Now they are deeply buried under younger formations. Those ribbons of sandstone are the Tyler formation, the source of most of the oil in central Montana.

The most conspicuous Pennsylvanian sedimentary rocks are a massive sandstone called the Quadrant formation in Montana and the Tensleep formation in Wyoming, which locally contains oil. In many areas, several hundred feet of that sandstone cover the red Amsden mudstone. It resists weathering and erosion, making towering cliffs and bouldery ridges in many parts of the northern Rocky Mountains. Like most sandstone ridges in the northern Rocky Moun-

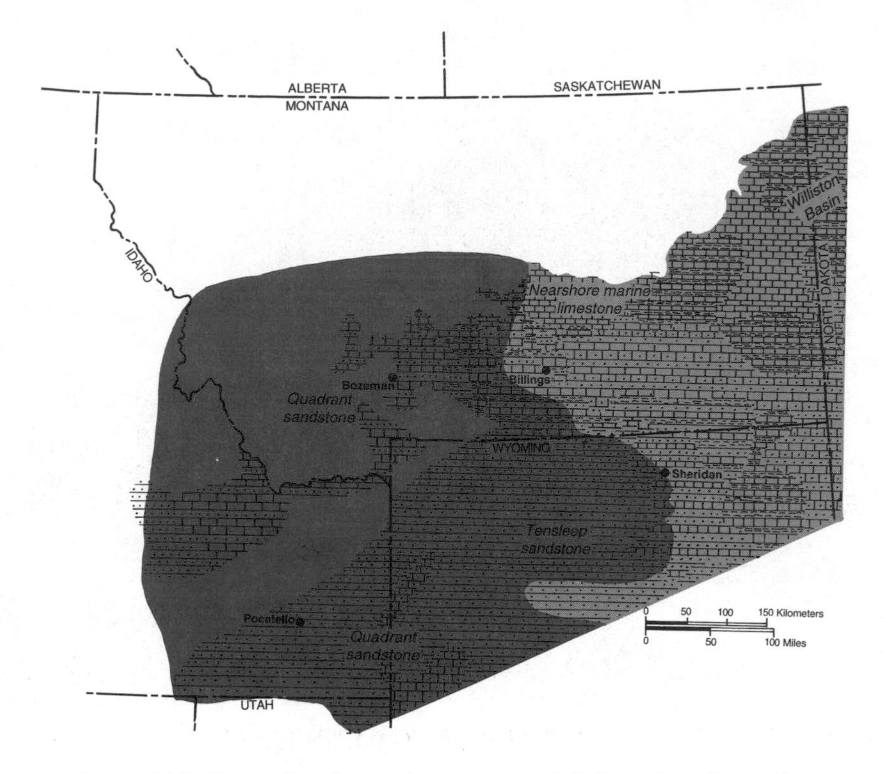

Where middle Pennsylvanian rocks are exposed. Paler colors show where they are buried under younger rocks. —Modified from Mallory, 1972; Skipp and Hall, 1980

tains, they generally contain enough water to support pine and spruce trees.

In large areas of Wyoming and locally in Montana, the Pennsylvanian sandstone contains the distinctive patterns of internal layering characteristic of sand deposited as wind-driven dunes. So far as we know, unmistakable dune sandstones did not accumulate anywhere until late Pennsylvanian time, when they were deposited in many parts of the world. The evidence from the rocks suggests that not until then did the wind first blow hard enough to move sand.

Permian Formations

Except in the Williston basin, that perennial sump for Paleozoic sediments, most of our region contains no early or middle Permian formations. It seems that it was then above sea level and eroding. Some sand dunes survive in south-central Wyoming, part of a dune

Crossbedding in Pennsylvanian sandstone in Wyoming. —W. T. Lee, U.S. Geological Survey

field that expanded southward. Throughout western North America, the early and middle Permian rocks appear to have been laid down in a desert.

In late Permian time, about 255 million years ago, much of the inland part of our region was again a shallow sea. Limestone and sandstone accumulated from northwestern Wyoming well into central Montana, red shale and gypsum farther east. In northern Utah, southeastern Idaho, and southwestern Montana, layers of limestone, sandstone, black shale, chert, and phosphate rock of the Phosphoria formation collected. The formation thickens from two to three hundred feet in western Wyoming to as much as 1,300 feet in southeastern Idaho. The Phosphoria formation rarely makes outcrops prominent enough to add much to the scenery. Geologists find it by looking on top of the Pennsylvanian Quadrant or Tensleep sandstone, which is easy to find.

As its name suggests, the Phosphoria formation contains phosphate rock, the source of mineral phosphate. Phosphate is used in fertilizers, in detergents, and in such weapons as tracer bullets, white phosphorous hand grenades, and incendiary shells. Demand for all those items seems certain to increase with the population. According to most estimates, the Phosphoria formation contains the world's largest reserves of minable phosphate rock. So far, the high shipping costs to major markets have limited development of the Rocky Mountain phosphate field to a few large mines, some of which have been operating since the 1920s. Many more mines will certainly start as more accessible phosphate reserves dwindle and demand increases. The Phosphoria formation will soon support a large mining industry.

The Phosphoria formation also contains two layers of waxy black shale about two or three feet thick. It is genuine oil shale; you can

The Phosphoria formation is exposed in southwestern Montana and eastern Idaho. Sediments of the same age farther east are shale, limestone, and gypsum. —Compiled from Mallory, 1972; Wardlaw, 1980; and Peterson, 1986

light it with a match and watch it burn with a smoky, yellow flame. It also contains an abundance of copper, nickel, and other metals. Similar black shales of the same age in northern Germany and Poland have supported mines for more than a century. Although the oil is not a significant petroleum reserve, it might provide enough energy to extract the metals.

The abundant organic matter in the Phosphoria formation is the primary source of much of the petroleum in the northern Rocky Mountains. In the Big Horn basin of northwestern Wyoming and south-central Montana, the Phosphoria formation abruptly blends eastward into Permian red shales and gypsum. Several oil fields produce petroleum trapped in the zone of transition from one kind of Permian rock to the other.

The Absolutely Horrible Permian Extinction

Paleozoic time began with the sudden flowering of the Cambrian fauna. It ended with the extinction of at least 90 percent of all the animals on earth during late Permian time. There is a certain symmetry in that neatly bracketing pair of opposing and unrelated events.

The late Permian extinction came perilously close to killing all the animals on earth, much closer than the more famous extinction that killed the dinosaurs and many other groups of animals 65 million years ago.

It now seems clear that great extinctions consistently happen at the same time that flood basalt provinces erupt. The Tungusska flood basalts of western Siberia erupted during the several million years in which the late Permian extinction happened. They cover a greater area than any other flood basalt province, so it seems appropriate that they correspond to the greatest extinction.

In addition to the phosphatic black shale with a large content of metals, the Phosphoria formation also contains thick sections of chert, a sedimentary rock that consists essentially of extremely fine-grained quartz. Many other formations of the same age deposited elsewhere in the world also contain phosphate rock, black shales, and chert—in some combination. Late Permian time was a period when unusual sedimentary rocks accumulated all over the world; presumably it was a time when unusual conditions prevailed worldwide. And assemblages of metalliferous black shale, phosphate rock, and chert are associated with other times of major extinction, including the one that occurred during late Devonian time, when the Jefferson dolomite was laid down.

Perhaps the enormous volcanic eruptions that produced the Tungusska basalts of western Siberia may also have produced enough carbon dioxide to cause a very strong global greenhouse effect. The high temperatures and shifts in climatic patterns that would result might spell the end for many land animals. And an atmosphere greatly enriched in carbon dioxide would also make surface water somewhat acidic, which might eliminate many kinds of animals that live in the seas.

Shattercones formed in bedrock by impact of an asteroid. Similar shattercones would have accompanied the impact that caused the great Permian extinction.

▶ Part 3 ◀

MESOZOIC TIME

The First Half

245 to 175 Million Years Ago

Mesozoic time began about 245 million years ago as the earth emerged from the massive extinction of late Permian time. It ended 65 million years ago, when another catastrophic extinction wiped out about 65 percent of the world's animals. The three periods of Mesozoic time that intervened—the Triassic, Jurassic, and Cretaceous—saw the earth acquire and support a distinctive population of plants and animals. This was the age of the dinosaurs.

The geologic periods of Mesozoic time.

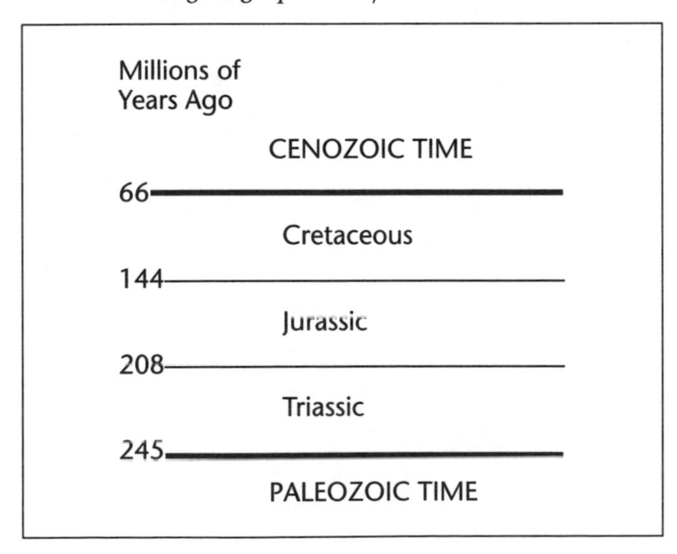

When Mesozoic time began, our region was still the partly submerged fringe on the western edge of the great supercontinent, Pangaea. Everything west of the old continental margin was open ocean, and large parts of the continental interior to the east were submerged under shallow seawater. When Mesozoic time ended, most of the continental interior was above sea level, and the inland sea was finally almost gone. Most of the rocks in the northern Rocky Mountains had formed and were almost in their present places, and large pieces of continental and oceanic crust had been added to the old continental margin. A raggedly embayed shoreline embraced much of the future Pacific Northwest.

Chapter 8

MESOZOIC TIME BEGINS

245 Million Years Ago

Relatively few vertebrate animals roamed the land in Permian time and those were spraddling sorts, amphibians and fairly primitive reptiles. Almost all vanished in the great Permian extinction, as did most of the enormously varied fauna of invertebrate animals that lived in the sea. Paleontologists who comb the preserved remains of Triassic beaches and shallow sea floors looking for fossil shells do not see the distinctive animals that flourished so abundantly in Permian time.

During early and middle Triassic time, the shallow inland sea that deposited the Phosphoria formation during late Permian time still flooded much of western Wyoming, southern Idaho, southwestern Montana, and the Williston basin. But its waters no longer laid down phosphatic sediments. Formations of mudstone and limestone accumulated to a thickness of as much as 4,000 feet in the western part of that area, and to about 500 feet in the Williston basin.

An ammonite from Jurassic rocks.
—G. Stanley

Meanwhile, red mudstones and sandstones accumulated on land in parts of central and eastern Montana and Wyoming. By late Triassic time, the inland sea had shriveled to a small area in southern Wyoming and northwestern Colorado, and red sedimentary formations were accumulating in a correspondingly larger area of our remote western end of the great continent of Pangaea.

Pangaea

During the latter part of the Paleozoic era, most of the earth's inventory of slowly drifting continental crust came together into the supercontinent Pangaea. For a few tens of millions of years, the earth had one hemisphere that was mostly continental, another that was almost entirely oceanic. The narrow Tethys Sea nearly divided the continental hemisphere into northern and southern land masses: Laurasia and Gondwana.

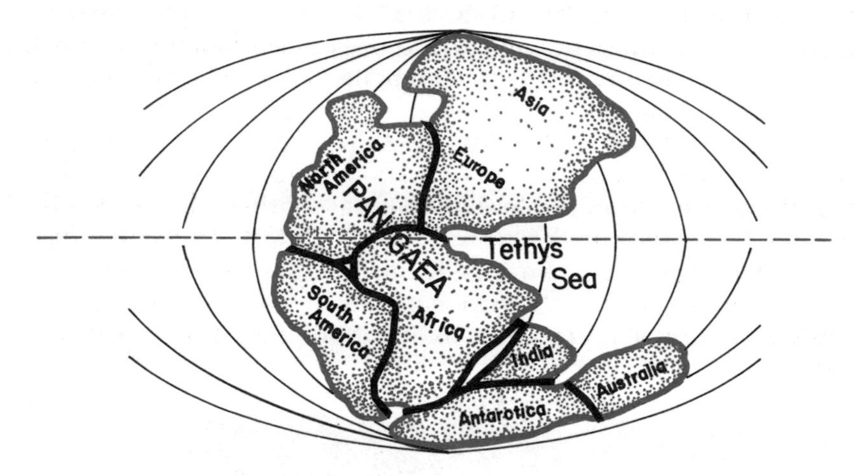

Pangaea as it was at about the end of Permian time.

You can reconstruct Pangaea by cutting the modern continents from a map and fitting their edges together as though they were the pieces of a jigsaw puzzle. The rocks match just as well as the map outlines. Small remnants of the Tethys Sea survive in the Mediterranean and Black Seas.

While Pangaea existed, our region lay on the western edge of Laurasia, which included all of what is now North America, Europe, and Asia. The great Paleozoic flood had largely receded, leaving most of Pangaea high and dry. The reptiles and amphibians of Permian time and the early dinosaurs of Triassic time were nearly the same from one end of that great continent to the other. An ambitious and footloose Triassic dinosaur could have started walking east from Montana and hiked all the way across North America, Europe, and Asia to the Pacific Ocean.

The Red Rocks of Pangaea

Permian time saw enormous amounts of red sandstone and mudstone accumulate on dry land throughout Pangaea; Triassic time brought a great deal more. At no time before or since has so much iron-stained sediment been laid down. So far, no one has convincingly explained why.

Many geologists argue that the climate of Permian and Triassic time was warm and dry. That makes sense because many of the

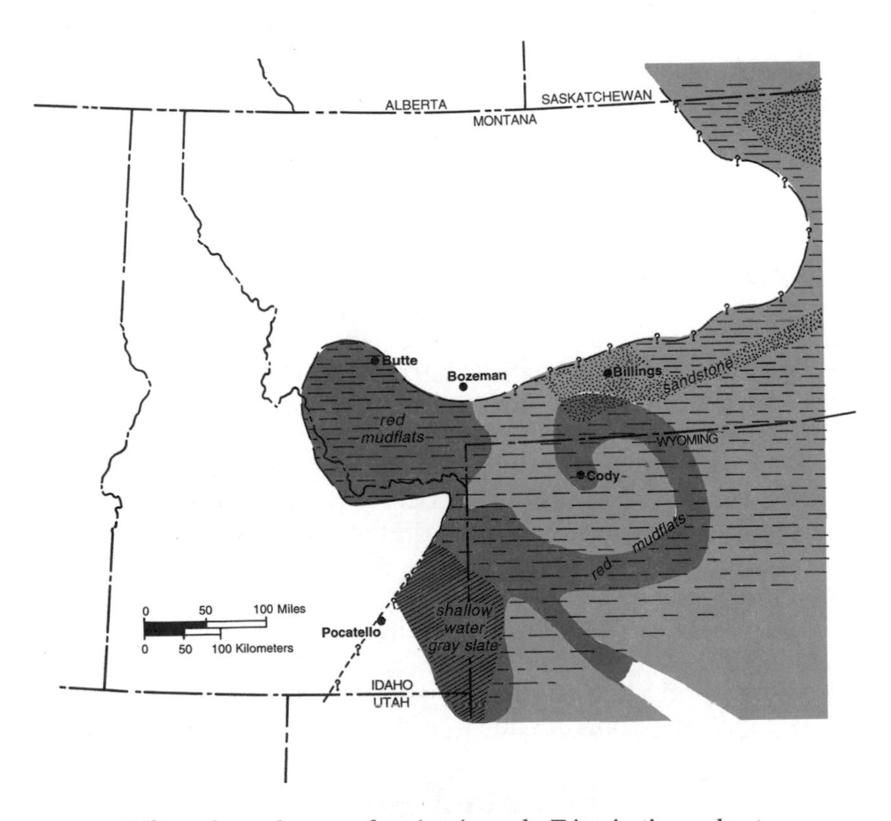

Where the rocks were forming in early Triassic time, about 240 million years ago. Paler shades in eastern Montana and Wyoming show where Triassic rocks are buried under much younger Cretaceous rocks. —Modified from Mallory, 1972

Triassic to Cretaceous sedimentary rocks in the Bighorn Basin of northern Wyoming. —W. T. Lee, U.S. Geological Survey

formations contain sand dunes, which fit nicely into a desert scene. And all those sedimentary formations deposited on land show that streams were not efficiently carrying sediment to the ocean, which also suggests that the climate was dry. Furthermore, a good many of those terrestrial formations contain salt and gypsum, typical desert sediments. But why was the climate so dry?

Some geologists point out that much of the vast interior landmass of Pangaea was far from the ocean, perhaps beyond the reach of its moist winds. Others suggest that abundant carbon dioxide in the atmosphere may have caused a worldwide greenhouse effect. Or perhaps our area was not far from the equator, and the trade winds were blowing off the dry interior of the continent. Of course, none of those explanations excludes the others. It would help if someone found a bottle full of Triassic air—or some way to measure the amount of carbon dioxide in ancient atmospheres.

Dinosaurs

As Triassic time began, the few amphibians and reptiles that survived the great late Permian extinction found themselves living in a world well supplied with plants to eat and lacking most of the competitors and predators that had plagued their ancestors. Theirs was a moment of vast opportunity: a food supply without a food chain. They soon established a new food chain as their numbers multiplied and their progeny evolved into new forms.

Within ten million years after the great Permian extinction, sometime during middle Triassic time, the surviving reptiles spawned the first dinosaurs, and the first mammals. The dinosaurs thrived mightily through the rest of Mesozoic time, while the mammals remained small, generally unimpressive, and probably scarce. Their turn would come when the dinosaurs were finally gone.

Dinosaurs were a new kind of animal unlike any that preceded or followed them. They obviously descended from reptilian ancestors, but with a novel and greatly improved architecture. Many paleontologists now believe that at least some dinosaurs were warm blooded—perhaps all of them were. If so, they were probably as active and alert as modern mammals and birds. They would prosper and dominate the continents for about 165 million years until they perished in the next overwhelming catastrophe, 65 million years ago.

The End of Pangaea

During Triassic time, Pangaea began to break into big pieces, the modern continents, which moved away from each other as new oceans opened between them. Gondwana, the southern giant continent, went first, as India separated from Africa and the combined mass of Australia and Antarctica detached from India and Africa. Then South America and Africa parted company. Those southern separations hardly affected our region.

Our region did feel the effects, however, when Laurasia, the northern giant continent, began to break about 205 million years ago, at the end of Triassic time. That was also the time of another great extinction, although it was not nearly so massive as the one that ended Permian time or the one that would later end the Cretaceous period. North America moved away from Eurasia and Africa, while the middle part of the Atlantic Ocean began to open between them. That started the series of events that added the Pacific Northwest to our region. It is a long story, and it begins with the growth of the Atlantic Ocean.

Reddish-colored crossbeds in Triassic sand dunes.

Chapter 9

PLATES SEPARATE AND COLLIDE

208 Million Years Ago, and Continuing

As Jurassic time began, about 205 million years ago, the inland part of our region was still a desert. The wind blew sand into large dune fields that survive as thick sedimentary formations in southwestern Wyoming and eastern Utah. In middle Jurassic time, a shallow inland sea advanced southward across Idaho, Montana, North Dakota, the northwestern half of Wyoming, and most of Utah. It left behind limestone and shale that are as much as 1,000 feet thick in southeastern Idaho and central Utah. Farther east, the Jurassic sediments include a few hundred feet of red shale, limestone, and gypsum. Meanwhile, beds of salt accumulated in that perennial inland sea, the Williston basin.

By late Jurassic time, about 160 million years ago, a broad area of beach sand separated land in Idaho and western Montana from the inland sea to the east. That sea laid down deposits of shale along with a bit of limestone in eastern Montana and Wyoming. Meanwhile, the evolving dinosaurs had long since produced the brontosaurus, that behemoth of the barrel body, long tail, and enormously long neck with a tiny lump of a head on the end. Feathered birds certainly flew through the air then, and presumably whistled in the trees—or perhaps they squawked.

That was the scene in our region as the middle part of the Atlantic Ocean began to open, starting a long train of geologic events along the western margin of North America. The Atlantic Ocean is still growing, still changing our region, thousands of miles away and 205 million years later.

An Ocean Grows

An oceanic ridge, the mid-Atlantic ridge, runs down the middle of the Atlantic Ocean. Its northern part rises above sea level in Iceland, the only place where an oceanic ridge is above sea level and the best place to see an ocean grow. A broad valley that cuts a dogleg course across the central part of Iceland is part of the rift that exactly follows the crest of the mid-Atlantic ridge; a rift follows the crest of every oceanic ridge.

Icelandic surveyors watch their carefully laid property lines change as that valley in the middle of the island grows about two inches wider every year. The long written history of Iceland records that every 200 years, more or less, a jagged fissure appears in the valley floor and basalt lava erupts in a large flow that covers as much as several hundred square miles. If the next fissure and lava flow are on schedule, the Icelanders can expect them any day.

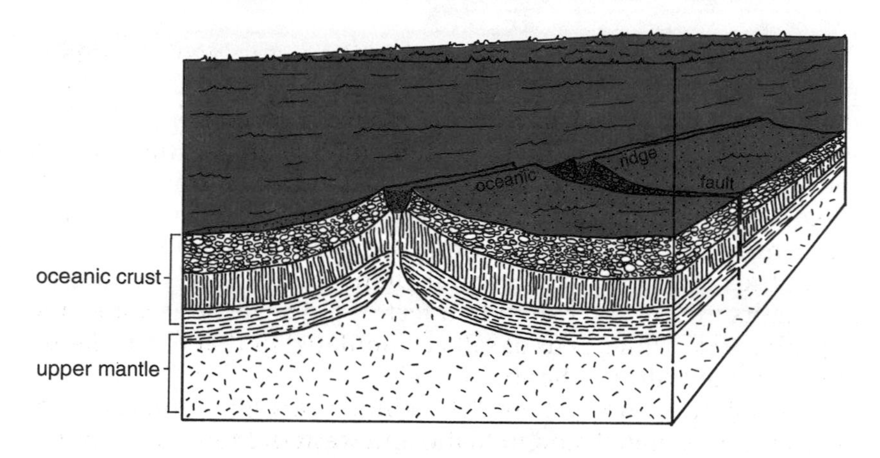

Section through an oceanic ridge.

It now seems clear that the submerged parts of the world's oceanic ridges behave like the valley in Iceland. Their rifts grow wider as the opposite sides pull away from each other. Fissures open in their floors, and great basalt lava flows erupt. Each flow adds a little more to the oceanic crust.

Some geologists have compared the growth of new oceanic crust at the crest of an oceanic ridge to two ice floes: Picture them pulling slowly away from each other with water welling into the opening gap between them. If water freezes onto the trailing edges of the floes at the same rate at which they separate, each will grow larger while the width of the gap between them remains constant. So it is that the oceanic crust grows as basalt erupts in the opening gap between two lithospheric plates that move away from each other at the crest of an oceanic ridge.

Ophiolites

Back in the 1950s, a group of eminent geologists formed an eminently informal club, which they called the American Miscellaneous Society. It may have been unique among professional societies in its lack of a formal list of members, official meetings, elected officers, regular dues, and boring publications. According to most accounts, it was basically a social group, but its occasional meetings did lead to a serious project.

That was an attempt to bore a hole all the way through the oceanic crust and into the mantle to see what kinds of rocks it might penetrate—the compositions of the oceanic crust and mantle were then a subject of much controversy. They made a serious attempt, but no hole ever reached the mantle. They would not have tried had they understood ophiolites.

Geologists knew about ophiolites long before they recognized them as slabs of oceanic crust misplaced into the continent instead of on the ocean floor. Somehow, no one knows exactly why or how,

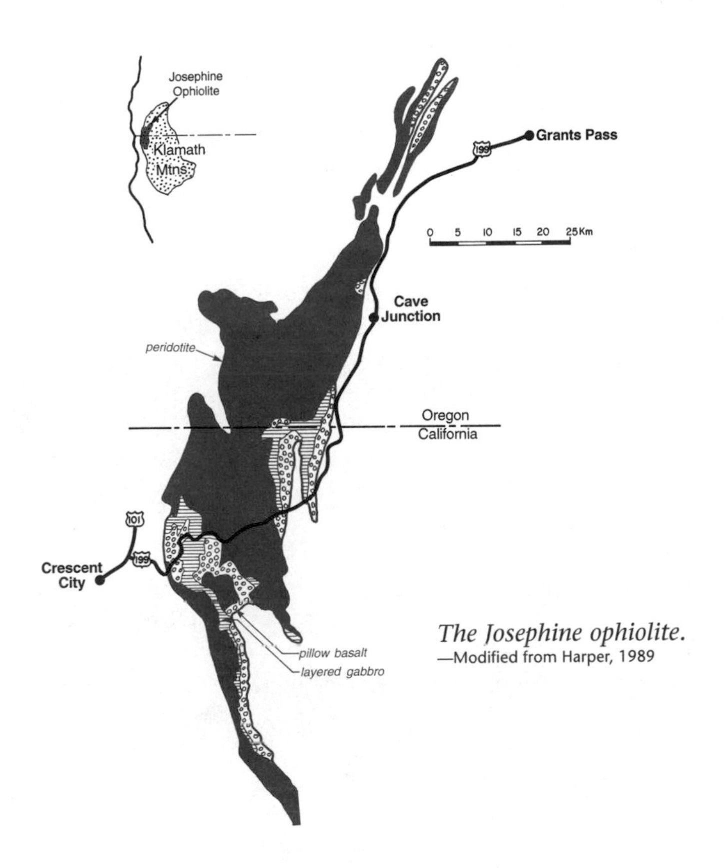

The Josephine ophiolite.
—Modified from Harper, 1989

those slices shear off into the stuff filling a trench instead of sliding smoothly down into the mantle, the fate ordained for most oceanic crust.

Our region contains a good many ophiolites, of which the Josephine ophiolite of southwestern Oregon and northwestern California is probably the finest example. Roadcuts along U.S. 199 through the Klamath Mountains expose nearly the entire oceanic crust, from the muddy brown oceanic sediments at the top down to mantle rocks at the base. You can see in a day all the rocks the distinguished geologists of the American Miscellaneous Society never managed to drill through. Watch the roadcuts for somber oceanic rocks in various nasty shades of darkly greenish gray.

The Josephine ophiolite apparently formed between an active volcanic chain and the continent during late Jurassic time, about 150 million years ago. Instead of sliding smoothly through the trench and down into the mantle, it sheared off and was shoved over the volcanic chain to the west. Perhaps the slab, consisting as it did of new oceanic crust, was still hot and therefore lighter than the older rocks in the volcanic chain.

The top of an ophiolite complex is its cover of sediments, especially muddy chert, deposited on the deep ocean floor. They lie on basalt lava flows, typically a few thousand feet of them, all erupted onto the floor of the rift valley in the crest of the oceanic ridge. Basalt that erupts underwater commonly breaks up into cylindrical

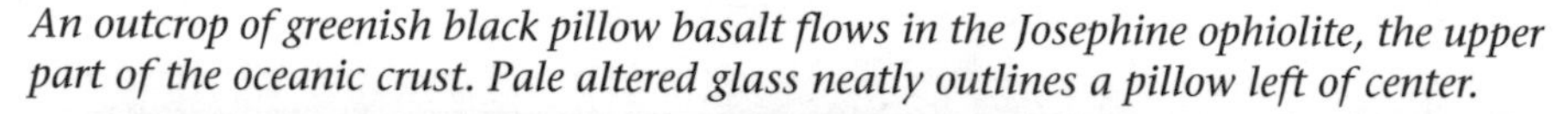

An outcrop of greenish black pillow basalt flows in the Josephine ophiolite, the upper part of the oceanic crust. Pale altered glass neatly outlines a pillow left of center.

masses that suggest a pile of pillows when you see them sectioned in a cliff or road cut.

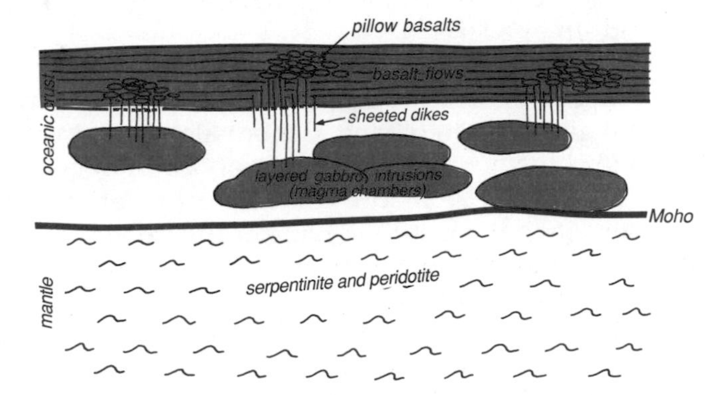

Section through the oceanic crust. Parts of such sections are exposed in several places in the Pacific Northwest.

Below the pillow basalts are several thousand feet of vertical basalt dikes packed one next to the other like the pages of a book standing on a shelf. The dikes are typically several feet thick. Each fills a fissure that erupted one of the pillow basalt lava flows into the upper part of the oceanic crust. It seems that the ocean grows wider one dike at a time.

Below the dikes lie several thousand feet of layered gabbro, dark gray rock that differs from basalt mainly in consisting of mineral grains large enough to see without a magnifying glass. The layered gabbro must have crystallized from pools of molten basalt that sup-

Layered gabbro in the Josephine ophiolite. It looks almost like a sedimentary rock.

plied the magma for the dikes and lava flows above. Large masses of gabbro are quite commonly layered, apparently because growing crystals settle out of the molten liquid like grains of sand settling to the bottom of a lake.

The layered gabbro lies directly on the dense greenish black rocks of the upper mantle, peridotite composed mainly of black pyroxene and glassy green olivine. Gabbro also contains those minerals, but dilutes them with a generous proportion of greenish white plagioclase feldspar. Peridotite contains only dark minerals, no feldspar. Although it constitutes more of the earth than any other kind of rock, peridotite properly belongs in the mantle, so you rarely see it exposed in a continent.

So the oceanic crust consists essentially of about five miles of basalt in various forms lying directly on the black and much denser peridotite of the earth's mantle. The peridotite in the outer 60 or so miles of the mantle differs from that beneath mainly in being cooler and therefore more rigid. Together, the crust and the relatively cool and rigid outer part of the mantle make an outer rind on the earth called the lithosphere. It is thin at the oceanic ridge but thickens as it moves away from the ridge and cools.

The Moving Continent Meets the Moving Ocean Floor

If North America is moving west, away from the mid-Atlantic ridge, then it must collide with the floor of the Pacific Ocean. Colliding lithospheric plates come as close as anything we can imagine

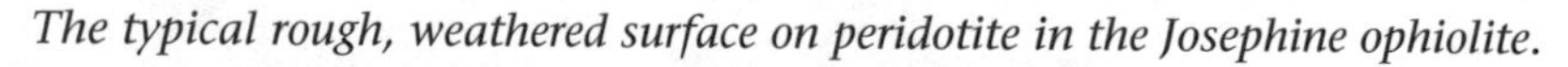

The typical rough, weathered surface on peridotite in the Josephine ophiolite.

Serpentinite, the hydrated and much-deformed equivalent of peridotite, exposed in western Washington.

to the proverbial encounter between an irresistible force and an immovable object.

In such cases, the earth invariably resolves its ponderous dilemma by letting oceanic lithosphere sink into its interior, ahead of the oncoming plate. Oceanic lithosphere sinks because it is cold and therefore denser than the much hotter but otherwise similar mantle rock beneath. The area of oceanic lithosphere that sinks each year into the earth's interior must match the area the oceanic ridges generate.

When the Atlantic Ocean began to open, the lithosphere broke along the old western margin of North America, creating a new plate boundary. During the 205 million years that new oceanic crust has been forming between them at the mid-Atlantic oceanic ridge, North America and Europe have separated the full width of the Atlantic Ocean. Meanwhile, the Pacific Ocean has shrunk as old oceanic crust slides beneath the advancing western edge of North America and, of course, the advancing eastern edge of Asia.

A long depression forms on the ocean floor where the sinking plate buckles and bends down to begin its long dive. That depression is an oceanic trench. The oceanic lithosphere slides down through the trench, then beneath the advancing continent, and on into the hot depths of the mantle. Most trenches swallow several inches of ocean floor every year.

Most sedimentary rocks are much too light to sink into the denser interior of the earth. Those that lie on the surface of the sinking oceanic plate scrape off into the trench or jam beneath the edge of the advancing continent. Continents and oceanic sediments refuse to sink into the earth's interior for precisely the same reason that the marshmallow refuses to sink into a cup of hot cocoa.

The sinking oceanic lithosphere soaks up heat as it descends into the mantle. The lithosphere, after all, differs from the rest of the mantle only in being the relatively cold and rigid outer rind of the earth. Heat it up, and it is neither cold nor rigid. That finally happens at a depth of one or two hundred miles. It simply changes into hot and soft rock in the same way that a hard lump of cold wax or modeling clay softens in the warmth of your hand.

Chapter 10

FOLDS AND GRANITE: THE KOOTENAY ARC

205 Million Years Ago

By Jurassic time, the western continental margin established sometime around 800 million years ago had existed for some 600 million years, no doubt with some modifications during the Antler mountain building event of about 350 million years ago. It had accumulated enough sediment during all those years to form a broad coastal plain and continental shelf. Those were the first hapless victims of the plate collision that began about 160 million years ago.

The Crumpled Margin

Most of those sediments were lying on oceanic crust and banked up against the margin of the continent as though they were carpets piled against a wall. When the oceanic crust finally began to move beneath them after having lain stationary since the Antler event, it mashed the fringe of sediments against the old western margin of

When the oceanic crust began to slide beneath the western edge of North America, it mashed the old coastal plain and continental shelf against the edge of the continent.

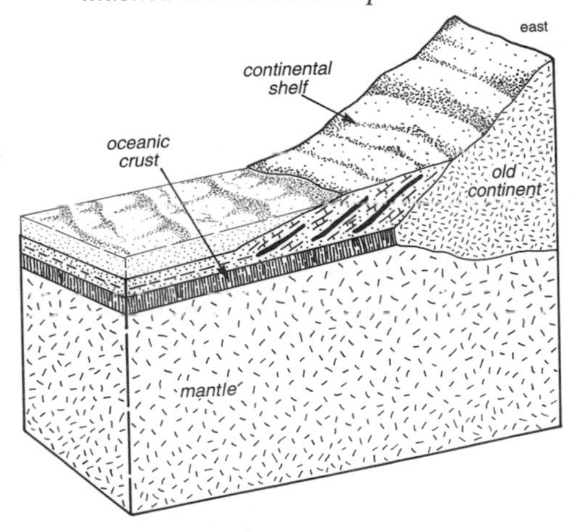

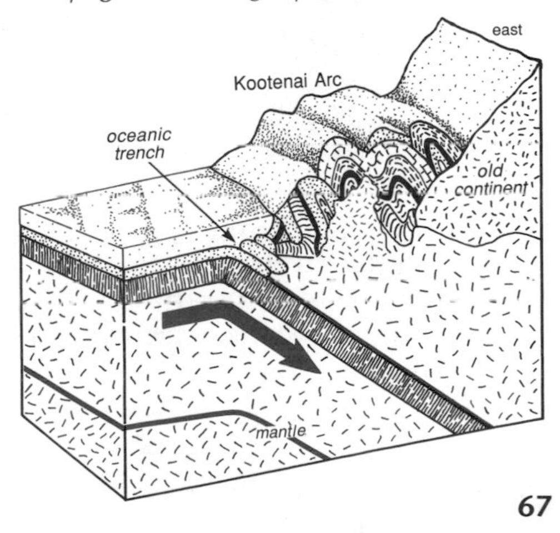

North America. Now they are tightly crumpled into folds, deeply telescoped along thrust faults, and full of granite intrusions.

Geologists in Washington and British Columbia call that belt of sedimentary rocks jammed hard against the former western edge of the continent the Kootenay arc. It includes coastal sedimentary rocks as old as the Windermere formations laid down during late Precambrian time, formations representing most of Paleozoic time, and Mesozoic formations apparently laid down on the open ocean floor shortly before the plate collision began.

We lose sight of the Kootenay arc just north of Spokane, where it disappears southward beneath much younger and absolutely opaque basalt lava flows. Or perhaps it disappears entirely. If the Kootenay arc continues farther south, the next place to look is in California, where the old continental margin emerges from beneath the younger volcanic rocks that cover so much of the Pacific Northwest.

Rocks in the Sierra Nevada of northern California also consist mainly of sedimentary formations that accumulated along the old

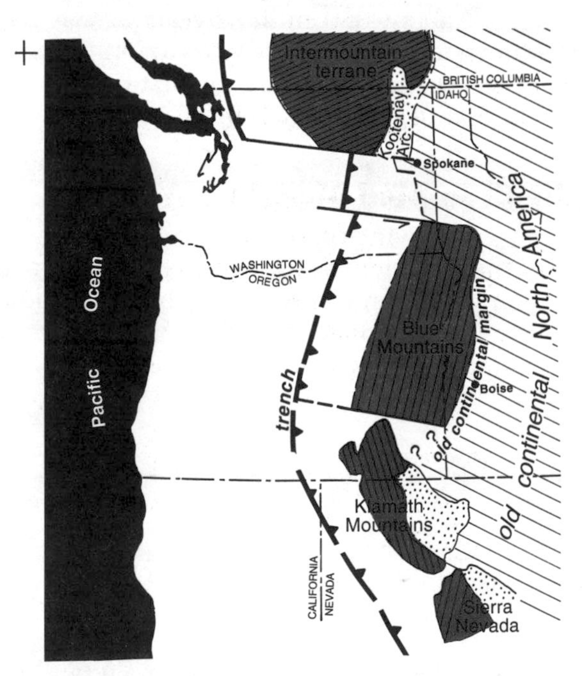

The old continental margin and the trench that formed west of it created the rocks in both the Kootenay arc and the Sierra Nevada.

continental margin, then were tightly folded and heavily injected by molten granite magma. They formed in essentially the same way as those in the Kootenay arc, and at the same time. The Klamath Mountains of northwestern California and southwestern Oregon appear to be the detached and displaced northern end of the Sierra Nevada.

Granite

The Kootenay arc and Sierra Nevada both contain granite. Its story begins at the oceanic ridge, leads down through the oceanic trench, and finally ends as the sinking oceanic lithosphere dives deep into the earth's mantle.

Cold seawater sinks through fractures near the crest of the oceanic ridge and into the red-hot peridotite in the upper mantle. The chemical reaction between water and peridotite creates a slippery and darkly greenish rock called serpentinite. Many people call it soapstone, because it feels a bit soapy. Souvenir shops are full of little greenish figurines and miniature boxes carved from serpentinite.

Much later, when the same rocks are finally sinking through the oceanic trench and into the mantle, the serpentinite soaks up heat. When it reaches a red heat, serpentinite reverts to the original peridotite, meanwhile releasing its water as steam—flaming red-hot steam, not the tepid vapor that whistles out of a friendly tea kettle.

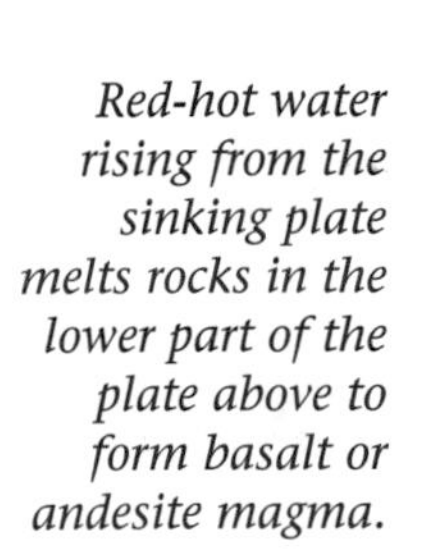

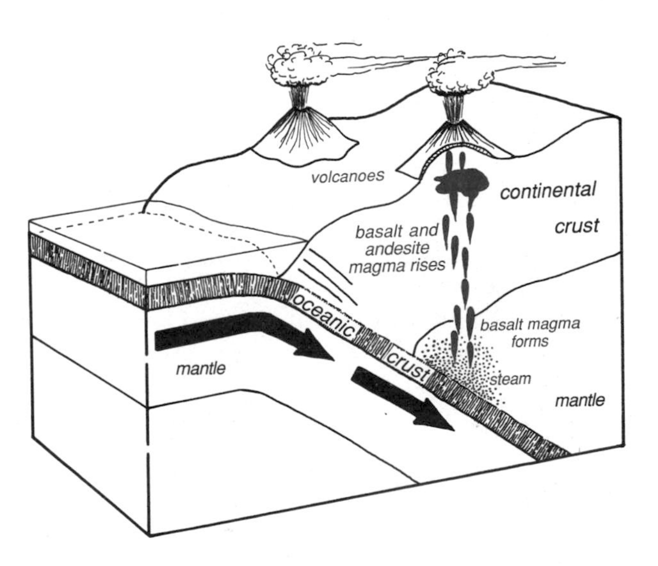

Red-hot water rising from the sinking plate melts rocks in the lower part of the plate above to form basalt or andesite magma.

That blazing water rises from the sinking plate into the peridotite above it, lowering its melting temperature—steam added to any rock makes it melt more easily. Melting the peridotite with steam produces basalt or andesite magma, which rises along a line parallel to the trench and 100 or so miles inland, building high volcanoes. The High Cascade volcanoes erupt such lavas.

Where molten basalt or andesite rises through old continental crust, or the sediments stuffed into a trench, it heats the rocks and in some places melts them. Large volumes of those higher rocks recrystallize to form metamorphic rocks, mostly gneiss and schist. Some may get hot enough to go beyond metamorphism and melt to form granite magma—granite because continental crust has the same average composition as granite.

It is a nice bit of recycling: Continental crustal rocks, mostly schist, gneiss, and granite, weather to form soils that erode to become sediments. Rivers carry the eroded sediments to the ocean. Eventually, the moving ocean floor sweeps those sediments into the depths of a trench, where they heat up and recrystallize to form a new generation of schist and gneiss, or melt to form masses of new granite. That is what happened in the Kootenay arc. Large volumes of molten granite magma invaded the folding and recrystallizing sedimentary rocks, creating a complex of granite and metamorphic rocks. The final result is, in effect, new continental crust, a new generation of basement rocks.

Chapter 11

NEW MOUNTAINS ALONG AN OLD COAST

205 to 100 Million Years Ago

On just the right kind of winter morning, the crisp snow breaks ahead of your shovel, and the pieces slide onto each other as you push them into a thickening ridge. The fracture surfaces on which those slabs of snow move are miniature thrust faults, and the moving slabs are thrust slices.

The collision between North America and the Pacific Ocean floor crushed the old western margin of the continent in much the same way as the shovel crushes the snow. Imagine the continental crust breaking along big fractures that become thrust faults as thick slices of rock above the faults pushed east over those below. Such stacking greatly thickened the western margin of the continent, raising a broad welt of early mountains. It was the first stage in the formation of the northern Rocky Mountains.

The Early Mountains

That early mountainous welt probably began to rise in middle Jurassic time and continued to rise into late Cretaceous time, at least until about 100 million years ago. Those early mountains probably resembled the modern Andes Mountains of South America, which owe their existence partly to thrust faults that stacked slices

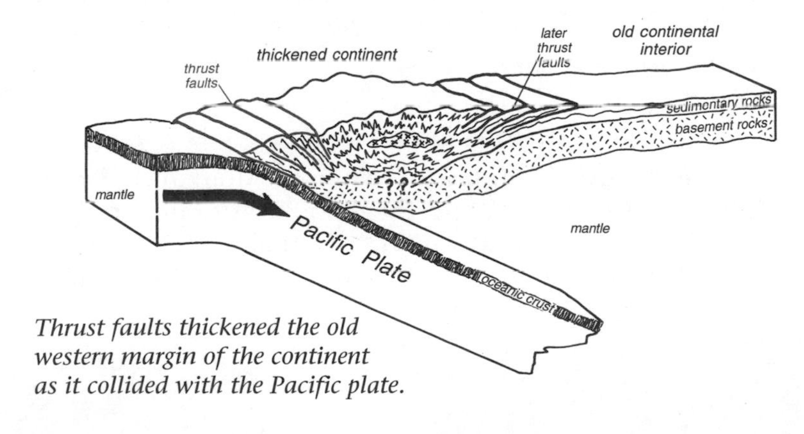

Thrust faults thickened the old western margin of the continent as it collided with the Pacific plate.

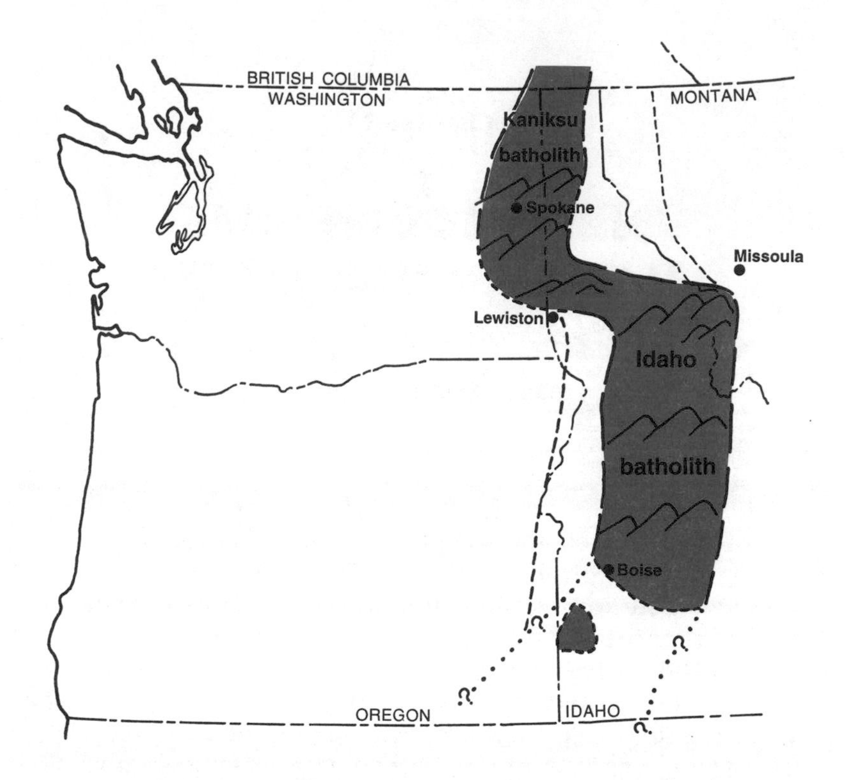

The approximate outlines of the early mountain welt along the western edge of North America. The color shows areas where older rocks metamorphosed and granite magma invaded them.

of rock along the western margin of a continent. For reasons that will emerge as our story unfolds, we think those first mountains were probably about 20,000 feet high.

Streams undoubtedly carved those early mountains as they rose, maintaining a rugged landscape. Nothing remains to tell whether the slopes were barren or forested, but their position along the old west coast makes it seem reasonable to imagine a cover of green forests, at least on their seaward slopes. The climate inland was certainly dry.

Inland Deserts

During late Jurassic time, the beaches of central Montana retreated eastward. By early Cretaceous time, the dinosaurs of Montana and Wyoming roamed across a nearly level plain that stretched east to the low coast of a shallow inland sea. That sea constantly changed its shape and size with variations in sea level or the elevation of the land. At times, it reached far into western Montana, leaving layers of marine sedimentary rocks as souvenirs of its incursion. At other times it retreated far to the east, and terrestrial deposits accumulated on the plain west of the inland coast.

Those terrestrial deposits include the Morrison formation, the nearest thing we have to a Jurassic zoo. It is the source of wonderful dinosaur skeletons displayed in museums all over the world. The Morrison formation is a variegated and colorful assortment of mudstone, sandstone, and limestone that surfaces here and there in wide areas of Colorado, Wyoming, Utah, and Montana. Morrison sandstones abound in parts of western Wyoming; gravelly sandstones dominate in westernmost Wyoming and southeastern Idaho. The streams that deposited them flowed east from highlands that stretched from central Idaho into Nevada. In some areas, the Morrison sandstones and shales contain large coal seams.

Rocks of late Jurassic time, about 160 million years ago.

—Compiled from Mallory, 1972; Imlay, 1980; and Peterson, 1986

Red shales laid down during Jurassic time. —W. T. Lee, U.S. Geological Survey

Many geologists interpret the Morrison formation as sediments laid down on river floodplains, but that is hard to believe. How could floodplains cover such an enormous area? On what kind of floodplain would limestone accumulate?

It seems more reasonable to think of the Morrison formation as sediment deposited in a desert. Similar sediments accumulate today in many of the undrained desert valleys of the west, such as Carson Sink in Nevada. Perhaps the dinosaurs of the Morrison formation roamed arid plains, dying in desperate groups around shrinking desert water holes where the next rains entombed their bones in mud.

The long drought seems to have continued as the new mountains rose in the west during early Cretaceous time. The Kootenai formation accumulated then in large areas of Montana and Wyo-

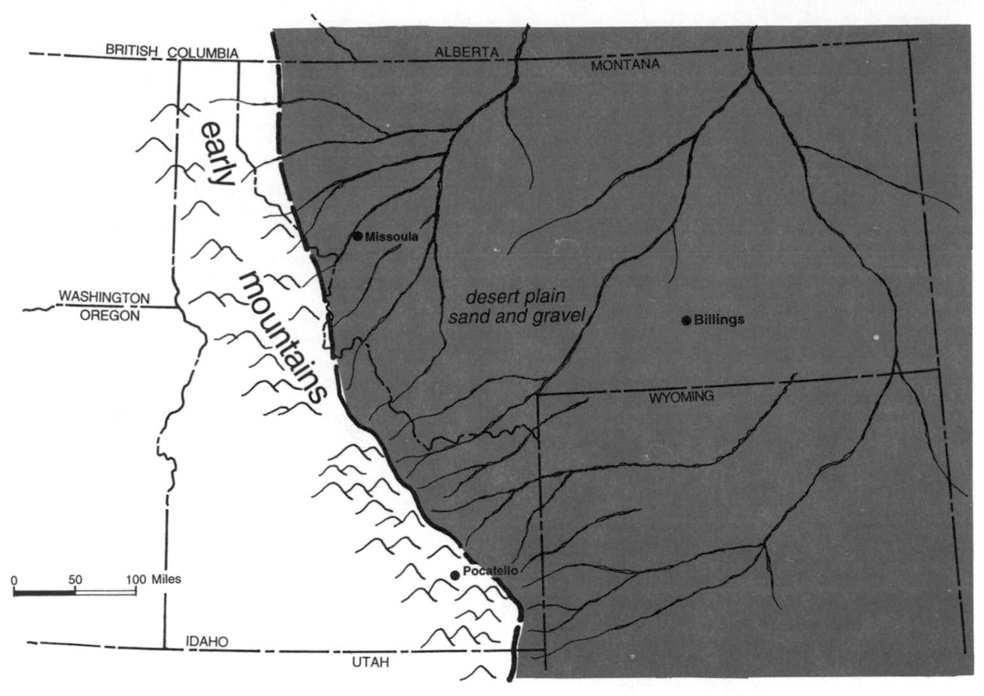

A desert plain sloped gently northeastward for hundreds of miles in early Cretaceous time, about 110 million years ago. —Modified from Eicher and others, 1969

ming, east of the rising early mountains. Like the Morrison formation, it also includes variegated assortments of colorful mudstone, sandstone, and limestone, and appears to be another desert deposit. The thin limestone beds, which spread over very large areas, contain shells of snails and clams that probably did not live in seawater, mixed with the bones and teeth of fish that probably did. It is hard to imagine their environment of deposition, perhaps large coastal lagoons nearly isolated from the inland sea to the east. The Kootenai formation contains fragments of dinosaur bones, but nothing like the magnificent skeletons that come from the Morrison formation.

Offshore and Underfoot

So those early mountains looked east across arid plains and shallow inland seas, west across an open ocean beyond the western edge of Idaho. Meanwhile, the continent was still moving west, overriding the floor of the Pacific Ocean, which was sliding through the trench offshore and sinking into the earth's hot interior.

The moving continent steadily advanced toward two large scraps of continental crust and several groups of islands out in the Pacific Ocean. They were destined to become the first major pieces of real estate added to the Pacific Northwest. And red-hot steam rising from the sinking slab was playing like a giant blowtorch on the underside of the continent. It would eventually demolish the early generation of mountains, transforming them into the northern Rocky Mountains.

First, the islands.

▸ Part 4 ◂

ADDING PIECES TO THE COLLAGE

175 to 65 Million Years Ago

When faced with a messy bunch of rocks, modern geologists generally begin by dividing them up into terranes, groups of rocks that appear to have a lot in common. Most terranes consist of rocks related in both origin and history; the boundaries of most are faults.

If a continental island, a microcontinent, is geologically complex, as most are, it will contain a number of different terranes. If it then joins a continent, the boundary between the former island and the continent will form another terrane boundary. How then do you find the outline of the former microcontinent? How do you tell the difference between terrane boundaries within the former island microcontinent and the boundary between it and the continent it joined?

In some cases, that can be very hard to do. In general, the youngest terrane boundaries are likely to mark the outline of the former microcontinent. But it is often hard to determine the age of a boundary. And in some cases, later movement can occur on an older boundary. This is not a business of certifiably cut-and-dried scientific fact.

So the story of the growth of the Pacific Northwest may have some of the outlines of microcontinents following the wrong terrane boundaries. Those details will change as geologists continue their work, but their ideas about the major boundaries have not changed radically over the last 10 or 15 years. And the main point of the story will certainly remain valid: The region is a geologic collage, a mosaic of terranes patched together into island microcontinents somewhere out in the Pacific Ocean and then patched onto North America. Or perhaps some of the terranes are slices detached from Asia or Australia before the Pacific Ocean existed.

Chapter 12

LANDING THE INTERMOUNTAIN TERRANES

175 Million Years Ago

Inevitably, your kite comes to your hand as you wind the string onto the reel. Just as inevitably, islands in the ocean come to a trench as it steadily swallows the intervening ocean floor, generally at a rate of a few inches every year. You can think of those islands coming in from the sea as though they were ships standing in to port. Or you can think of the continent moving west to join the islands in the ocean. Both views are correct.

The Intermountain terranes.

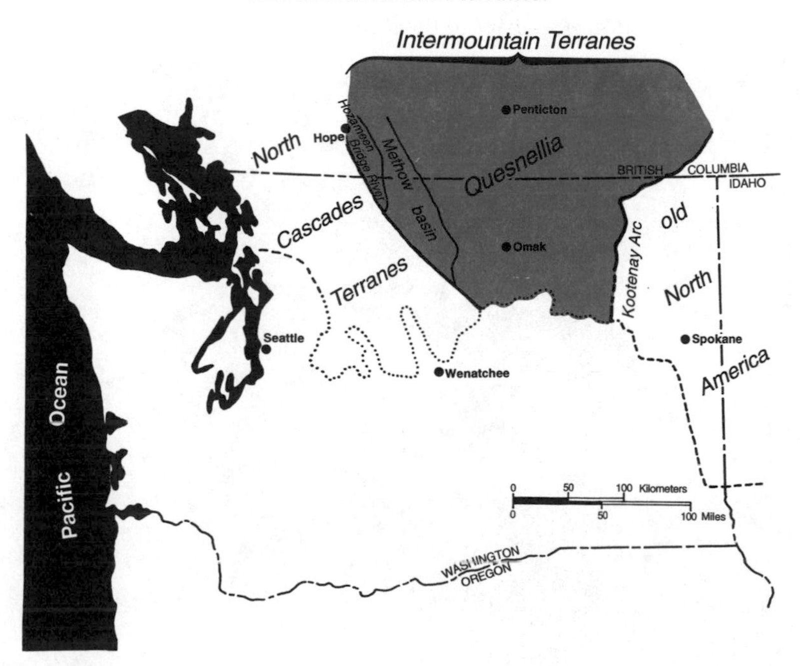

The Intermountain Terranes

We will call the first and largest single addition to the Pacific Northwest the Intermountain terranes. Most of that was a single very large terrane called Quesnellia, after a small town in British Columbia. It was once an island microcontinent out in the Pacific Ocean which then lodged against the Kootenay arc. It is now a landlocked former microcontinent that extends from north-central Washington into far-northern British Columbia.

The geologic map of the Intermountain terranes is about as complex a mess as any you will see for an area of its size. Its abstract patchwork of map colors shows a number of separate terranes that apparently assembled into a microcontinent before they became part of North America. The region contains areas of schist and gneiss that appear to be fairly standard basement rocks plus an assortment of sedimentary and volcanic rocks. Some of those rocks formed while the microcontinent was still out in the Pacific Ocean, others after it joined North America.

The eastern boundary of the Intermountain terranes is the contact with the Kootenay arc. That much is easy; the other boundaries are still debatable. The western boundary is in western British Columbia and north-central Washington; the southern boundary is somewhere near the northern edge of the Columbia Plateau.

Sheared gneisses in the Intermountain terranes.

The largest of the Intermountain terranes, Quesnellia, makes up all but the western fringe of the former microcontinent. It includes the Okanogan Highlands, an area east of the Okanogan Valley dominated by granite and metamorphic rocks. The Okanogan Highlands are now a huge arch, of which the crest dropped like a keystone to make a long trench that filled with volcanic rocks during Eocene time.

The original rocks of Quesnellia were sandstones and muddy sandstones laid down on the ocean floor between Devonian and Triassic times, from about 380 to about 180 million years ago. If we can judge from the compositions of some of the sand grains, the sediments were eroded from a chain of volcanoes that probably erupted along a line parallel to a trench. Those rocks are now metamorphosed almost beyond recognition, recrystallized into schists and gneisses. That happened deep within the continental crust, probably as the terrane was squashed against the Kootenay arc during middle Jurassic time, about 185 million years ago.

Much later, about 65 million years ago, much of the terrane was engulfed in granite magma. That was about when enormous volumes of molten granite formed large batholiths in the Coast Mountains complex of western British Columbia, in central Idaho, and in the Sierra Nevada of California. All the granite magmas melted above the slab of oceanic lithosphere, which was descending through the trench, and under the western margin of North America.

The Trench Moves West

The first oceanic trench that formed when North America began moving west must have followed a line very close to the old continental margin. It swallowed the ocean floor that separated the old island microcontinent of the Intermountain terranes from North America, probably at the usual rate of a couple of inches every year. Before Jurassic time ended, all that ocean floor had gone down the trench, and the Intermountain terranes had joined North America.

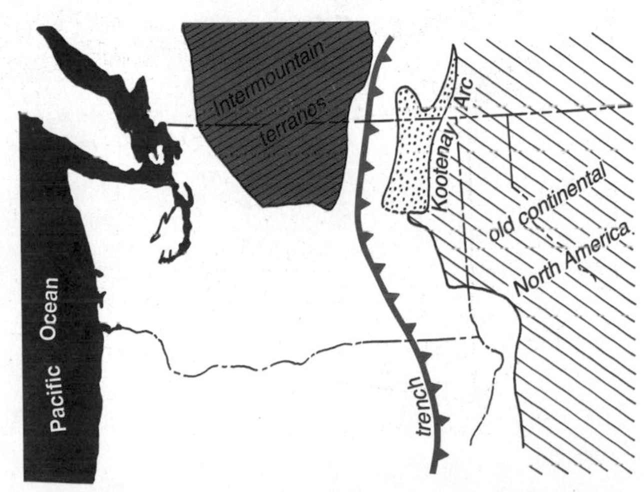

The trench before the Intermountain terranes landed.

North America began to move west at the beginning of Jurassic time, about 205 million years ago. The Intermountain terranes collided with North America just 30 million years later. If the trench swallowed the ocean floor at the rate of two inches per year, that works out to 31 miles every million years, or 930 miles in 30 million years. That is probably about the width of the ocean that separated the Intermountain terranes microcontinent from North America when Jurassic time began.

Landing the Intermountain terranes microcontinent.

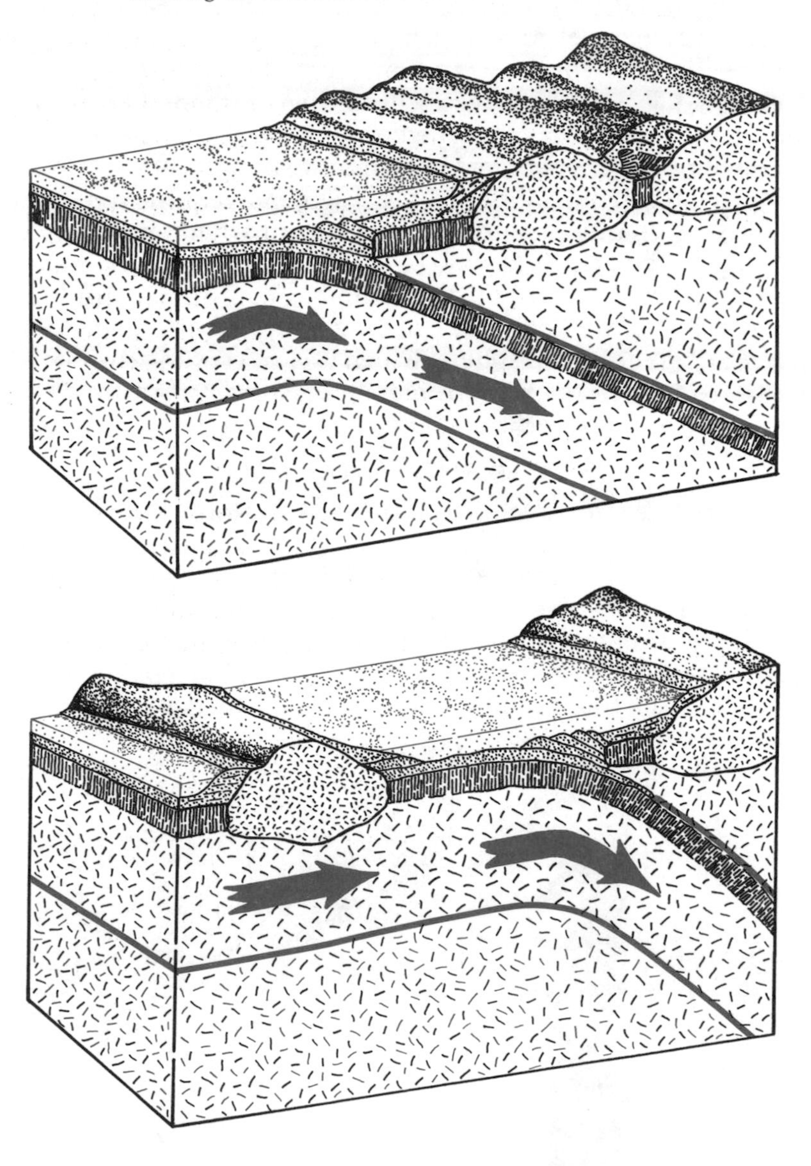

An oceanic trench is simply a place where the oceanic lithosphere bends down as it starts to sink into the earth's mantle. If a large mass of continental crust, such as a microcontinent, arrives at a trench, the light continental crust refuses to sink, and that kills the trench. It seems that the Intermountain terranes microcontinent killed the trench when it lodged against the Kootenay arc during Jurassic time.

But North America was still moving west, and the floor of the Pacific Ocean still had to make way for it, so a new trench formed off the western margin of the landed microcontinent. Or, more likely, a trench already existed there. If so, that would explain the ages of some of the igneous rocks within the Intermountain terranes, and also why the microcontinent apparently landed gently instead of jamming hard into the Kootenay arc.

The part of the trench south of the Intermountain terranes microcontinent must have remained in its original location because more wandering islands were yet to land against the old continental margin. So the trench was now in two segments: one west of the newly docked microcontinent and the other still hard against the old western edge of North America. Something had to connect them.

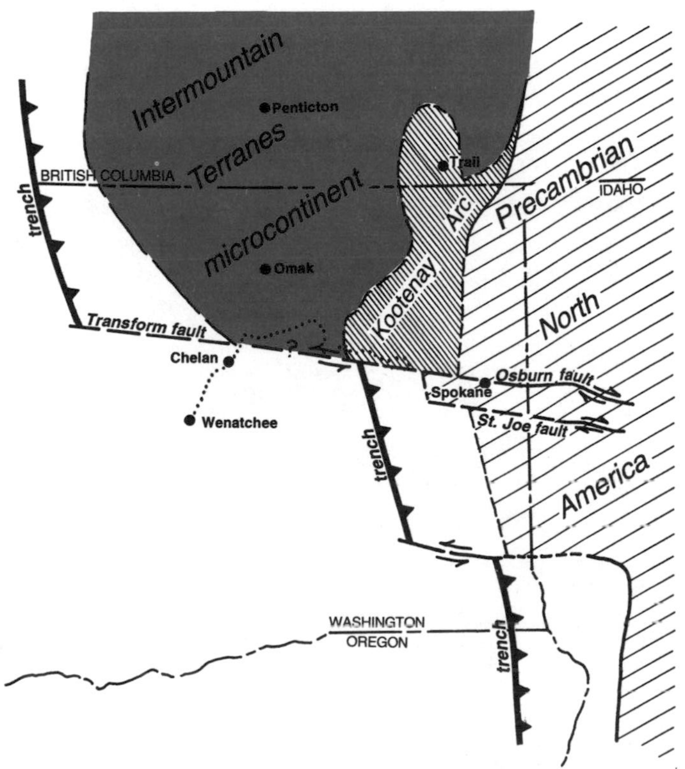

The hypothetical transform fault that may have joined the two segments of trench after the Intermountain terranes microcontinent landed.

A Hypothetical Transform, and Its Even More Hypothetical Continuation

A trench cannot simply end. Something has to define the boundary of the oceanic lithosphere sinking into it. Commonly that something is a sliding plate boundary, a transform fault. A sliding transform plate boundary that trended west may have connected the two segments of the trench after the Intermountain terranes docked onto western North America.

The transform boundary may have continued southeast across the northern Rocky Mountains along the line of the Osburn fault zone in Idaho and its continuation in the Lewis and Clark line of western Montana. That is a fairly racy suggestion because most transform faults on the ocean floor do not continue onto a continent.

The swarm of faults that make up the Osburn fault zone pass through the Coeur d'Alene mining district and continue into western Montana as the Lewis and Clark line. Interstate 90 follows it between Spokane and Missoula. Geologists have debated almost every aspect of that fault zone for years: which way it moved, how far, and when. It remains a puzzling problem.

Chapter 13

THE SEVEN DEVILS AND THE BLUE AND WALLOWA MOUNTAINS

100 to 90 Million Years Ago

Near the border between Oregon and Idaho, the Snake and Salmon rivers eroded their deep canyons through a thick section of basalt lava flows, opening a window into deeper rocks. The lower parts of those canyons cut through tightly folded sedimentary and volcanic rocks that were oceanic islands in Permian and Triassic times, then became part of North America during Cretaceous time. Those are the rocks of the Seven Devils and Riggins complexes. West of them, beyond a broad expanse of concealing lava flows, rise the Blue and Wallowa Mountains, which also contain the remains of old oceanic islands. They are probably continuous with the Seven Devils and Riggins rocks. Addition of all that rock to North America forced the southern segment of the trench to shift west, shortening the sliding transform plate boundary that linked it to the northern segment.

The Seven Devils Islands

The strongly deformed rocks of the Seven Devils and Riggins complexes lie along the border between central Idaho and Oregon. The best places to see these tortured rocks are along U.S. 95 between Riggins and Whitebird, Idaho, in the Salmon River Canyon, and in Hells Canyon, just a few miles to the west. They do not resemble the rocks in the Kootenay arc, and nothing resembling the Kootenay arc separates them from the old western margin of the continent.

The Seven Devils complex consists mostly of volcanic and sedimentary rocks that formed during Permian and Triassic times, around 250 million years ago. The Riggins complex is harder to read because most of its rocks are metamorphosed into gneiss and schist. Most geologists now think that the Riggins complex rocks

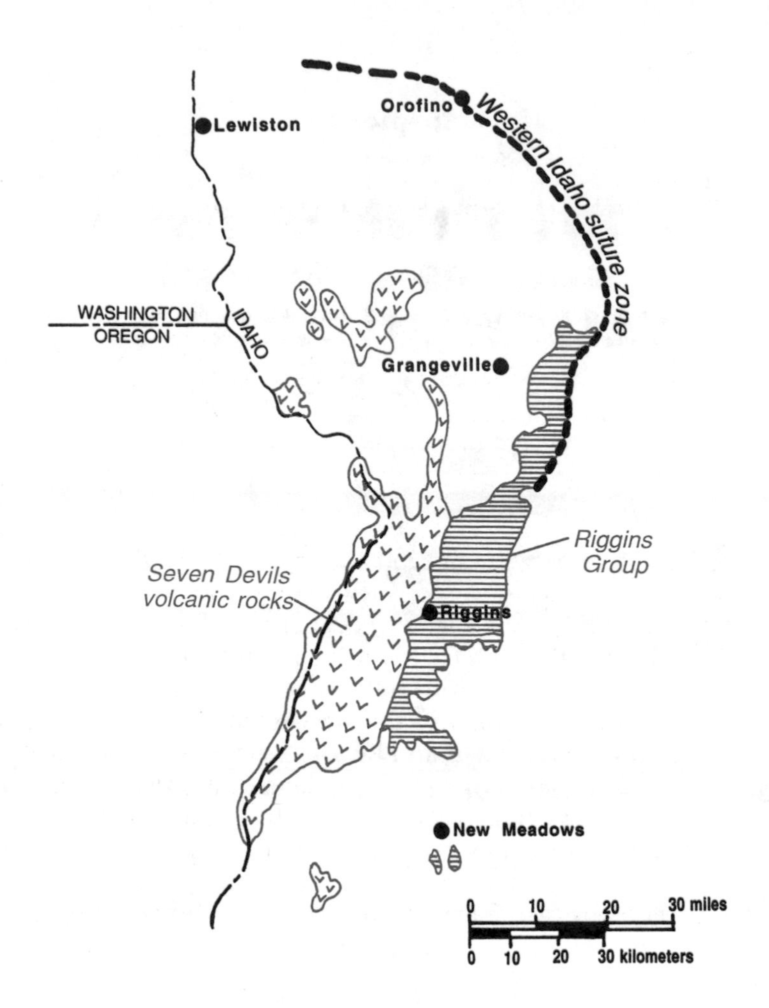

The Seven Devils and Riggins rocks.

are probably more of the same stuff, only more recrystallized and at much greater depths.

The better-preserved Seven Devils rocks are clearly the squashed remains of a group of volcanic islands. The volcanic rocks resemble those that typically erupt in oceanic volcanic chains, such as the Aleutians. Most of the sedimentary rocks are exactly what you would expect to form as such islands eroded and the debris collected on their lower slopes and around their coasts. Some of the sedimentary rocks are massive limestones that many geologists interpret as the remains of reefs that fringed the islands.

Fossils in the reef limestones vary greatly from place to place, as they would if they were the remains of animals living around a group of islands. If that mass had been one big island, its animals probably would not have varied so much within so small an area. If we can judge from the preferences of their modern descendants, the fossils are the remains of animals that lived in very warm water. And they more closely resemble the remains of animals of the same age in eastern Asia than those in North America. Wherever they came from, the Seven Devils rocks are a long way from home.

Studies of the orientation of the fossil magnetic fields recorded in those rocks suggest that they formed somewhere within about 20 degrees of latitude, north or south. That would place their original home somewhere south of Guatemala and north of Peru, which may help explain why the fossils appear to be the remains of animals that lived in very warm water.

The boundary between the islands and North America is known exactly, and masses of igneous rocks that were intruded along it reveal its age. Igneous intrusions emplaced along the boundary are between 110 and 90 million years old. Evidently, the islands were in contact with North America by 110 million years ago. Intrusive rocks more than about 100 million years old are sheared into gneiss, which must mean that movement was still continuing along the seam after they formed. Igneous rocks younger than about 80 million years old show no deformation, which must mean that the weld was firmly solid before they arrived in late Cretaceous time.

The Blue and Wallowa Mountains

The darkly forested slopes of the Blue and Wallowa Mountains rise like islands above the level basalt plains of northeastern Oregon. Parts of those mountains were indeed a group of islands far out in the Pacific Ocean long before they become part of North America. Like the Seven Devils and Riggins complexes, they lodged against the western edge of North America as the ocean floor that once separated them from the continent sank through the trench.

Bedrock in the Blue and Wallowa Mountains is a complicated and thoroughly confusing mess of different rock types. The usual way to cope with such a mess is to lump the rocks into terranes that appear to share broad similarities of origin and history. Even though different geologists may see different terranes, they generally tend to arrive at a similar view of the big picture. Once that emerges, the details begin to make sense.

Many geologists include most of the Blue and Wallowa Mountains within three large terranes: the Wallowa terrane in the north,

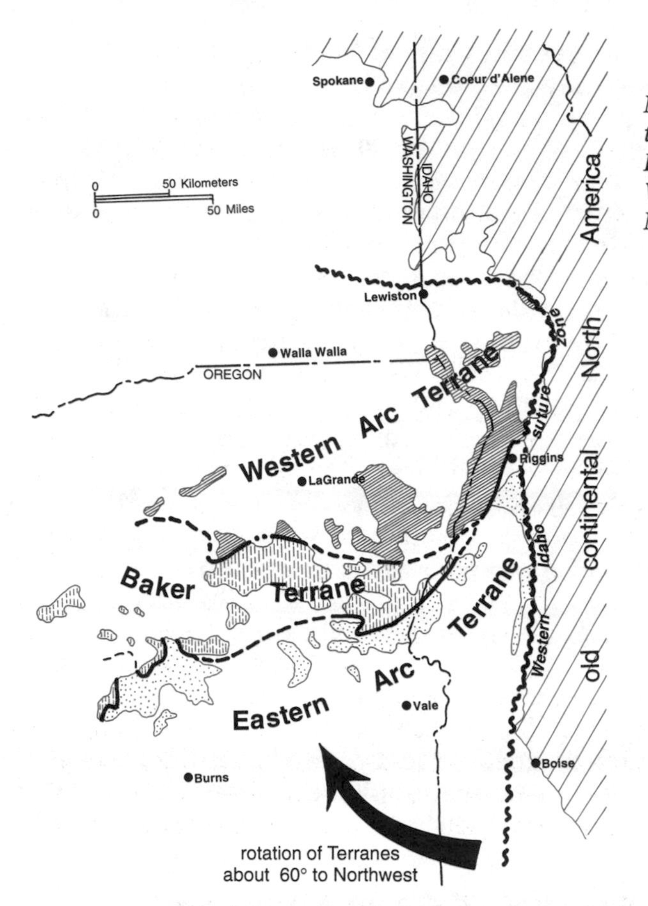

Major rock terranes in the Blue and Wallowa Mountains.

the Baker terrane south of it, and the Eastern Arc terrane south of that. All contain wild assortments of rocks that share a generally oceanic origin.

The Wallowa terrane makes a bold patch on the map, most of which is imaginary. It consists basically of the rocks in the Seven Devils and Riggins complexes, some rather similar rocks in the northern Blue and Wallowa Mountains, and rocks that probably exist beneath the intervening cover of basalt lava flows, but which may not. The boundary with the old edge of North America is exposed in western Idaho, and the boundary with the Baker terrane is in the northern Blue and Wallowa Mountains. The rest of the outline is just dashed in on the map. A large mass of granite, the Bald Mountain batholith, intruded the boundary with the Baker terrane in late Cretaceous time, so those two must have joined sometime before then.

The Baker terrane includes great slices of oceanic crust and muddy sandstones of the kind that accumulate on the deep ocean floor. Most of those rocks are about the same age as those in the Seven Devils complex—Permian and Triassic—having formed some millions of years before and after about 250 million years ago. They were probably the ocean floor bordering those volcanic islands before they were all jammed together at an oceanic trench.

The Eastern Arc terrane is southeast of the Baker terrane and generally a little younger. It contains an assortment of late Triassic and Jurassic volcanic rocks, along with a variety of sedimentary rocks composed of debris eroded from them. Some of the volcanic rocks are andesites, which almost certainly erupted from a chain of volcanic islands. In the western part of the Eastern Arc terrane, oceanic sedimentary rocks like those that normally accumulate in very deep water accompany the volcanic rocks. The youngest rocks in the terrane are limestones deposited during late Jurassic time, presumably in the shallow waters around the volcanic islands.

Oceanic muddy sandstones in the Baker terrane.

Another Segment of the Trench Shifts West

The trench choked on the great mass of the Seven Devils and Riggins complexes and the Blue and Wallowa Mountains; it could not swallow all those light rocks, which would not sink. So the segment of the trench that had landed those rocks shifted to their western margin, where it continued to swallow the ocean floor. That further shortened the sliding transform plate boundary that connected the two segments of the trench.

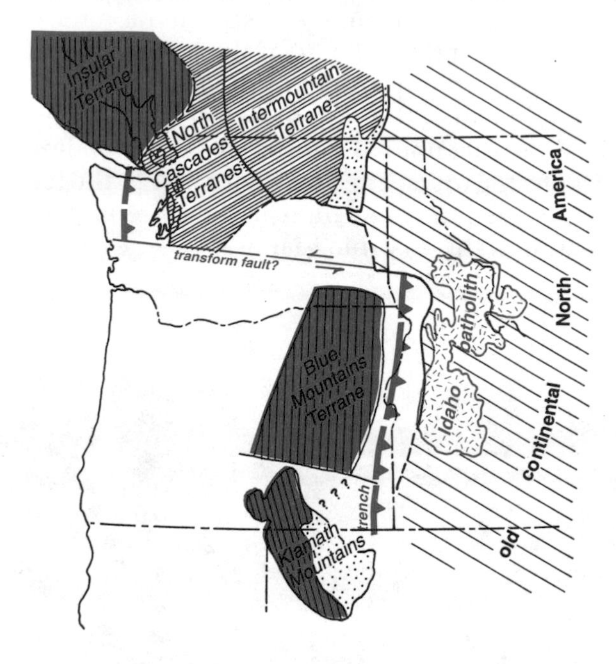

The probable position of the trench just before the Blue Mountains microcontinent docked to become part of the Pacific Northwest.

Evidence that the Seven Devils rocks joined North America about 100 million years ago is probably firm. But it does not tell us when the trench finally shifted to its new position. It may well have worked its way slowly westward, as it added one big piece of the Blue Mountains after another to the collage. The trench probably did not reach its next stable position until Eocene time. That was when it finally spawned the chain of volcanoes in the Ochoco Mountains of central Oregon.

Chapter 14

THE OLD CONTINENTAL MARGIN

100 Million Years Ago

The old western margin of North America still survives, in the rocks if not in the landscape. In most areas, geologists know where it is in a general way, but not precisely. Tracing it tends to mean separating the effects of the Antler mountain-building event from those of the much later Mesozoic events, and that is hard to do. But some parts of that old boundary are exactly known and are plotted on the map.

The Western Idaho Mylonite

About 80 miles of the trace of the old continental margin is clearly preserved in western Idaho. It appears most obviously in a change in the compositions of the rocks. Granitic rocks east of the western Idaho mylonite are genuine granite: very pale gray rocks that contain quite a bit of quartz and relatively few dark minerals. Rocks

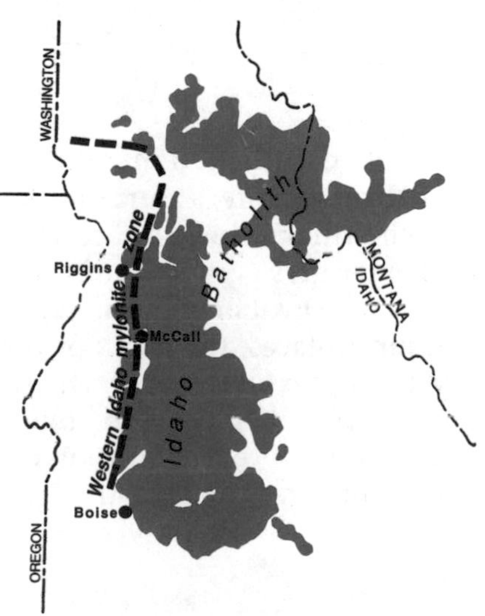

The western Idaho mylonite traces the old continental margin through part of western Idaho.

west of the western Idaho mylonite that look granitic are mostly diorite: medium to dark gray rocks that contain very little quartz, much plagioclase feldspar, and an abundance of dark minerals. The composition of the granites is typical of continental crust, that of the diorites more like that of oceanic rocks. Watch for that change in the roadcuts along U.S. 12 east of Orofino, Idaho.

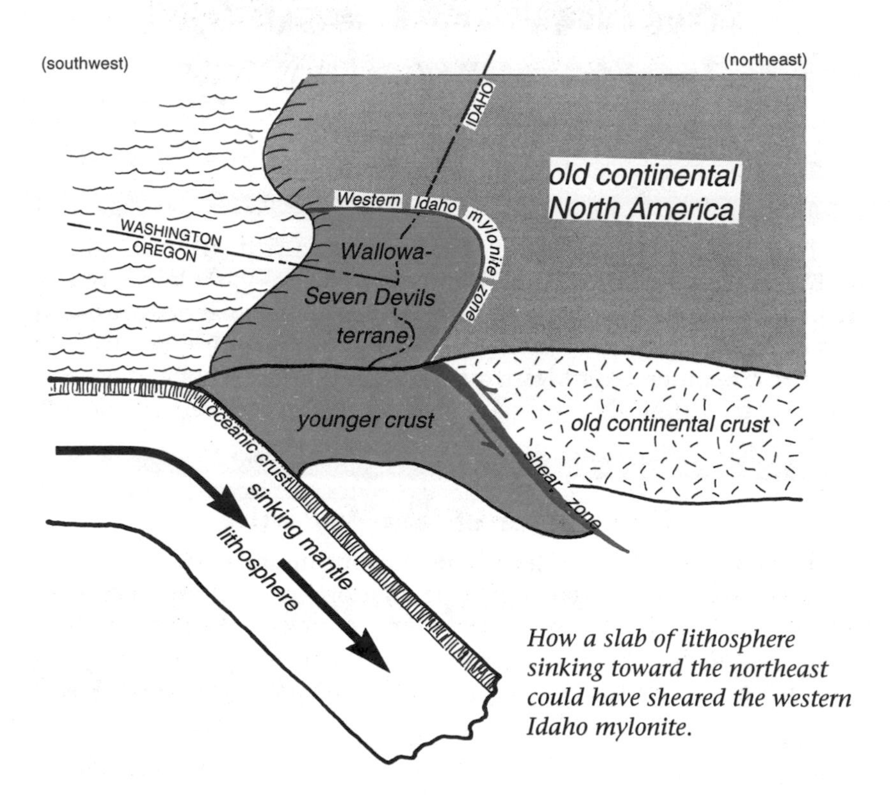

How a slab of lithosphere sinking toward the northeast could have sheared the western Idaho mylonite.

The old continental margin appears more precisely in the western Idaho mylonite, a zone of intensely sheared rock approximately one mile thick. The most convenient place to see those rocks in good exposures is near Orofino, Idaho, especially along the short road below Dworshak Dam.

In most places, the rocks of the western Idaho mylonite have a very strong textural grain, which points down to the northeast at an angle of 55 degrees, the angle of the shearing motion that created the mylonite. If you look closely at the surfaces of the slabs, perhaps turning them in the light to get the shadows to fall just

Sheared diorite gneiss in the western Idaho mylonite near Dworshak Dam, north of Orofino, Idaho.

right, you will see sets of parallel lines faintly grooved on them. Those lines also point northeast.

That faint northeast lineation almost certainly records the direction of the encounter between North America and the floor of the Pacific Ocean: The ocean floor was sinking northeastward beneath North America. And that presents a problem. We drew the transform that connects the two segments of the trench along a northwest trend for two reasons. The southern margin of the Intermountain terranes trends northwest, and it seems to align with the trend of the Osburn fault zone and Lewis and Clark line. A transform is a sliding plate boundary, so it must trend in exactly the direction of plate movement. A transform that trends northwest is incompatible with the lineations that trend northeast.

We can suggest no simple solution to this dilemma of our own making. Perhaps the old continental margin has been rotated out of the alignment it had back in middle Cretaceous time. If so, evidence of that movement will eventually emerge. Or perhaps our idea of the transform boundary is wrong.

Is a Piece of Idaho Missing?

Why is the Kootenay arc missing between the Seven Devils complex and the old continental margin? Could that peculiar jog in the trace of the old continental margin through western Idaho be more than a casual irregularity? Some geologists think it outlines the scar where a large piece of the Kootenay arc was torn off sometime before the islands of the Seven Devils and Riggins complexes came ashore.

If a piece of western Idaho is indeed missing, it must still exist, somewhere. Some geologists argue that it is in the Cassiar Mountains of northern British Columbia. That range consists of rocks that would fit comfortably in the western edge of Idaho. If the Cassiar Mountains are indeed a relocated chunk of Idaho, when did it move so far north?

Certainly not since the islands of the Seven Devils and Riggins complexes landed. The faintly drawn lines in the mylonite along the old continental margin seem to show that the ocean floor was heading northeast as it sank through the trench 90 or 100 million years ago. Any straight northward movement must have happened before then, perhaps in late Jurassic and early Cretaceous time.

Many geologists think that Idaho lost a piece where the old continental margin jogs east.

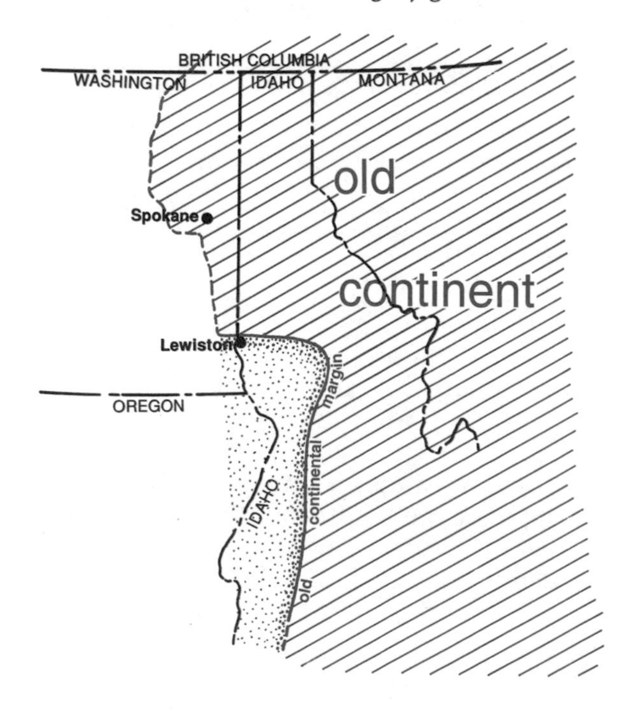

Chapter 15

THE NORTH CASCADES COLLAGE

100 to 90 Million Years Ago

After the microcontinent of the Intermountain terranes docked in middle Jurassic time, about 175 million years ago, the trench along its western coast continued to reel in the Pacific Ocean floor and all that was riding on it. Or, if you prefer, the ocean floor was sinking beneath North America as it moved west.

Meanwhile, the North Cascades terranes were still offshore, probably as a group of islands. The North Cascades are now a collage of at least six separate terranes, each with its own distinctive rocks and geologic history. They all jammed into the west coast of North America at about the same time, with about the effect you might

The collage of North Cascades terranes.

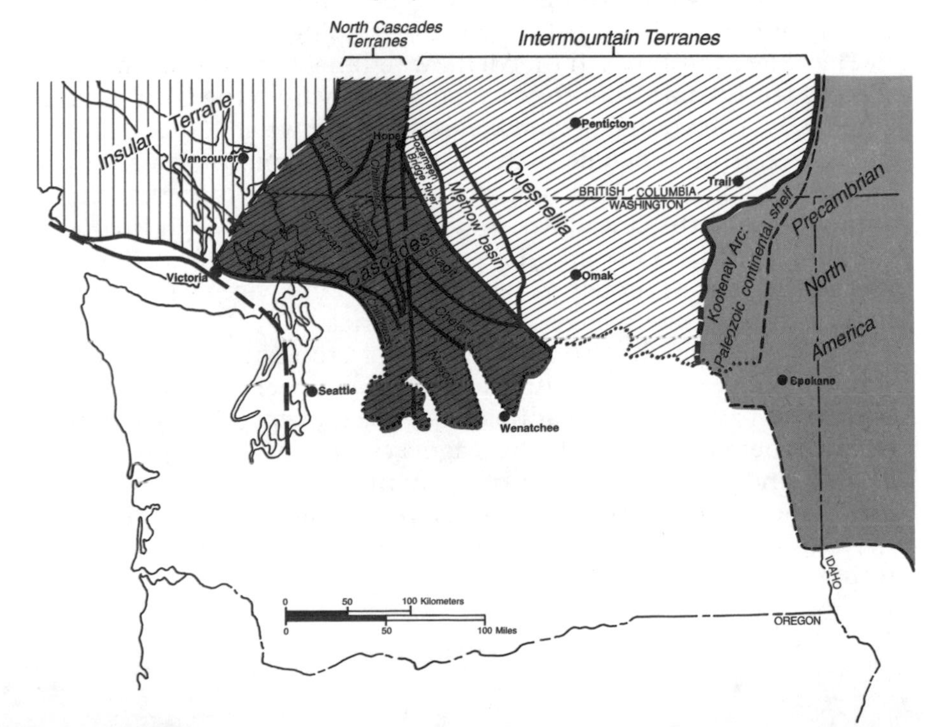

expect if Indonesia were to jam into southern China. That happened during middle Cretaceous time, about 100 to 90 million years ago. The collision squashed the North Cascades terranes and the Methow basin, the last expanse of ocean floor between the approaching islands and North America. As before, the collision killed the trench, and a new trench formed farther west.

After they jammed against the western edge of the continent, the former islands of the North Cascades terranes were a patchwork of narrow belts that trended generally northwest, each about 20 miles wide. Then a series of more or less horizontal faults cut them into thin slices that moved west and stacked up on each other like a deck of cards. Millions of years later, in Eocene time, a set of vertical faults again sliced the North Cascades into long strips that trend north. The rocks west of each of those faults moved north.

Finding the Eastern Boundary

The Methow terrane along the western edge of the Intermountain terranes contains part of the evidence that locates the boundary between the North Cascades and the Intermountain terranes microcontinent. The Hozameen and Bridge River terranes contain the rest.

The Methow terrane includes ocean floor basalts that erupted during Triassic time, and sandstones and shales deposited on it during Jurassic and Cretaceous time. The sandstones and shales have the general look of material that accumulates near a coast. They probably were laid down off the west coast of the Intermountain terranes microcontinent before and after it joined North America.

The older sandstones in the Methow terrane consist of sediment that evidently came from the east, presumably from the Okanogan Highlands. The youngest sandstones, those between 120 and 105 million years old, include material that came from both eastern and western sources. That influx of western sand tells us that the approaching North Cascades islands were then close enough to contribute sediment to the pile. Then they squashed the rocks of the Methow terrane as though between the jaws of a vice as they collided with the western edge of North America.

The Methow terrane also includes parts of a thoroughly scrambled slice through the oceanic crust and into the upper mantle. Thick sections of basalt flows are at the top of the sequence. The gabbros beneath them were probably the magma chambers that erupted the basalts. Under those are scraps of peridotite and its altered equivalent, serpentine—heavy, dark greenish rocks from the upper part of the mantle. All were shuffled into the trench in great slices, a confusing mess.

The Hozameen and Bridge River terranes, at the western edge of the Methow terrane, contain kinds of sedimentary rocks that accumulate only on remote reaches of the deep ocean floor. Their fossils show that they were laid down between Permian and middle Jurassic times. In places they include basalt, scraps of peridotite from the upper mantle, and parts of a chain of volcanoes. They rode into the trench on the moving oceanic crust, and now you see them crushed against the coastal sandstones of the Methow terrane.

It seems fantastic to stand in the Methow Valley now, near Winthrop or Twisp, look east to the Okanogan Highlands and west to the North Cascades, and realize that an ocean once separated them. But the rocks in the Hozameen and Bridge River terranes leave little doubt. How else could deep oceanic rocks arrive in such a place? They were stuffed into the nearby oceanic trench in middle Jurassic time, about 180 million years ago, just before the Intermountain terrane collided with North America.

The Shuksan Thrust Fault

The Shuksan thrust fault is the dominant member of the stack of nearly horizontal thrust faults in the San Juan Islands and the western part of the North Cascades. Rocks above it moved west or northwest over younger rocks as a nearly horizontal sheet. It lies beneath virtually all of the western 40 miles of the North Cascades. Some geologists argue that it extends as much as 40 or 45 miles farther east under the Skagit and Methow terranes, but that is largely conjecture. The Shuksan fault is best exposed along Highway 20 between Sedro Woolley and Marblemount.

When did the Shuksan fault move? It cuts rocks as young as 100 million years in British Columbia, so it must be younger than that.

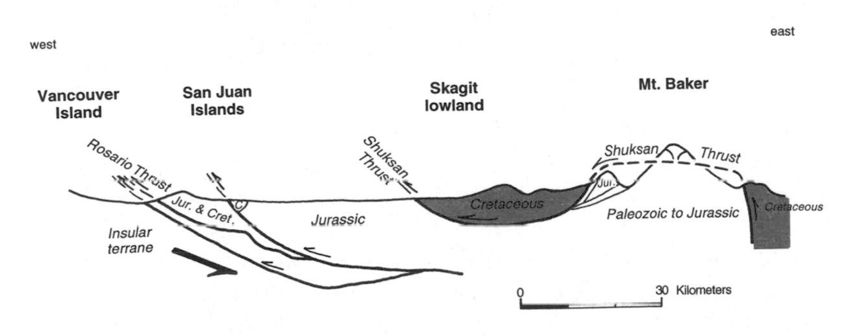

Section from Vancouver Island across the North Cascades with the movement on later faults removed. —Modified from Brandon, 1989; McGroder, 1991

Age dates on metamorphic rocks associated with the fault place its time of movement at about 100 to 90 million years ago. Those rocks had reached the surface by 90 to 70 million years ago, when sedimentary formations of the Nanaimo group were laid down on top of them on eastern Vancouver Island. And the Shuksan fault does not break the Chuckanut formation south of Bellingham, which was laid down across it during Eocene time, about 50 million years ago.

The San Juan Islands

The San Juan and North Cascades system of thrust faults is another stack of nearly horizontal slices beneath the level of the Shuksan thrust fault. Its western edge is in the San Juan Islands at the northern end of Puget Sound. Each slice moved west over the rocks beneath it along a nearly horizontal thrust fault.

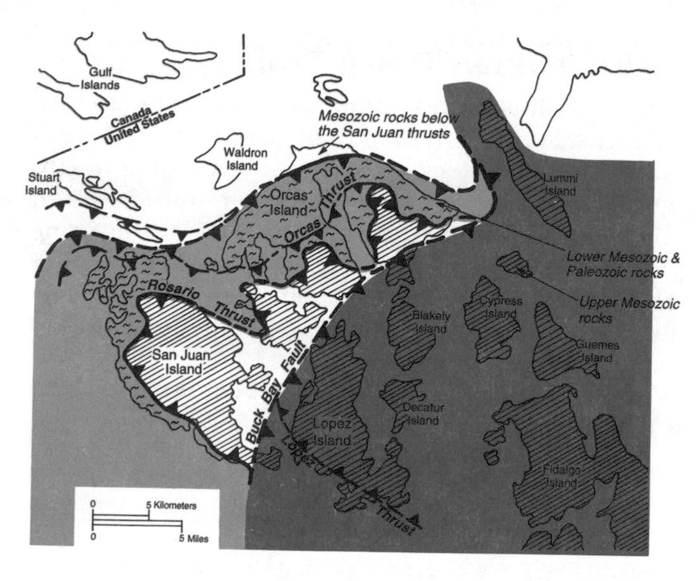

The thrust faults in the San Juan Islands moved generally northwest. —Modified from Brandon and others, 1988

Igneous rocks on the San Juan Islands include pillow basalts that erupted on the ocean floor, gabbro, and quartz diorite, along with andesites like those that erupt from the Cascade volcanoes. Sedimentary rocks include muddy sandstones laid down on the ocean floor during late Paleozoic and Mesozoic time, and minor amounts of chert and limestone laid down on the deep ocean floor as siliceous and calcareous oozes. The resemblance to the Franciscan rocks of the California Coast Ranges is so startling that some geologists have suggested that they came from there.

Pillow basalt on the San Juan Islands.

Geologists who study the stacked thrust slices conclude that if they could put everything back where it started, the North Cascades would be about 200 miles wider from east to west. Of course, the movement that stacked those slices was relative. You can imagine, as we just have, the terranes above moving westward over those below; or you can equally well imagine those below moving east beneath those above.

North Cascades Terranes

Long after the terranes were shuffled into a stack of nearly horizontal slices, a new swarm of faults that trend north cut the former islands of the North Cascades into long slices. The Straight Creek fault is the largest of the swarm. It moved far enough to separate the North Cascades into quite different halves.

The Skagit, Chelan, and Nason terranes are east of the Straight Creek fault. The easternmost is the Skagit gneiss, the crystalline core of the range. It consists mostly of metamorphic rocks, light and dark gneisses, and dark schists made of much smaller mineral grains. The original rocks recrystallized between about 90 and 60 million

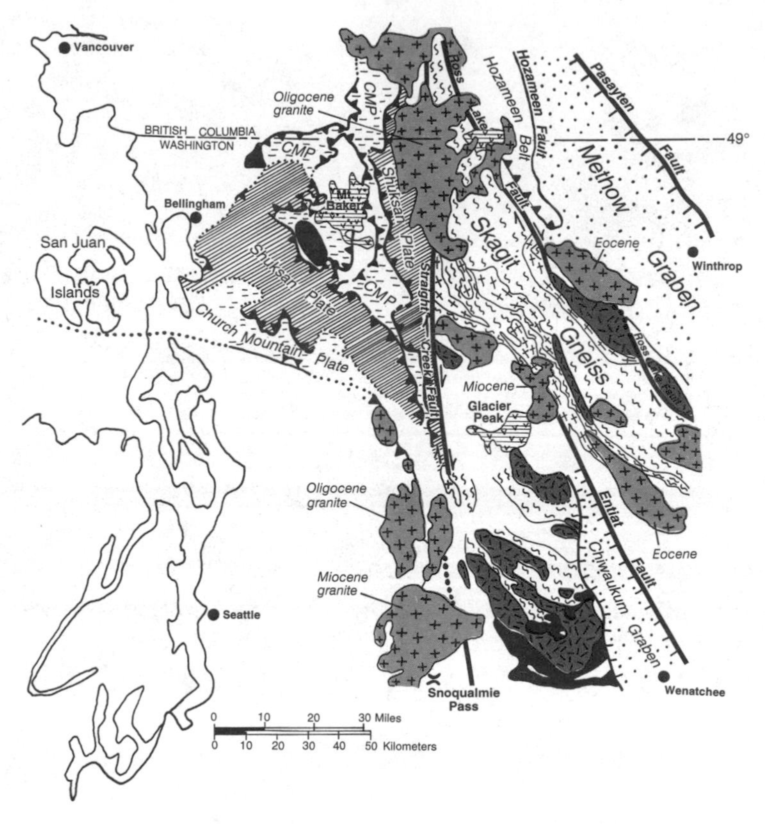

Complex metamorphic rocks and granite east of the Straight Creek fault are the core of the North Cascades. Eocene and much younger granites that cut the fault show that it has not moved since Eocene time. —Modified from Misch, 1988; Tabor and others, 1989

years ago; they appear to have been volcanic ash, muddy sandstones, and serpentinite, all dumped into an oceanic trench. Some geologists have concluded that at least 75 miles of the Skagit gneiss must be stuffed under the edge of the Methow terrane to the east.

The Chelan and Nason terranes are southwest of the Skagit gneiss. They probably began their careers on the ocean floor between Devonian and late Jurassic times, about 380 to 150 million years ago. The original rocks were a volcanic chain, and sediments eroded from it. They were metamorphosed during late Cretaceous time into a great variety of gray and dark gray schists and gneisses. The terrane also includes an igneous rock called quartz diorite that looks a bit like granite but is much too dark to be granite. Numerous age dates show that the rocks rose from the depths and cooled during late

Skagit gneiss.
—M. H. Staatz, U.S. Geological Survey

Veins of white granite in dark diorite gneiss in the Chelan complex near Chelan.

Cretaceous time. The same dates indirectly suggest that these terranes may have joined at that time. Watch for these dark rocks with numerous veins of white pegmatite along U.S. 2 across Stevens Pass and U.S. 97 between Wenatchee and Chelan. Those with prominent light and dark bands probably began as sedimentary rocks.

The Shuksan, Chilliwack, and Harrison Lake terranes are west of the Straight Creek fault. Their rocks are only slightly metamorphosed, not recrystallized enough to obliterate their original appearance. Some still contain recognizable fossils, which show that a few of the sedimentary rocks were laid down on the ocean floor in late Paleozoic time, but most were deposited during the early and middle parts of Mesozoic time.

The Shuksan terrane consists of schists and scraps of serpentinite that date from late Triassic and early Jurassic times. They are dark rocks, finely crystalline, probably old oceanic crust and muddy sand-

The Shuksan thrust plate.

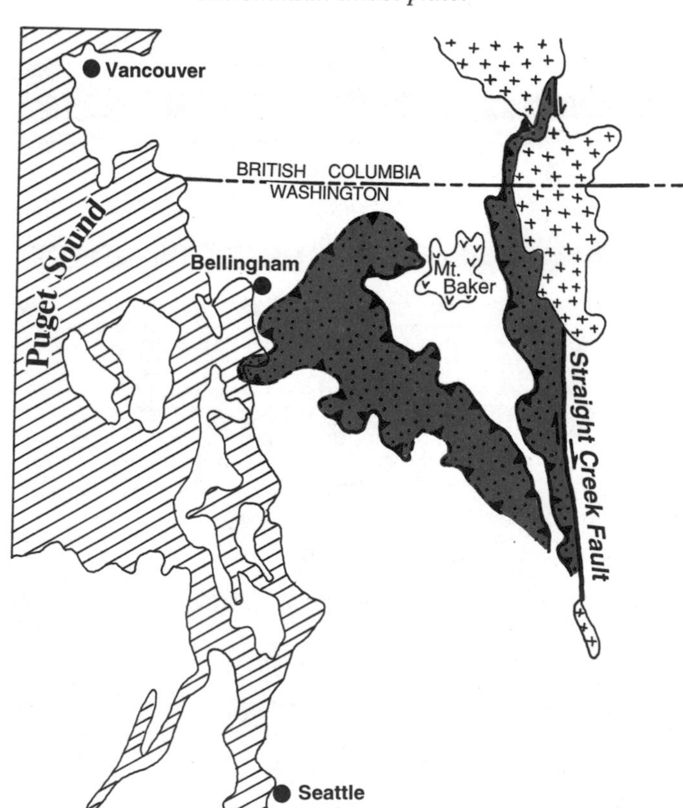

stones that were stuffed into the trench during Jurassic time. Many of the sand grains are fragments of volcanic rocks that must have eroded from a nearby chain of volcanoes.

The Shuksan rocks contain the distinctive metamorphic minerals lawsonite and aragonite. Laboratory experiments show that those minerals form when feldspar and aragonite recrystallize under pressures that correspond to depths of at least 12 miles. Evidently, the rocks of the Shuksan terrane were dragged that far down before they came back to the surface. Age dates on the metamorphic minerals show that most of the action happened during Cretaceous time.

The Chilliwack terrane includes sedimentary and volcanic rocks that date from Devonian to early Jurassic times. The volcanic rocks are the kinds that erupt in a volcanic chain like the modern High Cascades.

The much smaller Harrison Lake terrane, near Mount Baker, contains quite an assortment of rocks, including basalt, chert, mud-

Folded schist of the Shuksan metamorphic rocks southeast of Bellingham.

stone, and shale. Fossils show that all were laid down in Triassic and Jurassic times. The chert is distinctively the kind that forms from siliceous ooze laid down on remote regions of the deep ocean floor too far from land to receive much terrestrial sediment.

All those oceanic terranes west of the Straight Creek fault are now tightly folded and much broken along faults, no doubt the result of their having been rudely mashed into a trench. Rocks of the Shuksan terrane rode west along a fault over the rocks of the Chilliwack terrane, which rode west over those of the Harrison Lake terrane. Erosion in the high North Cascades has bitten down through the upper thrust slices, opening windows that expose those below.

Chapter 16

THE INSULAR TERRANE ARRIVES

100 to 90 Million Years Ago

Meanwhile, the segment of trench north of the transform link in Washington was reeling in yet another island microcontinent. It was once called Wrangellia, but geologists now call it the Insular terrane. It includes Vancouver Island and the west coast of British Columbia.

The Insular terrane jammed into and under the western edge of the North Cascades between 100 and 90 million years ago, about when the Seven Devils islands were coming ashore in western Idaho. The Insular terrane appears to have landed while the stacked thrust

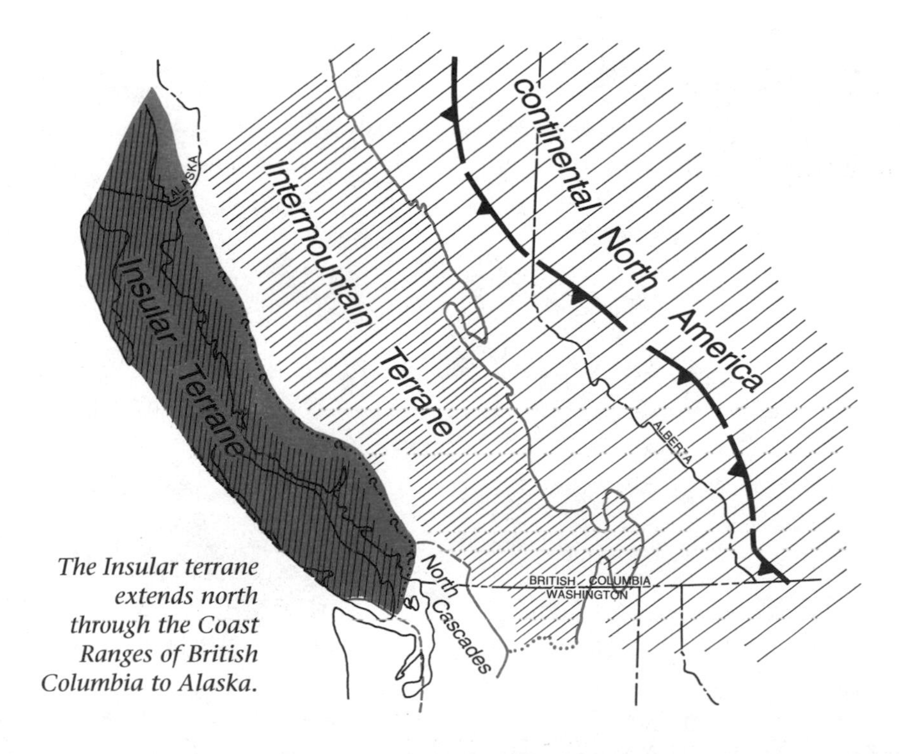

The Insular terrane extends north through the Coast Ranges of British Columbia to Alaska.

sheets in the North Cascades were moving, about 100 to 90 million years ago. At about the same time, between 110 and 60 million years ago, the Skagit gneiss was recrystallizing deep in the core of the North Cascades. And big masses of granite magma were invading many parts of the North Cascades 96 to 90 million years ago. Some geologists associate all of those events with the landing of the Insular terrane microcontinent. If so, it happened so soon after the arrival of the North Cascades islands that the two events were almost simultaneous.

Another Collage of Terranes

Its bold and irregular splashes of color make the geologic map of the Insular terrane look a bit like an old-time patchwork quilt. Each splash represents a body of rock, a product of a long and complex geologic history, much of which remains to be unraveled and told. Most of those rocks formed in late Paleozoic or Mesozoic time, long before the Insular terrane became part of North America; a few formed since then.

The Insular terrane and the Seven Devils complex could be parts of the same island group that rode into the trench on separate plates, with a transform boundary between them.

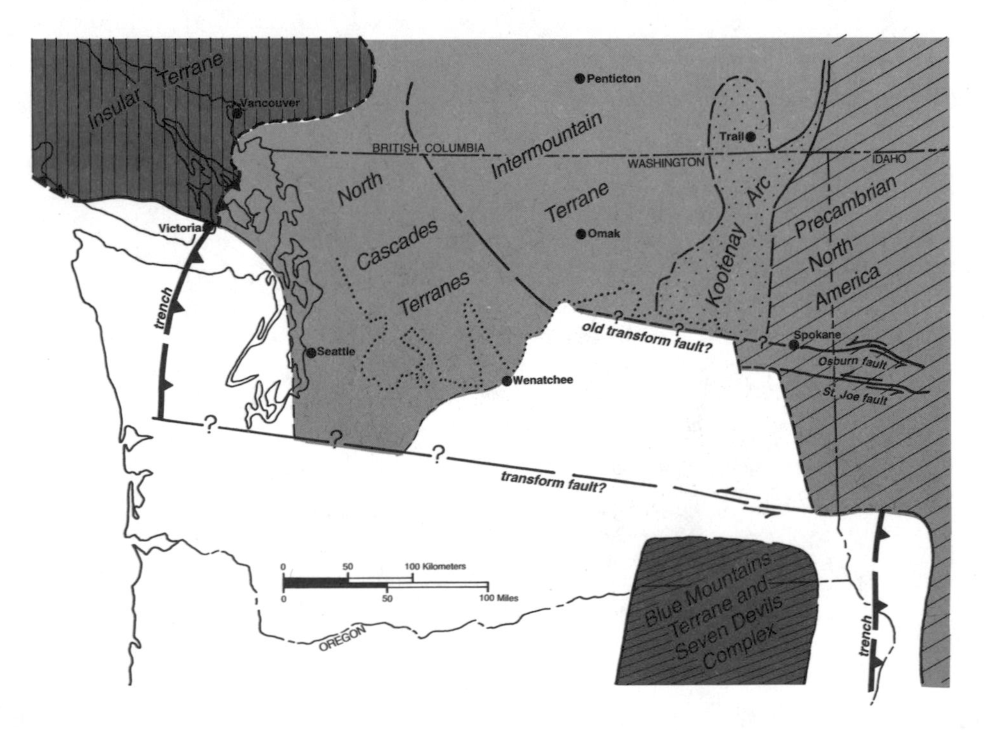

Much of the Insular terrane consists of volcanic rocks that erupted in a chain of volcanic islands during late Paleozoic and early Mesozoic time. The terrane also contains sedimentary rocks derived from the volcanoes and limestone that probably formed in fringing reefs. The magnetic fields in the volcanic rocks incline at an angle that suggests they erupted at equatorial latitudes. The limestones contain fossils of animals that probably lived in warm seawater. And the fossils more closely resemble fossils of the same age in southeast Asia than those in North America. In nearly every respect, the rocks of the Insular terrane resemble those of the Seven Devils complex of western Idaho.

In fact, the resemblance is so close that some geologists think that the former islands of the Insular terrane were part of the same group as those in the Seven Devils complex. That is certainly conceivable. The transform plate boundary that may have linked the segments of the trench could have split a group of islands into two parts that then came to the western edge of North America on opposite sides of the transform, at the two segments of the trench.

The eastern boundary of the Insular terrane is hopelessly lost in the Coast Plutonic complex of British Columbia, a mess of enormous granite batholiths that fill the coastal mountains all the way to Alaska. These rocks crystallized deep within the continental crust, presumably deep in the roots of mountains that rose as the Insular terrane collided with North America.

More Terranes

Like those in the North Cascades, the rocks in the Insular terrane are now shuffled into a stack of slices, each of which moved west over the rocks below on a nearly horizontal thrust fault. The process continues as the floor of the Pacific Ocean sinks through the oceanic trench off the modern coast, which continues to add pieces to the western edge of North America and shuffle them into slices.

The Modern Trench

Docking the Insular terrane onto North America killed the old trench, again because it could not swallow such light rocks. A new trench immediately formed west of the Insular terrane, or perhaps a trench already existed there. In any case, the trench west of the Insular terrane still exists off the modern west coast, still greedily swallowing the floor of the Pacific Ocean.

Reefs like this one in the present-day South Pacific surrounded some of the islands that collided with North America.

Chapter 17

NORTHERN CALIFORNIA TAKES SHAPE

100 Million Years Ago

Rocks along the southern edge of the Klamath Mountains so nearly match those along the northern edge of the Sierra Nevada that most geologists feel compelled to conclude that the two were once continuous. You need to shift the geologic map about 60 miles to find the best match, so it seems that the crustal block of the Klamath Mountains moved about that far west. Just when the Klamath block moved is most uncertain; it was probably during early Cretaceous time, before 100 million years ago.

The Klamath Block and the Seaway

What sawed that big piece out of the Sierra Nevada and moved it 60 miles west? We think that the Klamath block must have moved

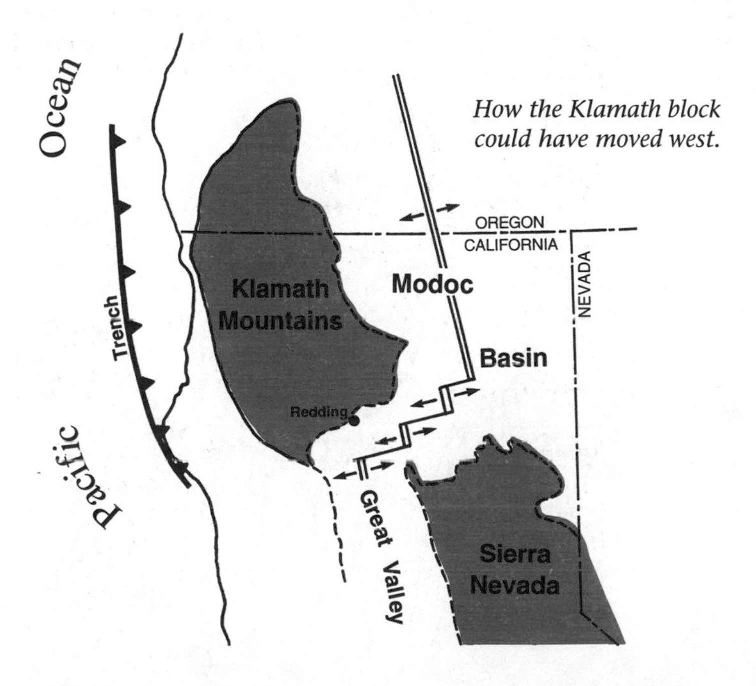

How the Klamath block could have moved west.

as a separate and very small lithospheric plate, a microplate. If so, it moved along some combination of the usual spreading, collision, or sliding plate boundaries. If a very short oceanic ridge developed east of the Klamath block, the offset between the southern margin of the Klamath block and the northern edge of the Sierra Nevada was a sliding transform plate boundary that connected the short ridge east of the block with the trench to the west. A similar boundary must have existed north of the Klamath block, but it is now completely buried under much younger volcanic rocks.

If this idea of how the Klamath block moved is correct, a miniature ocean about 60 miles wide opened east of the Klamath block sometime around 100 million years ago. It probably stretched from the broken northern end of the Sierra Nevada north into southwestern Oregon—to the northern transform plate boundary, wherever that was. Such a narrow ocean basin virtually surrounded by land must have received enormous volumes of sediment, presumably from all directions.

You can see the sediments. They appear as soft and shaley rocks of the Hornbrook formation in a number of road cuts along Interstate 5 between Medford, Oregon, and Yreka, California, along the eastern margin of the Klamath block. Parts of the Hornbrook formation contain fossils of animals that lived in seawater during late Cretaceous time. They are certain evidence that an open seaway then existed east of the Klamath block. The Hornbrook formation

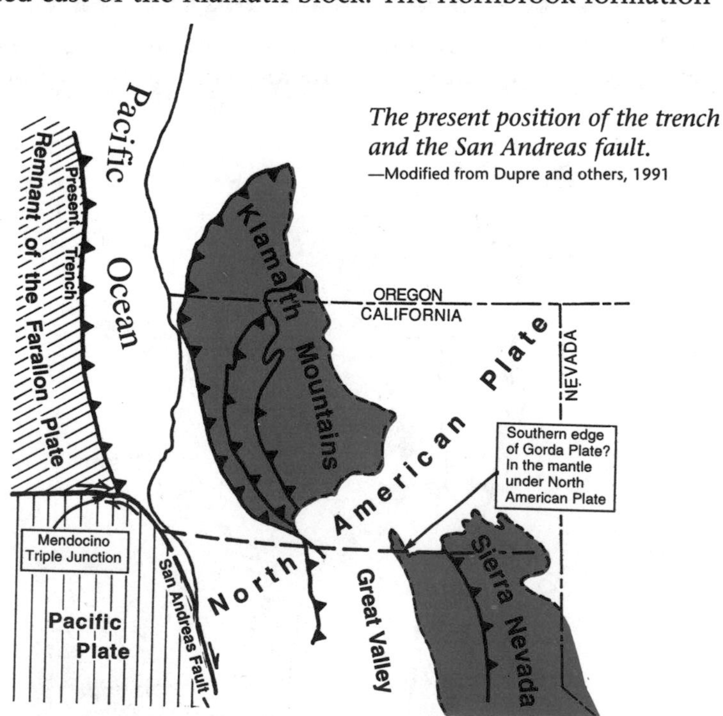

The present position of the trench and the San Andreas fault.
—Modified from Dupre and others, 1991

disappears eastward beneath sediments deposited on dry land during Eocene time, and beneath still younger volcanic rocks. The terrestrial sediments show that at least the southern part of the seaway was full by Eocene time, some 50 million years ago.

No solid evidence reveals how deep that pile of Cretaceous marine and Eocene terrestrial sediments may be. If our inference that an oceanic ridge existed east of the Klamath block is correct, the Cretaceous marine sediments must rest on oceanic crust. If so, their total thickness may be two or three miles: the average depth of the ocean near an oceanic ridge plus an allowance for subsidence beneath the weight of the accumulating sediments.

The earth's inventory of late Cretaceous sedimentary rocks deposited in seawater contains far more than its share of oil and gas. For some reason, that period was especially favorable for petroleum formation. That is precisely the age and kind of rock that probably fills the old seaway east of the Klamath block. Those sediments may well contain oil.

The Basic Blueprint for Northern California

If you could carry yourself back to late Cretaceous time to stand on the old Sierra Nevada coast, you would gaze west across an open ocean and north to the mountains of the Klamath block. With a few preliminary flaps of their great sails the pterodactyls soaring high along the coast could easily glide from the Sierra Nevada to

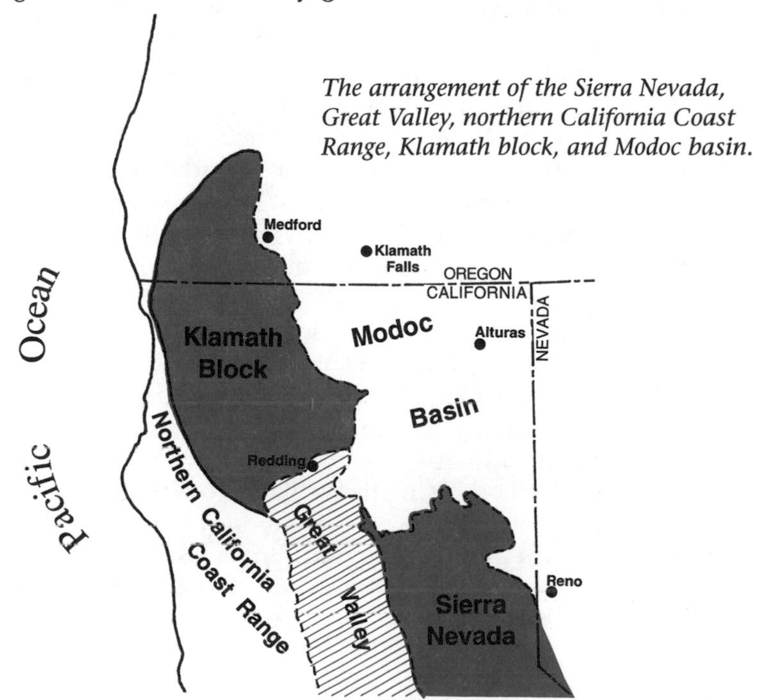

The arrangement of the Sierra Nevada, Great Valley, northern California Coast Range, Klamath block, and Modoc basin.

Greenstone knockers in black mudstone at Crescent City, California. Some of the sediments and scraps of ocean floor stuffed into the trench are thoroughly scrambled.

the Klamath block, across the narrow seaway that connected the open Pacific Ocean to the inland seaway east of the Klamath block.

Before the Klamath block moved west, the oceanic trench followed a line near the eastern base of the Sierra Nevada. As it moved west, the trench moved with it to a new position along the line of the California Coast Range. That shift left a long strip of oceanic crust stranded between the line of the old trench and that of the new trench. It became the bedrock floor of the Great Valley.

The Great Valley is a filled seaway between the Sierra Nevada and the Coast Range. Its width neatly measures the distance between the trench that shuffled the Sierra Nevada together and the one that assembled the California Coast Range. If we could expose the filled seaway east of the Klamath block by peeling the younger rocks off its surface, its width would match that of the Great Valley—about 60 miles.

For the better part of 50 million years after the trench jumped, the land of California ended somewhere along the western side of the Sierra Nevada, which had not yet risen into the high mountain range you see today. Meanwhile, the sinking ocean floor dragged some sediment underneath the oceanic crust that lay beneath the future Great Valley, raising it. We have no way of knowing whether it rose above sea level the way its modern equivalent has along the coast of Oregon and Washington. But it seems safe to assume that at least its edge rose a bit. That would have defined the future Great Valley as an enclosed basin separate from the open Pacific Ocean.

As soon as that rising ridge formed an edge along its western margin, the future Great Valley became an extremely efficient sediment trap. The thousands of feet of muddy sediments that now fill its floor began to accumulate, and the oceanic crust beneath that pile began to sink under its weight. Meanwhile, the trench beyond its western margin was consuming the floor of the Pacific Ocean, scraping its cover of sediments into itself. Each successive slab of sedimentary rock crammed in behind the last one, so the accumulation grew outward from east to west. From time to time, great slices of oceanic crust sheared off and jammed into the filling trench.

When sinking stops at an oceanic trench, the depression on the ocean floor rises, carrying with it all the sediments that were stuffed into the trench. That finally began to happen to the northern California Coast Range segment of the trench about 15 or 20 million years ago, when it began to swallow the offshore oceanic ridge. The San Andreas fault grows as the trench and ridge destroy each other.

In the next few million years, the trench and oceanic ridge north of the San Andreas fault will continue to meet and to destroy each other, extending the San Andreas fault northward. As the trench that is now swallowing the ocean floor off the coast of Washington and Oregon progressively disappears, the sediments now stuffed into it will rise. The contents of the trench will become the new Coast Range, and its rising will tilt the eroded slab of oceanic crust that is now the coastal range steeply down to the east. A similar tilted slab of oceanic crust now follows the eastern side of the California Coast Range.

Oceanic sediments stuffed into the oceanic trench make up the Coast Range.

Avalanches of sand and mud deposited these sediments in the oceanic trench off northern California.

► Part 5 ◄

THE NORTHERN ROCKY MOUNTAINS

100 to 65 Million Years Ago

The western edge of North America first collided with the floor of the Pacific Ocean at the beginning of Jurassic time, about 205 million years ago. Great thrust faults cut the western margin of the continent into slices and jammed them together to make a towering range of mountains. That was the first step in the making of the northern Rocky Mountains.

The second step began as steam and hot basalt magma rose above the sinking slab of ocean floor and melted the rocks in the lower part of the continental crust. Enormous volumes of molten granite magma rose into the early mountains, making them mechanically weak. They sheared off into great slabs that moved tens of miles east into western Montana. Those displaced rocks, and the deep rocks they exposed as they moved off, are the northern Rocky Mountains.

Granite of the Boulder batholith, east of Butte, Montana.

Chapter 18

THE IDAHO BATHOLITH

90 to 70 Million Years Ago

A batholith is simply a lot of granite. Any expanse of the stuff that covers more than about 40 square miles qualifies for the name. Most are many times that size. Granite tends to come in quantity.

For historical reasons, geologists have long called all the granite in central Idaho the Idaho batholith. The area actually contains two large batholiths and an assortment of smaller ones. Nevertheless, the phrase is useful, so we will use it.

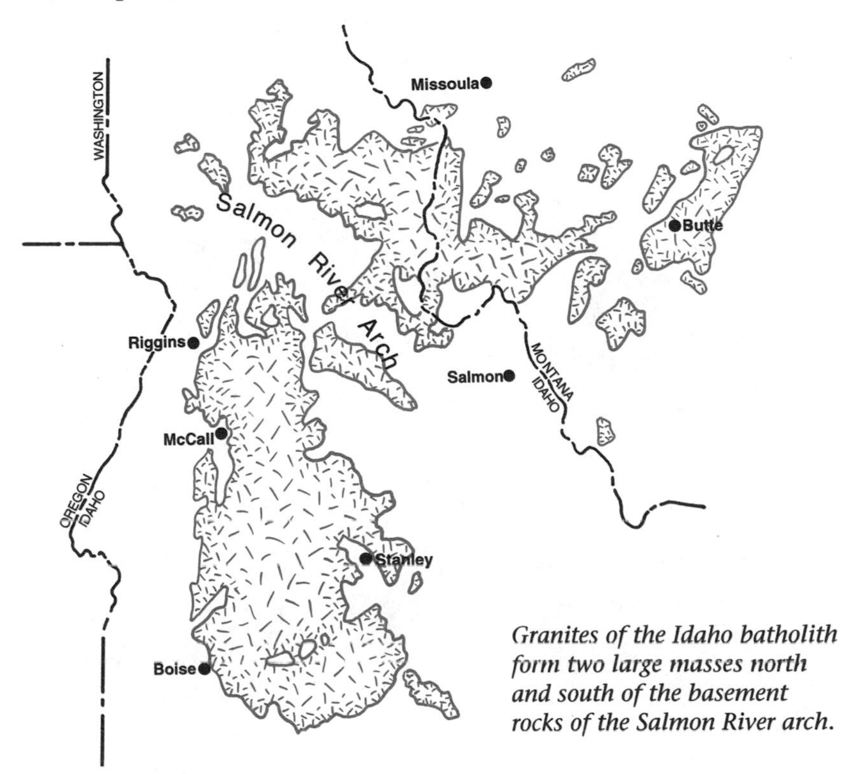

Granites of the Idaho batholith form two large masses north and south of the basement rocks of the Salmon River arch.

Making Granite Magma

Anything as common and abundant as granite must form fairly easily. The recipe calls for starting with any rock that has about the same overall chemical composition as granite, melting it, then letting the magma rise until it freezes. Most continental basement rocks have about the right chemical composition, as do many volcanic rocks. So do muddy sedimentary rocks like those that fill the lower part of the Belt basin. Melt any of those rocks and you get granite magma.

Granites of the Idaho batholith contain microscopic grains of zircon, a zirconium silicate mineral. They did not melt with the rest of the original rocks because the melting temperature of zircon is higher than that of granite. Age dates on the zircon crystals show that they are about 1,700 million years old. Since they are relics of the rocks that melted to become granite magma, their age must be the age of those rocks. They are older than the oldest known Belt rocks, so it seems likely that continental basement rocks melted to make the magma.

By sometime between 80 and 90 million years ago, flaming-hot steam rising from the sinking slab of oceanic lithosphere had soaked the mantle rocks above that sinking slab of lithosphere and the continental rocks above them. The rising steam added heat to those rocks and lowered their melting points. Add steam to any rock and you lower its melting temperature.

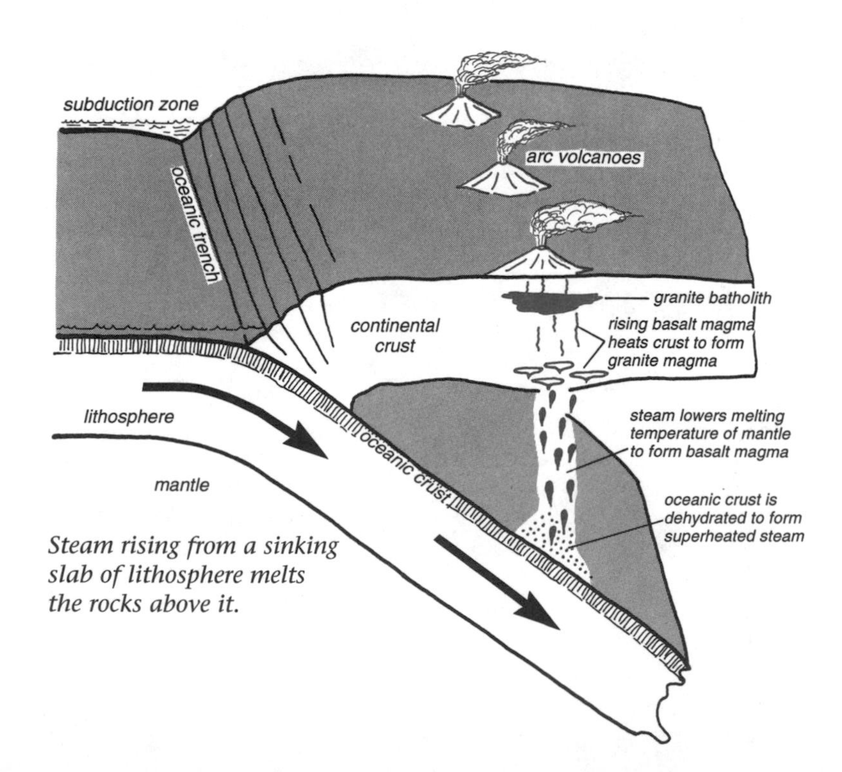

Steam rising from a sinking slab of lithosphere melts the rocks above it.

Rocks in the mantle above the sinking slab of lithosphere were probably the first to melt, making basalt magma that rose into the continental rocks above. Basalt melts at a temperature of about 1,300 degrees centigrade, some 500 degrees above the melting point of granite magma. Basement rocks in the lower part of the continental crust are normally very hot, though not hot enough to melt. Injecting large volumes of molten basalt and red-hot steam into them produces enormous volumes of molten granite magma. That magma was to dominate the development of the northern Rocky Mountains for the next 10 to 15 million years.

Experimental studies of granite magma consistently reveal that a heavy charge of steam causes the melting temperature of granite to increase with a decrease in pressure. That means that as steaming granite magma rises through the crust into regions of lower pressure, the temperature at which it will crystallize into solid rock steadily rises. When the magma reaches a level where its rising melting temperature matches its dropping actual temperature, it immediately crystallizes into solid granite. Masses of granite crystallize because steaming magma rose into a region of lower pressure, not because it cooled. Granite crystallizes first, then cools more slowly.

Things happen oppositely if the molten granite magma does not contain much steam. Then its melting point drops instead of rising as the magma ascends through the earth's crust. Such magmas remain molten until they reach the surface, where they erupt to become rhyolite, a volcanic rock that is in every way the fine-grained equivalent of granite.

The Deep Granites of Central Idaho

Imagine the rocks deep in the continental crust melting to form masses of magma that rose like bubbles into the rocks above. The rising magma dragged along large masses of Belt rocks that were hot enough to recrystallize into schist and gneiss, but not hot enough to melt. Those became the broad zones of metamorphic rocks that surround most of the Idaho batholith. The rising magma also dragged along masses of unmelted basement rocks.

Finally, their rising brought the masses of molten magma to a level where the pressure was low enough to make them crystallize. Experimental studies show that the minerals that make granite can form under a wide range of pressures, so they tell us little about the depth of crystallization. But the metamorphic rocks that surround the Idaho batholith do contain minerals that register pressure. They show that the northern part finally crystallized at a depth of about

Intricately folded schist near the Idaho batholith.

Granite of the Idaho batholith intrudes dark schist.

10 miles, the southern part at a depth of about seven miles. Such deep granites must have crystallized from magmas heavily charged with steam.

Granite in the Early Mountains

If you had trekked through central Idaho during late Cretaceous time, say about 80 million years ago, you would have clambered across Belt sedimentary formations as you crossed from one high snowfield to the next in mountains close to 20,000 feet high. Meanwhile, thousands of feet underfoot, molten granite magma was quietly invading the deep core of the mountains.

Tens of thousands of cubic miles of granite magma rose into the interior of the early mountains. It is possible to imagine it displacing huge masses of the Belt formations, penetrating fractures, and

Granite of the Idaho batholith in the southern Bitterroot Range of Montana.

injecting bedding planes. Meanwhile, its intense heat baked the muddy Belt formations enclosing the magma into glittering gneiss and schist.

The mechanical effect of all that molten magma permeating the interior of those high mountains was like that of spreading far too much soft icing between the layers of a large cake. Those early mountains eventually collapsed, and the pieces moved east, basically because everything sloped down that way. That transformed the high early mountains into the lower and much broader spread of the northern Rocky Mountains.

Chapter 19

THE EARLY MOUNTAINS COLLAPSE

80 to 70 Million Years Ago, Approximately

As large volumes of magma permeated the early generation of mountains, great slabs detached from the top of the pile and moved east, sliding on a sole of granite magma that put a lubricating banana peel under their passage. At least four enormous slabs of the upper continental crust sheared off the collapsing early mountains and moved tens of miles east into western Montana.

Some geologists think that those detachments unroofed the still-crystallizing granite of the Idaho batholith during late Cretaceous time, between 80 and 70 million years ago. Other geologists believe that continental stretching finally unroofed the batholith during Eocene time, about 40 or 50 million years ago. The available data does not seem to resolve the issue either way.

Whether it happened in two stages or in one, collapse of the early mountains certainly left the region with a much broader and correspondingly lower range of mountains. That, combined with all the igneous activity, established the basic architecture of the northern Rocky Mountains.

The Slippery Question of Lubrication

How is it possible for a large mass of rock to move along an almost horizontal slip surface? That question has troubled geologists ever since such overthrust faults were first recognized about a century ago. What force moves the rocks? Why does friction along the fault surface not stop them? Why do the moving slabs of rock move, instead of just crumpling into folds? The debate continues.

One large and deeply convinced school of geologic thought stoutly maintains that some force must shove a moving slab of rock from behind. Another large group contends that the moving mass glides down slope under the pull of gravity. Both groups have a serious problem in explaining why friction along the slip plane does not hold the mass of rock firmly in place, preventing its movement. No large mass, however propelled, can move down a gently sloping fault surface without good lubrication. What greased the skids?

Many geologic maps of areas in the northern Rocky Mountains show the faults that carried the big detached slabs of rock east disappearing into granite intrusions. Careful field work commonly reveals that the mass of granite is a sill, a mass shaped like a great board emplaced in the fault. Some are as much as several thousand feet thick. Molten granite magma could well have lubricated those faults, permitting the rocks above to glide down a gently sloping surface under the unrelenting pull of gravity. If those who prefer to think of a force shoving the slabs from behind are right, the same lubricant could have been at work.

The Sapphire Detachment

Several large blocks sheared off the top of the Idaho batholith, and moved east into western Montana. We will discuss only one of them, the Sapphire detachment. Although the deep granite of the Idaho batholith and its surrounding metamorphosed Belt formations lie west of the Sapphire detachment slab, they once were beneath it. The shallow sedimentary rocks to the east are the detached slab. They include Belt, Paleozoic, and Mesozoic formations. Tightly folded sedimentary rocks full of big thrust faults make up the mountains that wrap around the leading edges of the detachment slab. They outline a horseshoe, its open end facing west. The Bitterroot Valley, across the open end of the horseshoe, separates the areas of deep and shallow rocks.

Granite of the Idaho batholith along the western edge of the Bitterroot Valley, the eastern face of the Bitterroot Mountains, is sheared into Bitterroot mylonite, a strikingly platy gneiss. Mylonite develops where a fault penetrates into rocks so deep and hot that they flow instead of break. Imagine trying to bend a block of paraffin wax that is cold on top and warm below, watching the cold part break while the warm part bends as it flows internally. That is what happens as a fault passes through the cold rocks near the earth's surface into the red-hot rocks miles below. Where the detachment fault cut deep into the hot and still partly molten granite, it became a shearing zone of internal plastic flow, which you now see as a mylonite.

Outcrops near the mouth of any of the deep canyons that cut the eastern face of the Bitterroot Mountains provide a close view of the slabby mylonite. Think of the slabs as cards in a deck, then imagine

The Bitterroot mylonite lies beneath the smooth surfaces between canyons on the east face of the Bitterroot Mountains of western Montana.

shearing the deck so that each card slides past the one beneath, each taking its share of the total movement. That is how the thousands of slabs in the Bitterroot mylonite slipped as the Sapphire detachment slab moved east about 50 miles. The surfaces of the slabs show a pattern of faint parallel lines. From one end of the Bitterroot mylonite to the other, about 65 miles north to south, those lines point slightly south of east. They record the exact direction in which the Sapphire detachment slab moved.

Metamorphic rocks around the northern part of the Idaho batholith crystallized at a depth of about 10 miles, so they must have lost that much cover. If you add the thicknesses of the Belt sedimentary formations in the Sapphire detachment slab, the total comes to about the same figure. Those two facts lead to the conclusion that the Sapphire detachment slab is about 10 miles thick. Shearing off the northern Idaho batholith, it moved about 50 miles east, its leading edge crumpling into folds, breaking along thrust faults, and bulldozing the rocks ahead of it. Slabs of granite in the thrust faults that break the tightly crumpled rocks around its margins crystallized from the magma that probably lubricated the movement.

The Bitterroot Valley is a gap behind the trailing margin of the Sapphire detachment slab. It probably assumed essentially its present shape during the final movements of Eocene time, about 50 million years ago. Several thousand feet of younger sediments lie beneath the flat valley floor.

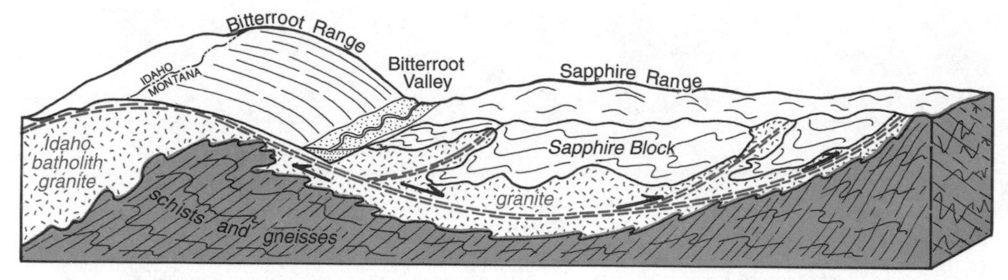

Section across the Idaho batholith and Sapphire detachment slab.

As the 10-mile thickness of the Sapphire detachment slab unloaded off central Idaho, the earth's crust there floated up to form a broad granite dome. At least some of that final movement, perhaps all of it, probably happened later, during Eocene time. Streams have since carved that dome into the rugged landscape of the Bitterroot Mountains. And glaciers added the final touches during the ice ages of the last two million years.

The Arithmetic of Trading Height for Breadth

Lift a can of soda pop from a floating air mattress and watch the depression it made flatten out. The dent will rise until enough water flows in beneath to equal the weight of the can. Precisely the same thing happens when something removes a load from the continental crust: The unloaded area rises until enough mantle rock flows in beneath to equal the weight that was removed from the surface.

Continental crust has an average density of approximately 2.7 times that of water. It floats on mantle rocks that are about 3.3 times as dense as water. It follows that if you take a load of rock off a continent and wait for denser mantle rock to flow in beneath, the compensating rise will restore the lost elevation in the ratio of 2.7 to 3.3, or approximately 82 percent. Conversely, if you load a slab of rock that has a density of 2.7 on the continental crust, it will sink about 82 percent of the height of the slab. Those are the basic terms of the exchange between height and width that created the northern Rocky Mountains some 70 or so million years ago.

The Sapphire detachment slab was probably the heftiest piece that came off the top of the Idaho batholith, about 50,000 feet thick. Assuming that the mountains of central Idaho were 20,000 feet high before it moved off and that the floating rise that followed amounted to 82 percent of 50,000 feet, or 41,000 feet, the mountains would have rebounded to an elevation of 11,000 feet. That is a little more than the present height of their highest peaks in the area that shed the Sapphire detachment. Write the difference off to erosion.

Meanwhile, the weight of the detached slab would depress the crust where it finally stopped. The amount of the depression in western Montana would be the same as that of the rebound in central Idaho. So the final height of the detachment slab would be 11,000 feet greater than the original elevation. Sedimentary formations deposited in shallow seawater during late Cretaceous time extend far into western Montana, so that region was then at an elevation very near sea level. The highest mountains within the Sapphire detachment slab are almost 10,000 feet high, about the height you would expect if a detachment slab 50,000 feet thick were loaded onto crust initially close to sea level and then eroded for 70 million years.

Footprints on the Map

Move a carpet and you leave the floor bare. Move big detachment and overthrust slabs and you expose older rocks that had been deeply buried. That is the broad pattern of the northern Rocky Mountains: Rocks exposed in the western part formed at consider-

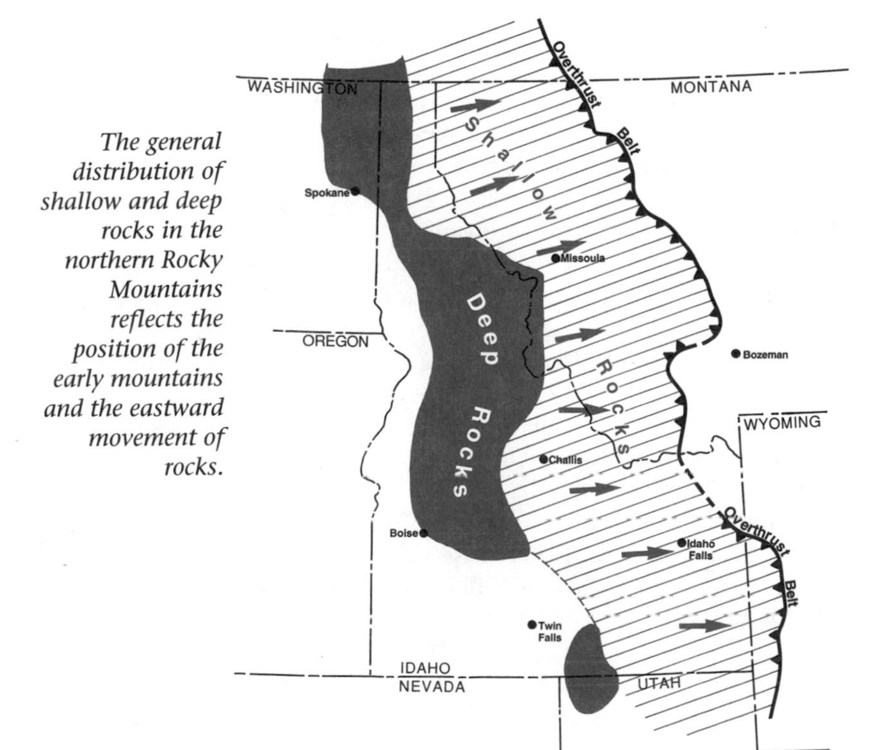

The general distribution of shallow and deep rocks in the northern Rocky Mountains reflects the position of the early mountains and the eastward movement of rocks.

able depth; they are the bare floor. Rocks in the eastern part were never more than a few thousand feet below the surface; they are the carpet that once covered the floor.

The deep rocks include the Idaho batholith, the metamorphic rocks that surround it, and the older Belt formations. They all contain minerals that crystallized under high temperatures and at pressures that correspond to depths of at least several miles. Their distribution outlines the early mountains that rose as thrust faults thickened the margin of the continent during Jurassic and early Cretaceous time. The shallow rocks farther east are the younger Belt rocks and the Paleozoic and Mesozoic sedimentary formations. None of them contain evidence of deep burial.

The Frontal Basins

The surface of a water bed sinks around you, as well as directly beneath you; the continental crust sinks around the margins of detachment slabs as well as directly beneath them. The stronger the continental crust, the farther the basin will extend beyond the margin of the detachment slab. Imagine a water bed topped with linoleum instead of with thin vinyl sheeting.

Many thousands of feet of late Cretaceous sedimentary formations lie along the eastern front of the northern Rocky Mountains, in deep basins along the eastern margins of the detachment slabs. Few of the sedimentary rocks appear to have accumulated in deep water, which probably means that the basins filled almost as fast as they sank. Much of the sediment is volcanic debris. It exists in immense volume, which testifies to a very high level of volcanic activity.

Andesite volcanoes tend to erode rapidly, primarily because the eruptions that build them are violent enough to produce great quantities of broken and incoherent rubble. Streams flowing east from the northern Rocky Mountains carried enormous quantities of volcanic sediment into the frontal basins and across the northern High Plains. Some of the formations in the frontal basins contain fossils of animals that lived in salty water, but most do not. The shallow inland sea evidently was well east of the mountain front during most of late Cretaceous time, and the water that flooded some of the frontal basins was not seawater. They may have held salty lake water, the kind that fills desert lakes.

Some of the frontal basins collected sediment eroded off the detachment slabs while they were still moving. In places, the leading edge of the Pioneer detachment rode over the Beaverhead gravel of southwestern Montana and nearby Idaho before it was cemented

Crushed cobbles of quartzite from the Beaverhead gravel.

into solid rock. The weight of the moving slab crushed and sheared the hard pebbles and cobbles, leaving them looking like abused hard-boiled eggs.

Volcanic Rocks, Absent and Present

For many years, most geologists assumed that a chain of volcanoes must have existed parallel to the trench off the old west coast, and it seemed reasonable to suppose that the granite of the Idaho batholith was emplaced in their roots. But where were the volcanic rocks? No late Cretaceous volcanic rocks are known to exist anywhere near the Idaho batholith. Now that clear evidence has emerged to show that the Idaho batholith crystallized at great depth, the absence of volcanic rocks makes sense. Those magmas never came within miles of the surface in Idaho. If any volcanic rocks had erupted in central Idaho, they would now be in western Montana, having ridden there on the big detachment slabs.

Late Cretaceous volcanic rocks do exist in considerable volume in western Montana, but not because they rode in on the big detachments. They erupted there. Those rocks are gray and brown andesites and white rhyolites the same age as the granite of the Idaho batholith, and with the same range of compositions. Most lie immediately east of the large detachment faults, and in some places beneath the detachment slabs. That position is important because it shows that the volcanic rocks erupted before the detachment slab stopped moving and that molten magma was indeed under the moving detachment slab, available to lubricate its passage.

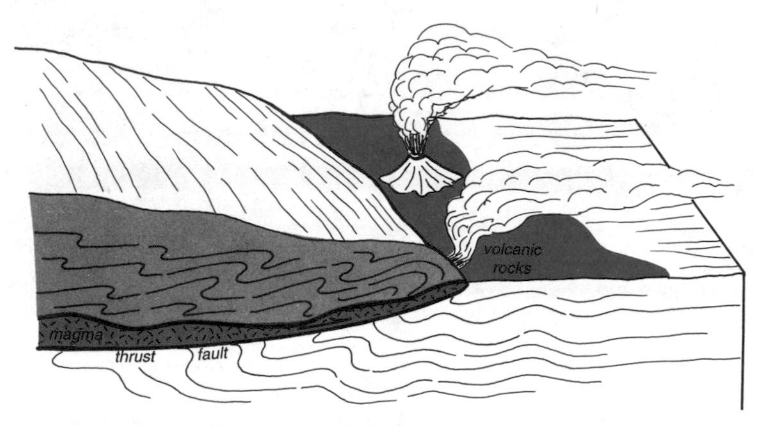

Magma erupting in front of a moving detachment slab.

Some of that magma crystallized into the large granite intrusions that lie in the detachment faults. In many places, those intrusions penetrate sedimentary formations that were very young then and almost at the surface. The rest of the magma made it to the surface and erupted at the leading edges of the detachment slabs to make big volcanic piles. The final result is a granite batholith beneath the detachment slab and a volcanic pile in front of it.

The Boulder Batholith

Two geologists who worked on the Boulder batholith during the 1960s raised quite a tempest in the teapot of academic geology when they compared it to a giant lava flow. They envisioned an extrusion of granite magma thousands of feet thick spreading very slowly across the surface as it continuously erupted its cover of volcanic rocks. They may well have been right. To judge from the age dates, that volcanic show probably lasted for several million years, centered around 75 million years ago—the same age as the Idaho batholith.

No place for a bicycle—a bouldery exposure of the Boulder batholith east of Butte, Montana. —W. H. Weed, U.S. Geological Survey

Unlike most granites, the Boulder batholith has no roof of older rocks. Its only cover is andesite of the Elkhorn Mountains volcanic pile, which is the same age as the granite and has about the same composition. The volcanic rocks do indeed appear to have erupted from the granite.

Why is the Boulder batholith more than 100 miles east of all that granite in central Idaho, which is itself farther east than anyone would normally expect? It is hard to imagine an independent source of heat that could melt granite magma so far east of the trench and in so restricted an area. It is much easier to suppose that the magma moved east from the Idaho batholith, perhaps along the detachment fault that carried the Sapphire detachment slab to the east.

The Boulder batholith is famous as the host of the big ore deposits at Butte, Montana, which claims to be the richest mining camp on earth. Mining started in 1864 with a rather limp placer gold discovery, the usual nineteenth-century story. Prospectors almost immediately found large quartz veins full of silver that looked a lot better than the gold in the creek. Other veins contained mainly copper, in great quantity, but copper hardly seemed relevant in the days before electricity.

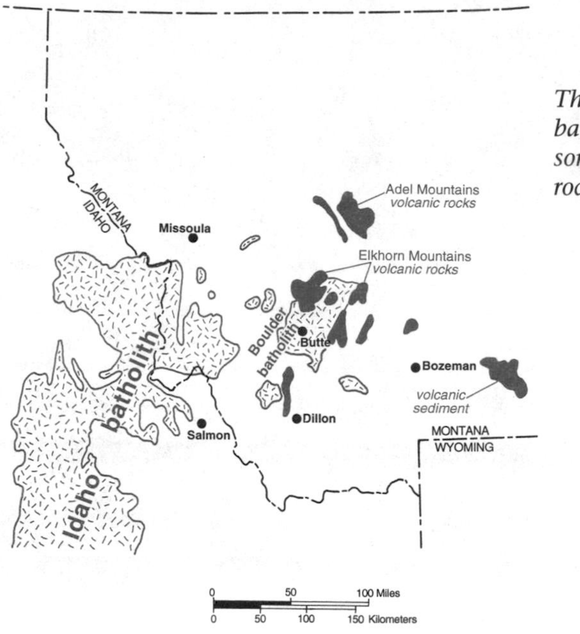

The Boulder batholith and some nearby rocks.

During the late 1880s, an emerging electrical industry created an enormous demand for copper. From then until about 1920, the mines of Butte supplied most of the copper for the wires that electrified America and connected its telephones. They continued to supply much of the national demand for copper for another six decades. But aluminum wires eroded the market for copper, and the electronics industry made communications technology wireless, leaving enormous quantities of scrap copper hanging from poles. By that time, the best copper ore was long gone, and foreign competition was stiff. Since 1983, the Butte mines have operated at only a low level. During all the years of its immense productivity, Butte did produce an amazing amount of gold and silver, primarily as by-products that sweetened the profits from copper.

The ores at Butte are the sort that develop beneath one of those volcanoes that quietly steams for thousands of years following their last big eruption. Surface water sinks deeply into the hot volcanic rocks, leaches minerals from them, and then deposits those minerals in veins as it cools on the way back to the surface. Generations of dinosaurs may have watched in dumb wonder as the plume of white steam rose hissing from the volcano. Then, 65 or so million years of erosion wiped the volcano off the landscape, exposing the steamed and mineralized rock underneath.

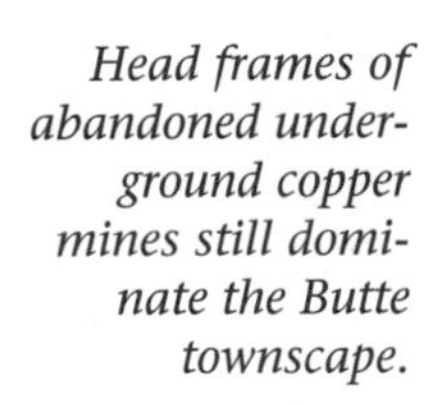

Head frames of abandoned underground copper mines still dominate the Butte townscape.

Ore and Climate

Geologists who study metallic ore deposits generally conclude that the circulating hot waters responsible for altering the host rocks and adding the ore minerals must be strong chemical solutions of one kind or another. Studies of modern systems in which volcanic heat is now driving a circulation of hot water and steam generally show that the water involved is local surface water. Where can you find surface water that is also a strong solution?

The shallow playa lakes that flood the floors of desert valleys are typically very salty. Most are strongly alkaline, and many contain other chemicals, such as Epsom salt or borax. Volcanic craters in arid regions commonly contain hot lakes filled with horrifying solutions, such as sulfuric acid. In a wetter climate, those craters would soon fill with water, their sides would collapse, and the whole thing would rinse clean.

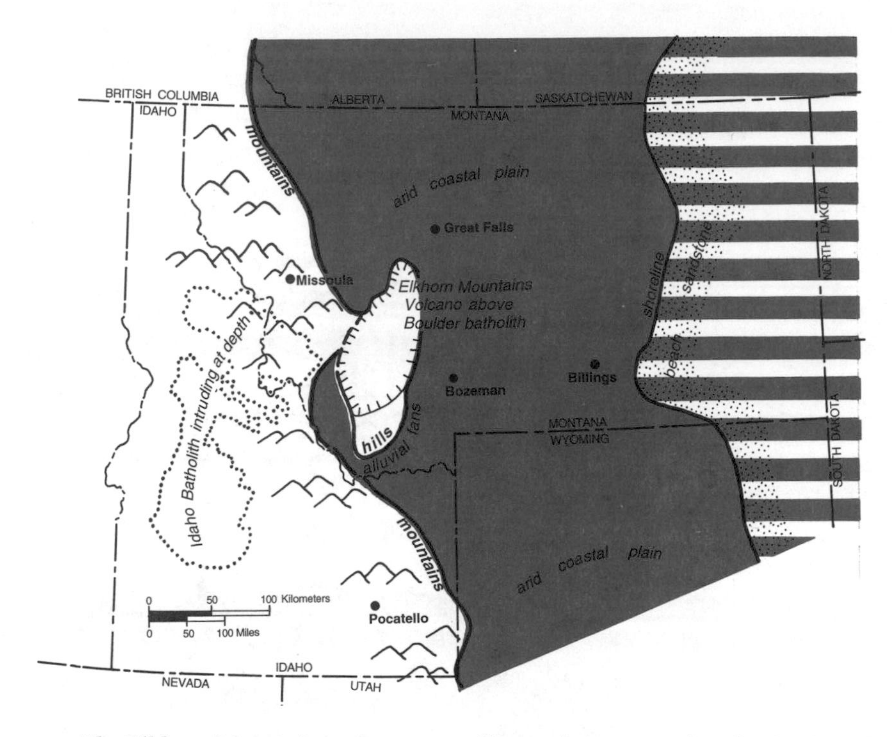

The Elkhorn Mountains volcano, standing on its magma chamber in the Boulder batholith, looked east across an arid coastal plain with scattered bands of grazing dinosaurs. —Modified from Mallory and others, 1972; Peterson, 1986

So it seems that the widespread development of mineral deposits associated with volcanic rocks requires an arid climate. That might explain, for example, why the late Cretaceous volcanic rocks in the northern Rocky Mountains contain many more ore bodies than the basically similar Eocene volcanic rocks. The climate of late Cretaceous time was arid; that of Eocene time was wet.

Those dinosaurs who watched the volcano steaming above the future site of Butte must have lived in an extremely dry setting. And the crater of that volcano may have contained a lake full of water at least as horrifying as that now filling the deep open pit of the old Berkeley Pit mine.

Chapter 20

THE OVERTHRUST BELT

70 to 55 Million Years Ago

The overthrust belt became famous during the late 1970s as the geologic feature that reaches from Alaska to Mexico, supposedly oozing oil all the way. In one form or another, a belt of overthrust faulting does indeed extend all that way, and parts of it certainly do contain oil and gas, most famously in the southern Canadian Rockies and southwestern Wyoming. But that long and complex geologic province differs so greatly from one segment to another that it is hard to imagine that it all contains oil.

The big detachment slabs in western Montana are one distinctive segment of the overthrust belt. The overthrust belt of northern

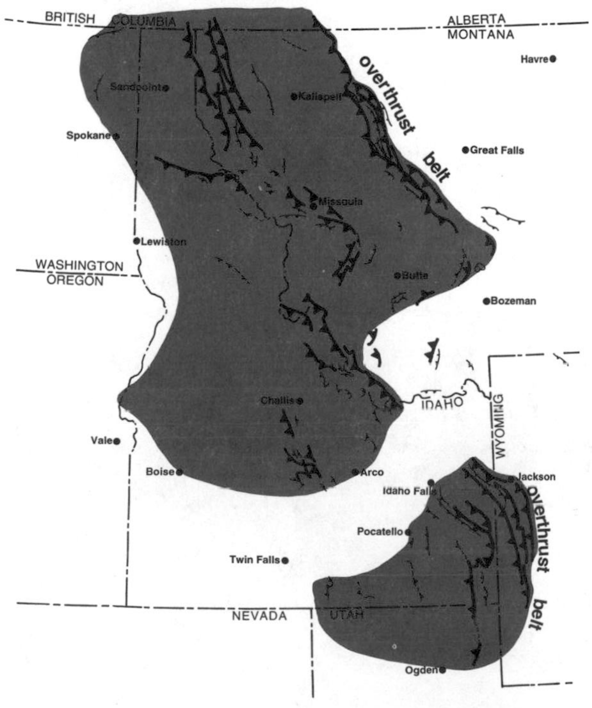

The overthrust belt defines the eastern margin of the mountains that formed during late Cretaceous time.

Montana and Alberta is another, equally distinctive. The two hardly resemble each other, except in being regions in which large masses of rock have moved east on nearly horizontal slip surfaces, or overthrust faults.

Overthrust Faults

Overthrust faults are gently dipping surfaces along which great slices of rock above moved over those below. The slices of rock that moved into the overthrust belt are much thinner than the big detachment slabs. Many overthrust slices a mile or so thick cover hundreds or even thousands of square miles, so their thickness is no greater in proportion to their area than that of a bargain carpet. No

The gently sloping line along the mountains is the Lewis overthrust fault in Glacier National Park, Montana.

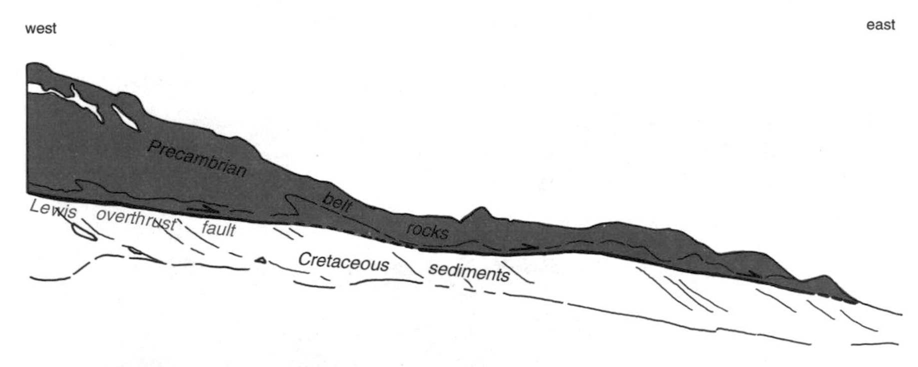

one yet knows exactly how far those thin slices moved, but it is not likely to have been less than 50 miles. That is probably a minimum estimate.

The problem in moving those slices, like that in moving the big detachment slabs, is friction. The weight of a slice of rock a mile or more thick creates tremendous friction along its lower surface. How is it possible for the slice to move on that surface, instead of just crumpling into folds? Most of the overthrust belt contains no granite, so magma did not lubricate the fault surfaces. How then could they move?

For many years, most geologists thought that extremely high water pressure within the rocks may have reduced internal friction, permitting them to slide past each other. That idea does make sense in some settings, but not in much of our region. All of the Belt formations and most of the Paleozoic formations in the northern Rocky Mountains part of the overthrust belt contain so little pore space that they cannot hold much water. Super-pressurized water can hardly have helped move them. The Paleozoic rocks probably moved by gliding on layers of soft shale, which can put a banana peel under a slice of rock almost as effectively as magma. But the Belt rocks include very little shale, and we are not sure how they moved.

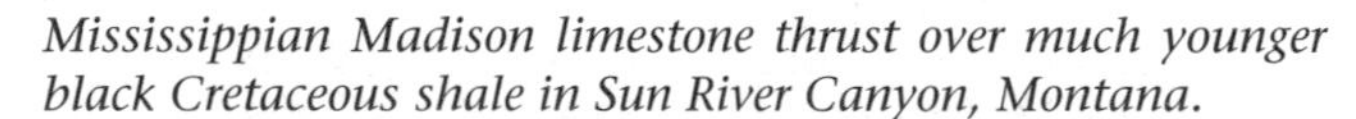

Mississippian Madison limestone thrust over much younger black Cretaceous shale in Sun River Canyon, Montana.

The Curious Architecture of the Overthrust Belt

Whatever lubricated them, the overthrust slices certainly rode east onto the High Plains, where they came to rest piled on the essentially undeformed plains. Each slice is stacked on the next slice to the east, imbricated like shingles on a roof, or scales on a fish.

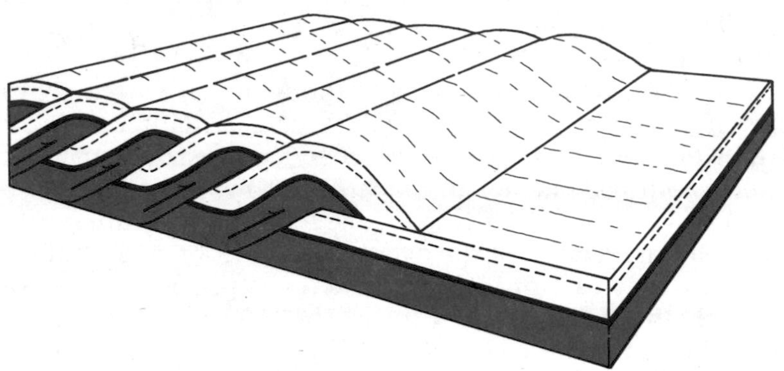

The imbricate pattern of fault-slice stacking in the northern Montana part of the overthrust belt. Movement along a thrust fault shortens the crust.

The youngest rocks are at the eastern edge of the overthrust belt, and the rocks in the fault slices become progressively older westward. So older rocks lie on top of younger rocks, precisely the reverse of the normal sequence. It was finding old rocks on top of younger ones that first convinced geologists of a century ago that overthrust faults exist.

When did the overthrust slices move? Clearly, the rocks had to exist before they could move or other rocks could move over them. Formations deposited during late Cretaceous time, less than 90 mil-

The northern Montana overthrust belt and the big fault blocks behind it.

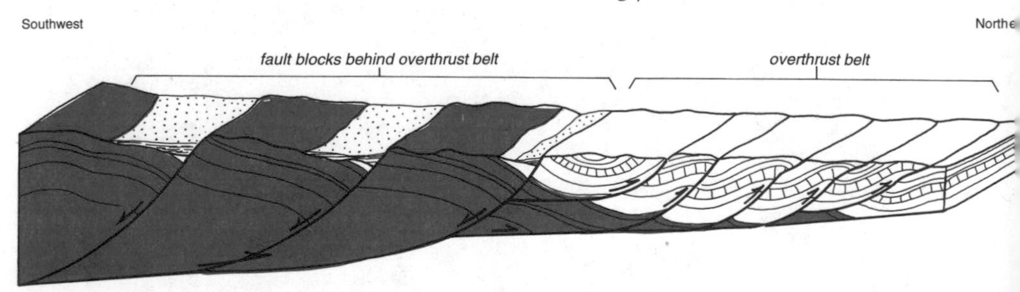

Big cliffs of Madison limestone thrust east over soft Cretaceous shales, now eroded into swales between the cliffs. Sun River Canyon, Montana. —W. C. Alden, U.S. Geological Survey

lion years old, moved and were overridden. In a few places, rocks deposited during Paleocene time, between 65 and 58 million years ago, were involved. Age dates on clays cooked in the warmth of overriding rocks show that they baked between 70 and 55 million years ago.

As the high mountains west of the overthrust belt grew, they squeezed thrust slices out to the east; as the thrust slices built into a thicker pile, they squeezed more thrust slices out farther east.

Oil in the Overthrust Belt

When the Mormon pioneers were on their way west in 1847, they found an oil seep in the southwestern Wyoming part of the overthrust belt. Brigham Young had them dig a well there to get more oil for their wagon hubs and to treat the sores on their livestock. That was probably the first oil well; the more famous Drake well in Pennsylvania was not drilled until 1859.

That oil seep left no doubt that oil existed in the southwestern Wyoming part of the overthrust belt. Nevertheless, hundreds of exploratory holes were drilled over a period of about 50 years be-

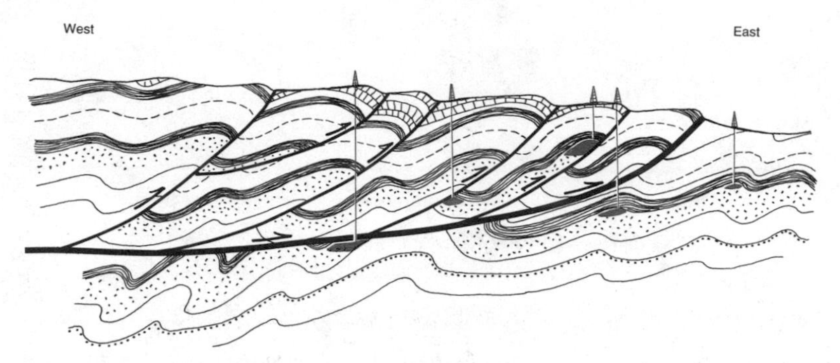

Exploration within and beneath the displaced slabs of the overthrust belt.

fore the first significant oil field was discovered. And finding several more oil fields required many more dry holes.

Exploration geologists who seek oil and gas in the intricately folded and broken rocks of the overthrust belt have a difficult task. They can look for reservoirs within the displaced overthrust slices, in the buried rocks of the plains beneath them, or both. A deep well may pass through several stacked overthrust slices before it finally enters the rocks in the plains beneath. It is extremely difficult to predict what the drill will penetrate in such extremely complex structures.

▶ Part 6 ◀

THE LATE CRETACEOUS WORLD, AND ITS HORRIBLE END

90 to 65 Million Years Ago

The northern High Plains contain thick accumulations of sedimentary rocks laid down during latest Cretaceous and earliest Tertiary times both on land and in the shallow sea that then flooded much of the continental interior. Those rocks, and the fossils they contain, provide a rich picture of the late Cretaceous world. The western part of our region contains virtually no geologic record of that fascinating time.

Then that world suddenly crashed, apparently in a very brief period of time. When the dust had settled, about 65 percent of all the animals then living were extinct. They included animals that roamed the land, swam in the seas, and flew through the air. The survivors emerged into a radically different scene, and eventually created a new world quite different from the one that vanished 65 million years ago.

The late Cretaceous Hell Creek beds, near Circle in eastern Montana, show little indication of the impending catastrophe. —C. E. Erdmann, U.S. Geological Survey

Chapter 21

THE LATE CRETACEOUS WORLD

90 to 65 Million Years Ago

The sedimentary formations deposited on the northern High Plains during late Cretaceous time contain the raw material for a portrait of those times. It was a rich world, a varied landscape full of animals that now seem very remote and of plants that would look quite familiar. Nothing in the geologic record of the time hints of impending doom.

The Shallow Inland Sea

Imagine the scene during the latter part of Cretaceous time when the northern Rocky Mountains were still rising and spreading in the west and none of the smaller mountain ranges that now stand east of the overthrust belt existed. The growing mountains of the overthrust belt looked east across a broad plain that shelved ever so gently down to the shallow inland sea. The land probably blended into the sea through a broad tract of coastal marshes. Every slight change in sea level or the elevation of the land shifted those marshy coasts long distances across the flat coastal plain, making it broad at some times, narrow at others.

During middle Cretaceous time, about 115 million years ago, the inland sea flooded west over much of Wyoming and Montana. A narrow belt of beach sandstone moved west with the shoreline from near the border between Wyoming and Idaho, and north through western Montana. Black shale accumulated offshore to the east, and peat that would become thin coal seams accumulated in coastal swamps.

Volcanic ash fell in that inland sea, probably as it drifted east from the big eruptions in front of the detachments in western Montana. Ash in seawater soon alters to bentonite clay, which swells when it is wet. Today, slimy surfaces of expanded bentonite make

Bold cliffs of Eagle sandstone in central Montana. —T. W. Stanton, U.S. Geological Survey

the unpaved roads of the northern High Plains as slick in the rain as in the snow. Bentonite is also the basis of oil-well-drilling mud, and it seals many ponds and irrigation ditches against leakage.

For several million years, the inland sea of middle Cretaceous time widened to the west. Then, about 105 million years ago, floods of sand reached the coast from a source in the rising mountains of southeastern Idaho as the coast moved east to southeastern Wyoming. By 90 million years ago, the sea again flooded west to central Wyoming as sand still came to the coast from southeastern Idaho. Offshore, sandstone and shale accumulated in a broad area of central and northeastern Montana, in a much narrower belt through central and southwestern Wyoming. Limestone was accumulating in the warm shallows farther east.

The next few million years saw the sea expanding west as far as the border between Idaho and Wyoming, then retreating to central Montana and central Wyoming by about 80 million years ago. The Eagle sandstone accumulated then along that migrating beach. Look for it in the bold cliffs of the Rimrock around Billings, Montana. Meanwhile, coal swamps were accumulating soggy messes of dark peat in western Wyoming.

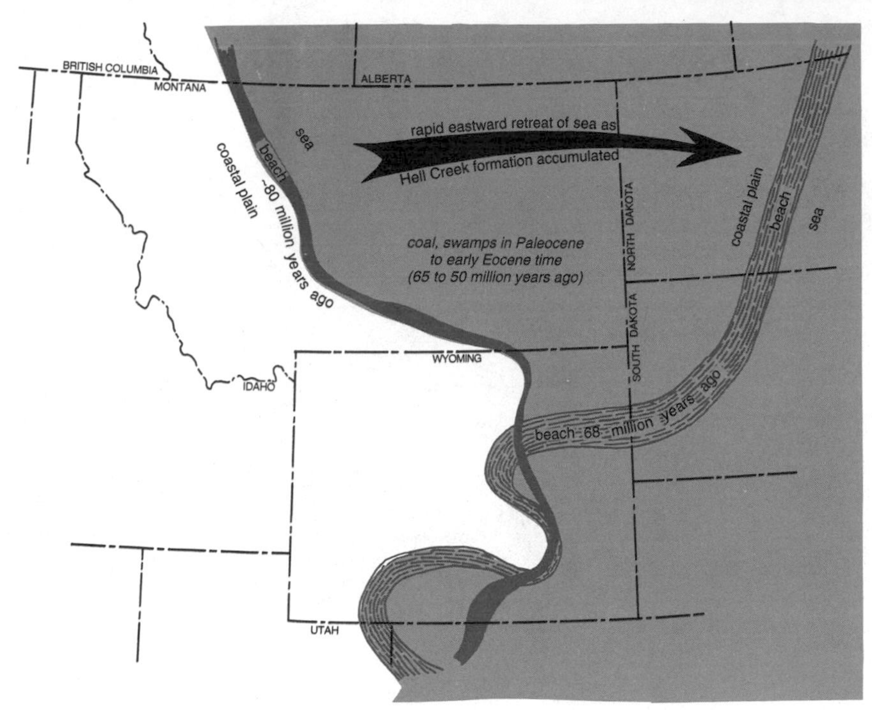

In the last few million years of Cretaceous time, the inland sea retreated rapidly eastward.

Toward the end of Cretaceous time, the inland sea advanced briefly into central Montana. Then it receded rapidly to the Dakotas, perhaps chased by the advancing overthrust belt. That retreat left a broad coastal plain that covered the eastern two-thirds of Montana and most of Wyoming. Sandstones and shales of the Hell Creek formation accumulated on that coastal plain.

That inland sea was shallow, probably no more than a few hundred feet deep anywhere. Banks of oysters lived so densely along the shore that some of the rocks it left behind are solid masses of oyster shells. Many of the rocks are full of broken fragments of the shells of inoceramus, a monster clam several feet long with a very thin shell. It was probably delicious, and that may help explain why the shells are always broken. Swarms of tentacled animals called ammonites swam about, looking a lot like the modern pearly nautilus. Great swimming reptiles, plesiosaurs and mosasaurs, chased the ammonites and the fish. The surface waters were full of minute floating animals that rained their microscopic shells on the bottom in such abundance that they made a distinctive kind of chalky limestone.

Sedimentary rocks of the Hell Creek formation in eastern Montana. —F. S. Jensen, U .S. Geological Survey

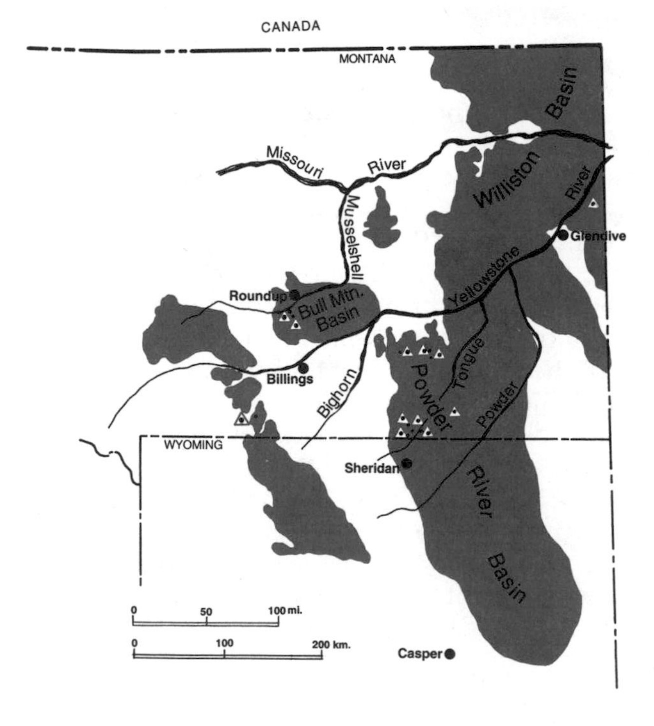

Distribution of coal beds in the northern High Plains. Triangles mark major deposits. —Montana Bureau of Mines and Geology

Coal

Coal forms from peat, a deposit of partially decayed plant matter. Bury peat shallowly and it turns to soft coal; bury it deeply, and it turns to hard coal. The late Cretaceous formations of the northern High Plains contain quite a lot of soft coal and some that is fairly hard.

Peat forms in all sorts of swampy or marshy environments. Some of the largest and most continuous peat environments in the modern world are in the coastal mangrove swamps of the tropics. Somewhat similar swamps probably existed along the shifting late-Cretaceous coast, extending landward along rivers that drained into the inland sea. The late Cretaceous coal seams do continue for long distances, and many of them are closely associated with sedimentary rocks that were laid down in seawater.

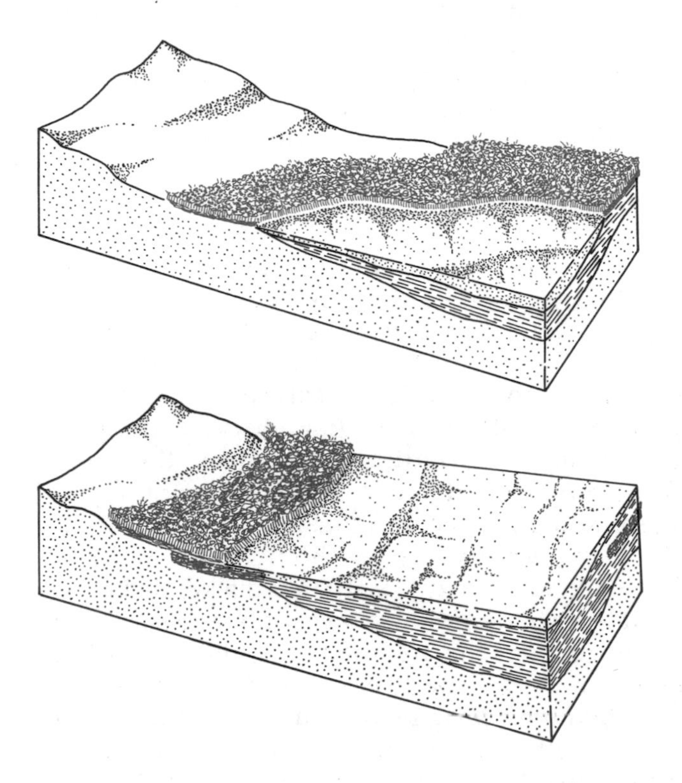

The shifting shoreline wove a tapestry of coal seams through the sedimentary formations accumulating on the northern High Plains.

Every time the coast of the shallow inland sea shifted to the west, the sediments accumulating in seawater buried the old coastal peat deposits. When the coast shifted to the east, the sediments accumulating on land buried the coastal peat. Either way, it was compressed into coal. Then a new deposit of coastal peat would begin to accumulate along the new coast, another future coal seam.

Those coal swamps must have been lush environments full of green vegetation, and presumably swarming with animals. But the land west of them was probably dry and forbidding. The juxtaposition of a lush coast against a dry land is not in any way contradictory. You can see that today along the east coast of northern Mexico, where deep deposits of peat accumulate in coastal mangrove swamps along the edge of a dry coastal plain covered with thorny bushes and small trees.

Dinosaurs in an Arid Landscape

During late Cretaceous time, layer after layer of sediment spread across the coastal plain between the rising mountains in the west and the shallow inland sea in the east. In the modern world, sedimentary deposits accumulate offshore in humid regions, but not onshore. If the climate is wet enough to maintain streams, they carry sediment to the sea. So the very existence of those late-Cretaceous coastal plain formations suggests a climate too dry to maintain streams.

And those formations contain buried desert soils full of calcium carbonate, caliche, and the clay minerals typical of arid soils. They contain bones, which must be buried in alkaline soils or sediments, the kind typical of dry regions, if they are to become fossils. They do not contain petrified wood, as you would expect if trees grew on that coastal plain. The dinosaurs of late Cretaceous time probably browsed on the harsh and scanty vegetation of arid plains. The sparse plants of dry regions are typically very nutritious because they grow in soil that has not lost its soluble fertilizer nutrients to the rain.

And the dinosaurs certainly did live in herds. John Horner, the great dinosaur paleontologist, estimated that a single bone bed in the Two Medicine formation just east of the northern Montana overthrust belt may contain as many as 10,000 duck-billed dinosaurs. They were suddenly buried under a layer of hot volcanic ash, which probably erupted from the area of the Boulder batholith. Duck-billed dinosaurs averaged some 30 or more feet long and probably weighed six to eight tons. Imagine that thundering great herd roaring and eating and defecating its way across the countryside. They must have migrated. It is beyond imagining that so much bone and flesh and scaly hide could feed itself in one pasture, even one full of highly nutritious plants.

Early paleontologists working in the late Cretaceous sedimentary rocks of the northern High Plains found many skeletons of duck-billed dinosaurs, triceratops, and even occasional scraps of fierce tyrannosaurs, those with a mouth full of viciously serrated teeth the size and shape of hunting knives. But they found neither eggs nor babies. During the 1980s, John Horner began finding nests, whole rookeries of them, full of dinosaur eggs and babies. The nests tell a vivid story of loving parental care and family life. It now seems clear that dinosaurs were not the coldly reptilian personalities everyone imagined until just a few years ago.

Those great squawking herds of dinosaurs that lived and nested along the flat coastal plain between the rising mountains to the west and the sea to the east surely saw and heard the volcanic eruptions. They watched the sky darken beneath plumes of ash drifting east on the wind, and they saw the streams run gray with volcanic ash. Despite those occasional hazards, the dinosaurs thrived mightily in late Cretaceous times. They must have been very numerous, to judge from the abundance of their bones, and they came in many varieties. Nothing in their bones, or the rocks that enclose them, forebodes their impending doom.

Chapter 22

THE GREAT CRASH OF 65 MILLION YEARS AGO

65 Million Years Ago

About 245 million years ago, the Permian period and Paleozoic time ended with the extinction of at least 90 percent of all the kinds of animals then living. Then Mesozoic time began. Sixty-five million years ago, the Cretaceous period and Mesozoic time ended with the extinction of about 65 percent of all the animal species then living. Then Tertiary time began.

Many kinds of animals abruptly vanished with the dinosaurs 65 million years ago. Ammonites were a varied and numerous tribe of shelled oceanic animals related to modern octopus and squid, and they disappeared, too. The plesiosaurs and mosasaurs, great swimming reptiles, also vanished. So did many kinds of sedentary reef-building animals, along with most of the microscopic animals that drifted in the surface waters of the ocean. It is hard to imagine that anything but a great catastrophe could exterminate so many different kinds of animals all at the same time.

Ours is a roundabout story about that horrible event of 65 million years ago. The tale meanders all the way to India and down any number of side avenues before it eventually comes back to the Pacific Northwest. It is a digression to be sure, and this view of it is controversial, but it illuminates the much later history of our own region.

Mass Extinction

Geologists have known for many decades of the sudden end of the dinosaurs, and have offered all sorts of explanations: The dinosaurs died from the climatic effects of changing sea levels, from the loss of too many eggs to small mammals with sharp teeth, from some sort of mysterious old age of the lineage, from failure of the urge to reproduce, from the effects of eating flowering plants, and

An eight-foot coal seam in the Paleocene Fort Union formation, near Glendive, Montana. —M. R. Campbell, U.S. Geological Survey

so on and on. Interesting as those ideas may be, they do not explain the simultaneous disappearance of all sorts of other animals, large and small, that lived on land and in the sea.

People who examine collections of late Cretaceous fossils and attempt to imagine the world as it then was see nothing that warns of impending catastrophe. The creatures destined for imminent extinction were thriving, apparently as well equipped to compete as those destined to survive. It is impossible to imagine that all those different kinds of animals that walked on the land, swam in the oceans, and flew through the air somehow failed for reasons of their own, all at the same time. Cretaceous time came to a terrible end.

What happened?

Boundary Clay

Whatever happened as the dinosaurs and their companions on earth died coincided with deposition of a layer of clay an inch or so thick, the boundary clay. In some places it is white, in other places pink or red, and in many places almost black. Whatever its color, the boundary clay neatly separates sedimentary rocks laid down in the Mesozoic world of the dinosaurs and the animals they knew, from those laid down in the much different world that followed. That precise separation of the rich late-Cretaceous fauna from the greatly impoverished one immediately above shows that the boundary clay was laid down as the great extinction happened. If the archives of catastrophe survive, they are in the boundary clay.

Exposures of the boundary clay have been found in more than 80 different places, all around the world, in sections of rock that were deposited on dry land, in lakes, in coal swamps, in shallow seas, and on the deep ocean floor. In all those settings, it is essentially the same material. That single thin layer of distinctive sediment must have fallen from the sky like winter snow. The High Plains of eastern Montana contain some of the best and most informative exposures.

The first solid evidence of what happened at the end of Cretaceous time appeared in 1980. Analyses of the boundary clay from a locality in Italy revealed amazing amounts of an extremely rare element called iridium, as much as 30 parts per billion. Such a minute concentration of iridium was amazing only because most sedimentary rocks contain none. It soon became clear that boundary clay everywhere contains equally amazing amounts of iridium and other platinum-group elements.

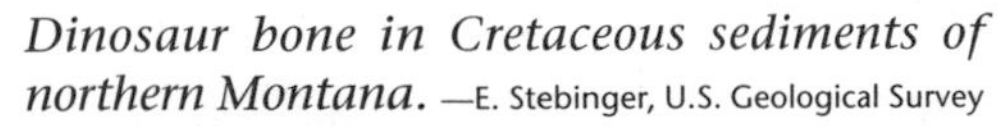

Dinosaur bone in Cretaceous sediments of northern Montana. —E. Stebinger, U.S. Geological Survey

How did those rarities get into the boundary clay?

The dark peridotites of the earth's mantle contain iridium and the other elements similar to platinum, and so do meteorites. The people who first found iridium in the boundary clay suggested that it came from a giant meteorite, an asteroid, that struck the earth and exploded, raising a cloud of dust that darkened the sky for weeks or months. The settled dust became the boundary clay.

Exploding Rocks

Asteroids are small bits of the solar system just large enough to see and track with an astronomical telescope. Most are about the size of a small mountain. Comets are masses of ice with rocks embedded in them. Asteroids and comets move through the solar system at speeds measured in tens of miles per second. Even the slowest asteroids and comets move many times faster than a rifle bullet. They all have tremendous energy of motion.

If you calculate the energy of motion of an orbiting asteroid or comet, you find that it contains far more energy in proportion to its weight than any ordinary explosive. Asteroids and comets weigh millions of tons, so they carry enormous energy of motion, thousands of times the explosive energy of all the world's nuclear arsenals combined. They are massive enough to pass through the atmosphere and hit the earth at nearly their full cosmic velocity. When that happens, part of their enormous energy of motion causes a tremendous earthquake, part blasts a crater, and part converts into heat as an exploding fireball.

So far, astronomers have found at least 200 asteroids in orbits that cross the earth's orbit, and they estimate that approximately ten times that many remain to be discovered. Any object orbiting in a path that crosses the earth's orbit is potentially able to collide with the earth. Any object large enough to see through a telescope is energetic enough to do a lot of damage. One of the many formulations of Murphy's Law holds that anything that can happen will happen, eventually. If an asteroid is in an orbit that crosses that of the earth, it may someday hit the earth. If hundreds of asteroids are in orbits that cross the earth's, sooner or later at least some of them will hit the earth. Our planet has no armor against the random violence of the universe.

The chances of a collision between the earth and a passing asteroid happening this year are much smaller than your chances of winning the grand prize in the state lottery on a single ticket. Bet against it. But do not bet too heavily against such a collision happening during the next ten or 20 million years. The odds improve with every lottery ticket you buy, with every year that comes around. The earth and the asteroids have all the years they need to wait for their rendezvous in space.

When the earth does collide with an asteroid or comet, the explosion causes a catastrophe beyond imagining. The first victims of those explosions are mineral grains unlucky enough to be near the point of impact. They keep a record of the experience.

Shocked Quartz

Quartz seen through a microscope is one of the dullest of minerals. Subject quartz to an extremely violent shock wave, as from a nearby nuclear blast, and it develops an intriguing pattern of fine fractures. Under the microscope, they look like sets of parallel lines faintly scribed across the grain. Such shocked quartz also exists in the rocks around small impact craters, such as Meteor Crater, Arizona. No known natural event except the shock wave from an asteroid impact can create shocked quartz.

Everywhere the boundary clay has been studied, its upper part contains abundant grains of shocked quartz. It seems that shocked quartz sifted out of the sky across our entire planet as the boundary clay was deposited. That shocked quartz provides direct and compelling evidence that an extraordinarily violent asteroid explosion brought Mesozoic time to an abrupt end.

We can picture that end with some confidence, reading part of the story directly from the evidence, the rest from the basic physics and chemistry of what must happen when an asteroid hits the earth.

The Violent End of Cretaceous Time

Somewhere, in the first instant of doomsday, a browsing dinosaur may have glanced up to see a dark object suddenly loom in the sky. A moment later, it blazed brighter than the sun for a few seconds as it consumed a small part of its energy of motion in the friction of its passage. That glare in the sky instantly incinerated our eyewitness. It was a merciful end, considering the occasion. In the few moments of its flaring, the fireball burned millions of tons of atmospheric nitrogen to form various nitrogen oxides. All of those chemicals would eventually combine with water in the clouds to form nitric and nitrous acids; they would later fall as a strongly acid rain.

But first came the impact.

Part of the asteroid's energy of motion blasted an enormous crater and caused an earthquake millions of times more energetic than any in the entire disorderly experience of the human race. Large parts of the earth almost certainly felt ground motion violent enough to hurl loose objects into the air. The shaking almost certainly continued for hours, as opposed to less than a minute for ordinary earthquakes. We doubt that any city on the planet would remain standing if such an impact explosion were to happen now. Earthquake ground motion undoubtedly raised enormous tidal waves, probably along every coast. Geologists have found direct evidence that giant tidal waves washed across the coasts of New Jersey, Texas, and the Caribbean region just as Cretaceous time ended.

But most of the asteroid's energy of motion was converted into heat. An enormous fireball rose from the crater as it opened, burning more atmospheric nitrogen to make more nitrogen oxides, more future acid rain. And a dark column of pulverized rock shot into the stratosphere, where it immediately spread into a broad mushroom cap. Dust from the mushroom cloud began to drift around the earth on the high altitude winds, casting a dark pall that converted day into night. It would eventually settle to become the boundary clay.

Everywhere, the lower part of the boundary clay contains enigmatic round spherules only faintly visible under the microscope. They appear to be the relics of white-hot droplets of molten rock that blew into the air with the fireball and then drifted on the wind. Nearly everywhere, the boundary clay contains soot, so it seems clear that enormous fires burned as Cretaceous time ended. The glare of the fireball was hot enough to ignite everything within view. A white-hot wind spreading from the impact site and those splattering droplets of molten rock probably ignited fires at great distances.

At least several weeks, perhaps months, would pass before the smoke and dust could settle enough to let the bright sun again shine through a clear sky. In the meantime, that dark cloud probably caused a long spell of very cold weather in dry regions, although humid regions may have remained reasonably warm. The darkness and to some extent the cold almost certainly slowed the growth of any plants that survived the fires. Many animals probably survived the explosion, the earthquake, and the fires, only to starve in slow misery during the dark weeks that followed.

Acid rain surely caused enormous havoc. Some geologists speculate about a marine environment turned suddenly acidic, a Strangelove Ocean in which animals died wholesale. Paleontologists who study fossil pollen and spores find evidence of all sorts of flowering plants in the deposits immediately beneath the boundary clay of eastern Montana but almost none within it. Nothing, it seems, bloomed while the boundary clay sifted out of the darkened sky. Immediately above the boundary clay, they find a rich abundance of fern spores, but still almost no pollen. Evidently, the ravished earth greened first with ferns. Then the pollen returned, recording the return of flowering plants, probably within a few months, or perhaps a year or so. But that was too late for the dinosaurs. Their bleached bones were dissolving in the acid rain.

A few plant species vanished in the northern hemisphere, none in the southern hemisphere. Some geologists interpret that pattern of plant extinction as evidence that the fatal meteorite struck in April. That is the month when plants of the northern hemisphere are in their most vulnerable stage of early-spring growth, while those in the southern hemisphere are entering their most resistant period of winter dormancy.

Where did the asteroid strike on that fatal day in April?

The Deccan Connection

Quartz abounds in continental rocks; the oceanic crust contains none. So the shocked quartz in the boundary clay leaves little doubt that the asteroid of 65 million years ago struck a continent. An explosion big enough to blanket the earth with an inch of boundary clay must have left quite a hole. Where is it?

Geologists have recognized about a hundred impact craters here and there about the earth. None are more than about 60 miles across. The explosion that killed the dinosaurs probably opened a crater substantially more than 60 miles across; most likely it is no longer an open hole because it filled with lava.

Rocks must expand if they are to melt into magma. Rocks in the asthenosphere part of the earth's mantle slow down and alter earthquake waves in ways that leave no doubt that they are already partly molten. A very large crater anywhere would relieve pressure on those hot rocks enough to allow them to expand, to partially melt into basalt magma. So it is reasonable to suppose that the site of the crater of 65 million years ago may have filled with molten basalt to become a lava lake, a large basalt volcanic field.

Age dates done by quite different methods show that the Deccan flood basalt province of western India is the same age as the boundary clay, within a probable margin of error of less than a half million years. It is the only large volcanic field exactly the same age as the boundary clay. Furthermore, it erupted within a continent, far from any plate boundaries that might otherwise explain its existence. It could well be the site of the impact explosion that ended Cretaceous time.

Many geologists argue that evidence of enormous waves in the Yucatan area of eastern Mexico at the end of Cretaceous time identifies that region as the site of the impact. Others have found evidence of enormous landslides that moved on the east coast of Africa 65 million years ago. So the Caribbean region was not the only one so disturbed. The earth's internal structure focuses seismic waves on the area antipodal to the earthquake, causing large ground motion there. At the end of Cretaceous time, India and the Caribbean were approximately on opposites sides of the earth. An asteroid impact in India would have caused large ground motion and giant waves in the Caribbean region.

In 1993, geologists restudying the limestones capping the impact breccia in the Yucatan crater found fossils of animals from late-Cretaceous time, some 25 million years before the end of Cretaceous time. If they are right, this impact predated the event that killed the dinosaurs.

▸ Part 7 ◂

EARLY AND MIDDLE TERTIARY TIME

65 to 17 Million Years Ago

Tertiary time began as the Mesozoic era ended, 65 million years ago. Most kinds of land animals and a few plants were gone, vanished in the aftermath of the impact explosion that ended Cretaceous time. Tertiary time continued until about two million years ago, when the first of the great ice ages began. Ever since then, our planet has been in the habit of staging ice ages.

The geologic epochs of early and middle Tertiary time.

Millions of Years Ago	
5.3	
	Miocene (late)
17	
	Miocene (early)
24	
	Oligocene
37	
	Eocene
58	
	Paleocene
66	
	MESOZOIC TIME

The catastrophe that ended Cretaceous time left the physical environment of our region fairly intact. Nevertheless, the first half of Tertiary time brought a series of new events. Some could have been foreseen in the pattern of earlier events; others could not. No one looking at the happenings of Cretaceous time would predict that new mountains would rise east of the overthrust belt, or that the inland parts of our region would see widespread crustal stretching and igneous activity.

In our part of the world, this first phase of Tertiary time was to last just 48 million years. It was destined to come to an abrupt end a little more than 17 million years ago, when the random impact of another vagrant asteroid set the pattern of events on a new course that no one could have predicted. That impact created the great flood basalt province of the Pacific Northwest. Nothing has been the same since. So we will divide our discussion of Tertiary time into before and after the flood basalts.

Chapter 23

TERTIARY TIME DAWNS

65 to 55 Million Years Ago

The northern High Plains contain one of the earth's best-known records of animal life during that critical period just before and after the Mesozoic world crashed. Sedimentary rocks continued to accumulate there through the catastrophe. They contain the fossil remains of the animals that perished at the end of Cretaceous time, and those of the survivors that multiplied and diversified during earliest Tertiary time.

Episodes

If gasoline were suddenly to disappear, so would the cars that use it. New companies would spring up to build new kinds of cars, and our roads would soon look quite different. Disaster for the old kind of cars would create opportunities for new kinds. The fossil record seems to show such a pattern. Most groups of animals remain stable for a long time with few new species appearing, presumably because species already exist to exploit every ecological niche. Under those conditions, organic evolution is a painfully slow process, and new species rarely appear. Then something decimates the old group of animals, and the survivors quickly evolve into new groups.

The holocaust that ended Cretaceous time left our earth with almost as many kinds of plants as ever, but many fewer animals: that is, with an excellent basis for a food chain but almost no food chain. The surviving animals quickly seized their opportunity, multiplied, and covered the earth. The toothy little mammals that had scurried through the underbrush of Mesozoic time soon spawned an amazing zoo of descendants: large and small, vegetarians and carnivores, swimmers and fliers, mammals for every occasion. That happened within a few million years.

It appears that one branch of the dinosaurs survived the crash of the Mesozoic world. The skeletal architecture of birds is remarkably similar to that of the predatory theropod dinosaurs, the group that spawned the awesome tyrannosaurs. Many paleontologists now regard birds as part of the theropod line, the only surviving dinosaurs. Think of that when you next see an ostrich stalking across its pen in the zoo, or the next time you eat fried chicken.

Did a Catastrophe Really Kill the Dinosaurs?

It seems that no one has ever found a dinosaur skeleton in the several feet of rock just below the boundary clay. Except in those few feet, the late Cretaceous rocks and those immediately above them generally contain the mineral calcite; it fizzes when you spray acid on the rock. Some geologists interpret the absence of calcite from the rocks just below the boundary clay as evidence of strongly acid rain falling in the aftermath of a great impact explosion, dissolving all the calcite out of the soil. The mineral matter of bones, calcium phosphate, also dissolves in acid. If acid rain dissolved the calcite out of those sediments, then it would have dissolved bones, too. That might explain why dinosaur fossils are so scarce just below the boundary clay.

Some paleontologists interpret the fossil gap below the boundary clay as evidence that the dinosaurs died out before the end of Cretaceous time, in the natural course of events, rather than in a great catastrophe. The late Cretaceous rocks of the High Plains contain many intervals at least as thick as the fossil gap below the boundary clay in which no one has ever found a skeleton. But no one has ever interpreted those other gaps as evidence of periods when the dinosaurs were somehow missing.

After much debate over the fossil gap, a group of paleontologists ransacked that interval in eastern Montana for scraps of bone and teeth—any evidence of dinosaurs. They found them, right up to the boundary clay. Evidently, the dinosaurs lived until the very end of Cretaceous time. Does that mean that they really did vanish as the boundary clay was laid down?

Not according to some paleontologists, who have found scraps of dinosaur bones and teeth in the rocks above the boundary clay. They interpret them as evidence that the dinosaurs survived into early Tertiary time. But such bits and pieces of fossils can easily erode out of older sediments and come to rest in younger deposits. When someone finds a connected skeleton above the boundary clay, everyone will agree that some dinosaurs survived into Tertiary time.

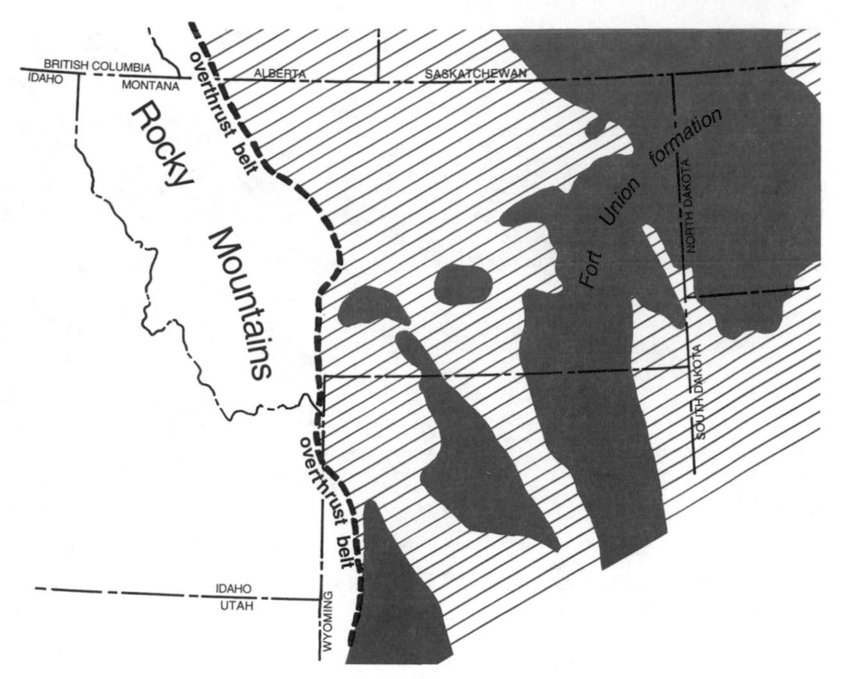

The Fort Union formation. The diagonal line pattern shows where it has been lost to erosion.

The Fort Union Formation

As Cretaceous time ended, a flat plain stretched east of the overthrust belt to the shallow inland sea that still flooded parts of the northern High Plains. All the territory from the overthrust belt to Minnesota was probably within a few hundred feet of sea level, one way or the other. Small changes in sea level or land elevation shifted the shoreline long distances across the gentle slope of the plain. The Hell Creek formation of eastern Montana accumulated on that flat coastal plain during latest Cretaceous time, the Fort Union formation during earliest Tertiary time. They generally resemble each other, but with important differences. That coastal plain evidently changed as Cretaceous time ended.

The Hell Creek formation contains many layers of clay that is dark with organic matter, and a few coal seams. The Fort Union formation contains one of the world's great storehouses of coal, much of it in immensely thick seams. Most parts of both formations are soft enough to dig with a shovel. Both contain pale mudstones and sandstones that come in pastel shades of gray, yellow, pink, and brown. Watch for them in the roadcuts and badlands of eastern Montana.

Sandstone in the Fort Union formation, near Roundup, Montana.

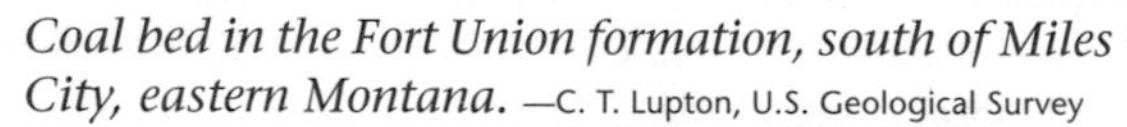

Coal bed in the Fort Union formation, south of Miles City, eastern Montana. —C. T. Lupton, U.S. Geological Survey

The Fort Union formation contains so much coal that estimates of its tonnage boggle the mind—as do measurements of geologic time, astronomical distances, and the national debt. It contains billions of tons of coal, enough to supply the nation's energy needs for centuries. And a kind providence arranged that the biggest and thickest coal seams are in the upper part of the formation, where they are easiest to mine. How did the northern High Plains acquire that rich benevolence of fossil fuel?

Imagine again that low and arid plain that stretched in endless monotony from the overthrust belt in the west to the shifting shoreline of the shallow inland sea on the east. Coal swamps fringed the coast and fingered inland along the rivers on a scale the world has rarely seen. Deep deposits of partly decomposed vegetation called peat accumulated in them while mud and sand were laid down on the plain that stretched west to the mountains. More mud and sand full of seashells accumulated on the shallow sea floor east of the coast.

As the shoreline shifted with slight changes of sea level or land elevation, muds and sands were laid down over the old deposits of peat, while new peat beds accumulated along the new coast. The weight of the sediments that buried them compacted the peat beds into the coal seams that weave a complex pattern through the Fort Union formation.

Fort Union coal was a great blessing to the early white settlers on the treeless northern High Plains, where buffalo chips were considered prime fuel. They began to mine it for domestic use, then for

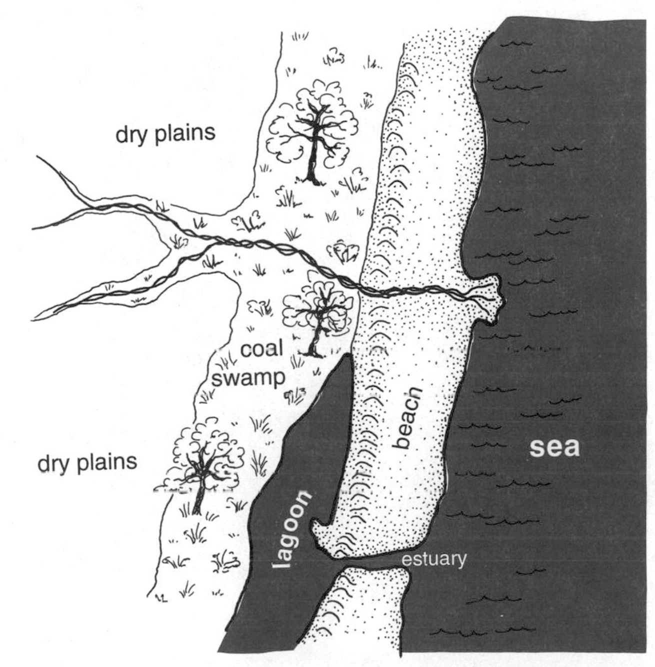

Typical arrangement of beach, sand dunes, and coal swamps.

the railroads. The coal mines prospered through the early decades of this century, until cheap petroleum and diesel engines replaced steam. Then the energy crises of the 1970s and the fear of rapidly dwindling world oil reserves renewed interest in coal.

The new generation of mines, like many of the early ones, are open pits that use enormous power shovels to uncover long sections of coal and then mine it. If not reclaimed, such mines leave long ridges of waste rock, stripped overburden, that parallel the deep coal trenches. Some of the early mines left behind hideous scars, a landscape that looks as though it had been turned by some sort of devilish plow with a blade that sliced two hundred feet into the ground. Reclamation laws now require mining companies to restore the landscape to its original shape and condition.

Ton for ton, Fort Union coal produces less heat than the hard coals of the Midwestern and Eastern states. Some Fort Union coal, however, especially in the southern part of the formation, contains remarkably little sulfur. That is important because sulfur burns into sulfur dioxide, one of the major causes of acid rain. Some large industrial consumers find it cheaper to ship Fort Union coal all the way to the Midwest than to scrub sulfur dioxide out of stack gases.

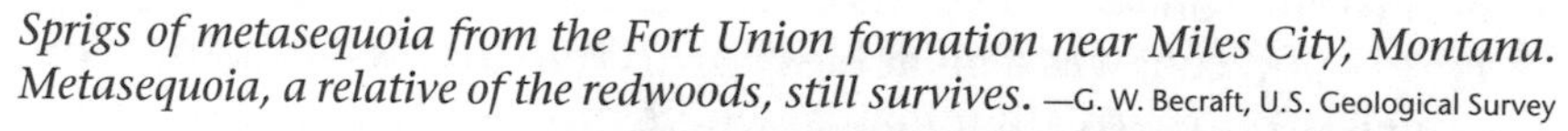

Sprigs of metasequoia from the Fort Union formation near Miles City, Montana. Metasequoia, a relative of the redwoods, still survives. —G. W. Becraft, U.S. Geological Survey

Chapter 24

MORE TERRANES

55 to 37 Million Years Ago

Eocene time brought the last terranes into the North Cascades and another to the Oregon and Washington coast. They completed the collage of geologic terranes that makes the Pacific Northwest. Eocene time also brought a brief episode of folding and faulting to the North Cascades. It may be related to the arrival of one of the new terranes.

The Pacific Rim Terrane Arrives

After the Insular terrane arrived, the Pacific Rim terrane collided with its western edge to complete the North Cascades. That probably happened during Eocene time. Remnants of that terrane extend in a thin belt across southern Vancouver Island and into the eastern lowlands of Puget Sound. They consist of volcanic rocks that erupted from a chain of volcanoes during late Triassic time. Muddy sandstones and chert deposited on the deep ocean floor during late Jurassic and early Cretaceous time cover them, along with some basalt lava flows.

The stacked thrust faults of the western half of the North Cascades and Insular terrane.

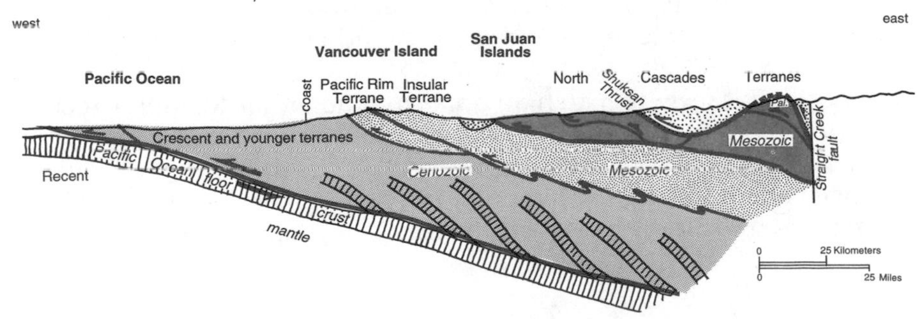

Jamming the Pacific Rim terrane into the western edge of the Insular terrane stacked more nearly horizontal thrust sheets in the western part of the North Cascades, thus continuing the tradition established in late Cretaceous time.

Slicing the North Cascades

During the early part of Eocene time, several large faults that trend generally to the north, cut the North Cascades into long slices. In every case, the slice west of the fault moved north. The largest are the Straight Creek fault, which becomes the Fraser fault in British Columbia, and the Ross Lake fault. The Straight Creek fault does

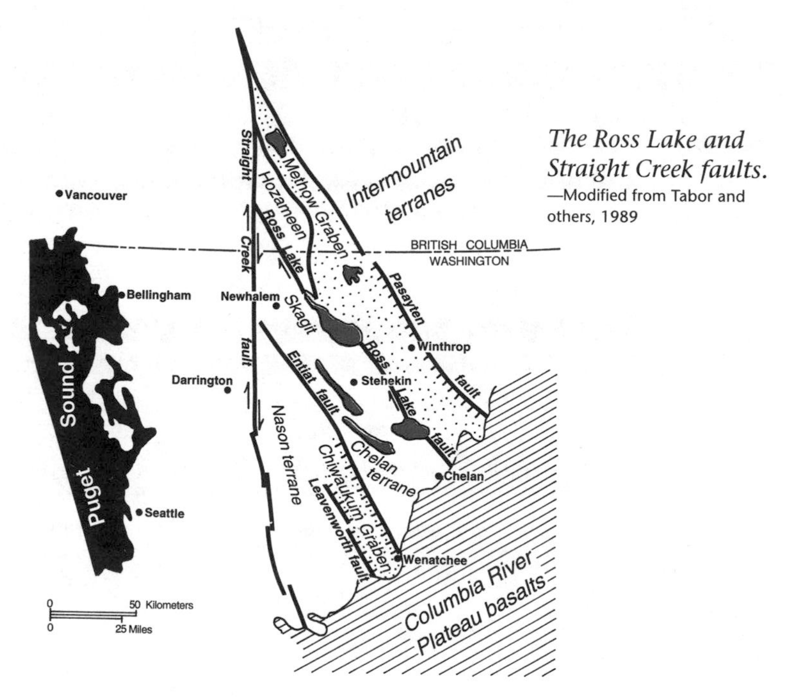

The Ross Lake and Straight Creek faults.
—Modified from Tabor and others, 1989

not offset granite intrusions that invaded it during early Oligocene time, about 35 million years ago, so it had stopped moving by then. It seems reasonable to suppose that the others have a similar history.

Rocks that seem to match on opposite sides of the Straight Creek fault in Washington and British Columbia suggest that it moved about 68 miles. A similar match suggests that the Ross Lake fault moved about 93 miles. Obviously, the present shape of the North Cascades and the terranes within them can hardly resemble the

Eocene continental sediments in the Chiwaukum graben.

islands that landed during Cretaceous time. Indeed, the Straight Creek fault neatly divides the North Cascades into distinctive eastern and western halves.

At its southern end, the Straight Creek fault swings to the southeast, bringing it parallel to the Chiwaukum graben, a dropped fault block filled with the Chumstick formation. It consists mostly of sediments, along with some volcanic rocks, all laid down during Eocene time. The Chiwaukum graben is parallel to the Methow graben, another fault slice that dropped at about the same time, but neither trends in the direction you would expect if the plate sinking through the trench offshore were providing the force. We have no good explanation.

Eocene Formations

Sedimentary rock formations that look like the battered remains of an old coastal plain exist along the eastern side of Puget Sound. Those formations, which geologists lump into the Puget group, consist largely of stream and delta sediments deposited during Eocene time. They contain fossils of broadleaf evergreens that suggest a tropical climate. Rather similar rocks in the general vicinity of Cle Elum lie farther inland and appear to have accumulated on land. In both areas, the rocks include layers of sandstone, shale, volcanic ash, and coal.

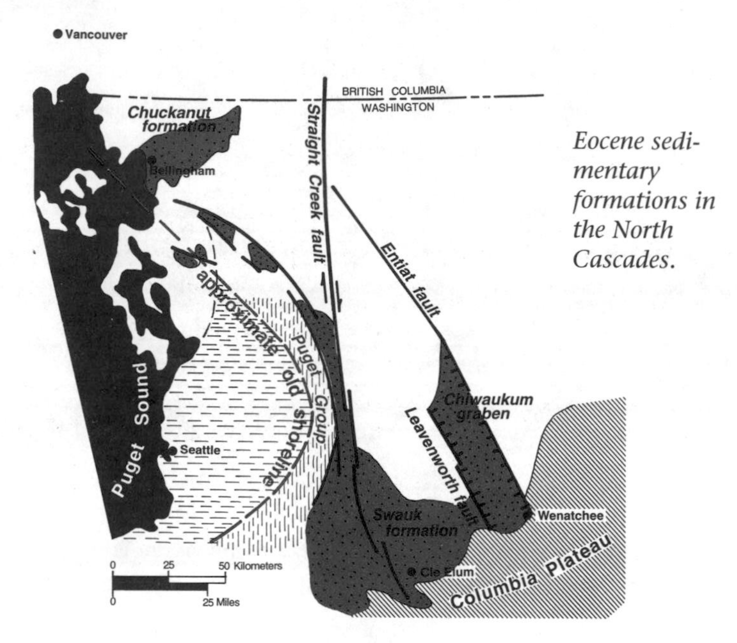

Eocene sedimentary formations in the North Cascades.

Mines produced coal from the Eocene rocks in both areas until cheap oil and diesel locomotives put them out of business. The mines worked underground because folds steeply tilted the coal seams, making open pits impractical. The need to work underground makes it seem unlikely that renewed demand for coal will reopen the mines anytime soon. Huge open-pit coal mines in the Centralia area south of Puget Sound do supply fuel to generate electricity.

The Swauk formation, a pile of sandstones, shales, volcanic ash, and coal, lies generally west of Wenatchee, along the eastern side of the North Cascades. It was deposited on land during Eocene time. The distribution of the rocks and the patterns of sedimentary layers in the sandstones indicate that the streams brought sediment from the northwest.

Folds and Faults

The Eocene sedimentary formations east of Puget Sound—those in the Cle Elum area and the Chumstick and Swauk formations—were all rumpled into folds during Eocene time. Folds in sedimen-

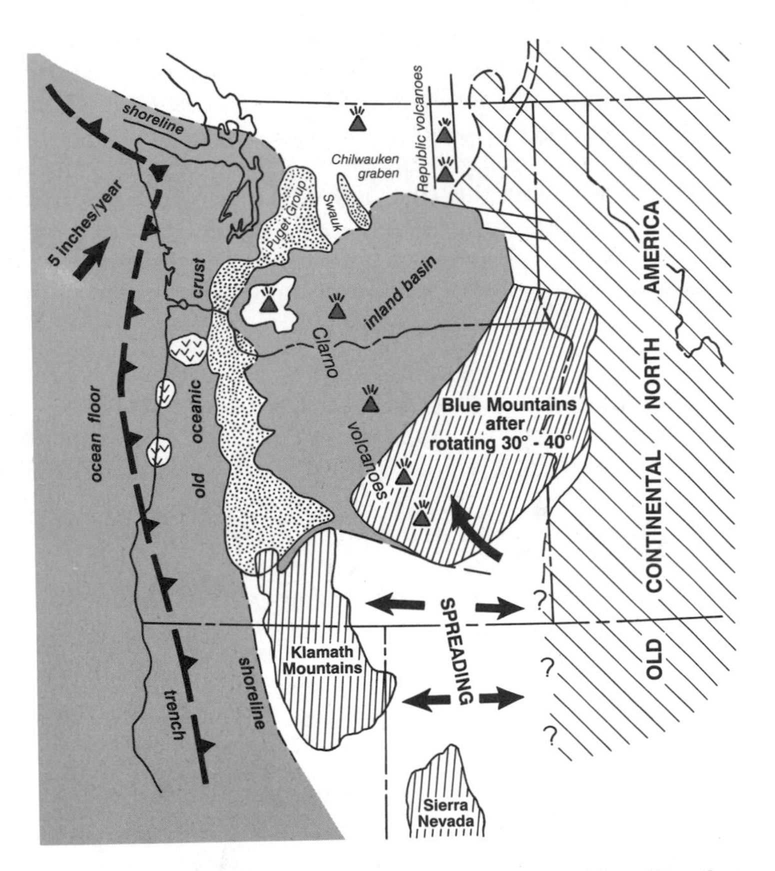

The Northwest in late Eocene to Oligocene times, 45 to 37 million years ago. —Modified from Christiansen and Yates, 1986; Wells and Heller, 1988

tary rocks, like folds in a crumpled rug, trend roughly across the direction of the movement that caused the crumpling. All these folds trend generally northwest, evidently because the force that made them was directed from the southwest.

Age dates show that the Teanaway basalt flow near Cle Elum erupted about 45 million years ago. It lies on tightly folded sediments of about the same age, but is itself only slightly folded. It appears that the Eocene stream sediments were deposited, folded, and then covered by the Teanaway flow within a short time around 45 million years ago. That was when the Straight Creek fault and its cohorts were chopping the North Cascades into long slices and moving them north.

What caused all that faulting and folding? The Farallon plate was sinking through the offshore trench and beneath the western edge of North America in a generally northeasterly direction. It is easy to imagine that its dragging beneath the North Cascades pushed them to the northeast, crumpling the Eocene sedimentary rocks, and driving the faults. But why did that action happen in a brief part of Eocene time and then stop? The Farallon plate has been sinking through the offshore trench ever since then. The deformation may have happened while the Pacific Rim terrane was jamming into the western edge of the North Cascades.

The Crescent and Siletz Terranes Arrive

The Crescent terrane is a great slab of oceanic crust that stands almost vertically where it wraps around the eastern side of the Olympic Peninsula. Farther south, it lies nearly flat beneath the Willapa Hills of southwestern Washington and the northern end of the Oregon Coast Range. The Siletz terrane is part of that slab of oceanic crust that lies almost flat beneath the Oregon Coast Range south all the way to the Klamath block.

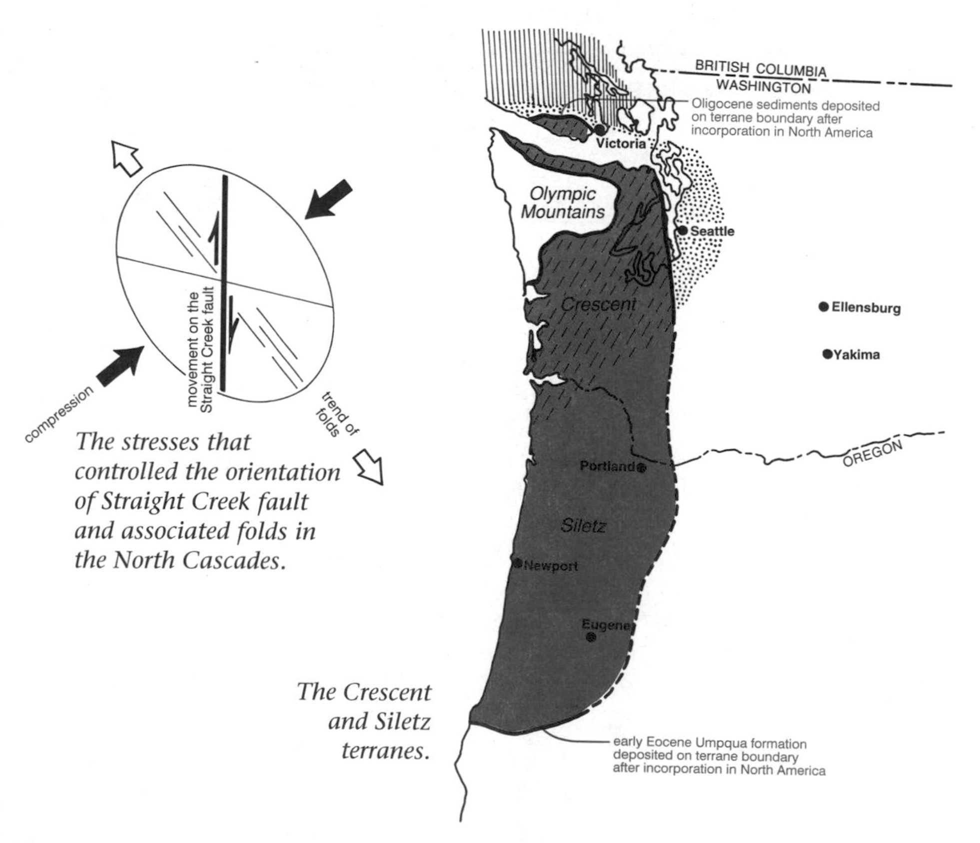

The stresses that controlled the orientation of Straight Creek fault and associated folds in the North Cascades.

The Crescent and Siletz terranes.

Age dates show that most of the basalt of the oceanic crust in the Crescent and Siletz terranes erupted during Eocene time, about 50 to 60 million years ago. Fossils in the oceanic sediments that cover them are the remains of animals that lived during Eocene and Oligocene times, as recently as about 35 million years ago. Those great slabs of ocean floor became part of the Pacific Northwest. Their eastern edges, which hold evidence of exactly when they docked, are mostly buried under Cascade volcanic rocks. The Umpqua formation, which dates from Eocene time buries the southern end. Younger sediments lap onto the contact in the southern end of Vancouver Island. Perhaps the Siletz terrane docked before the Crescent terrane.

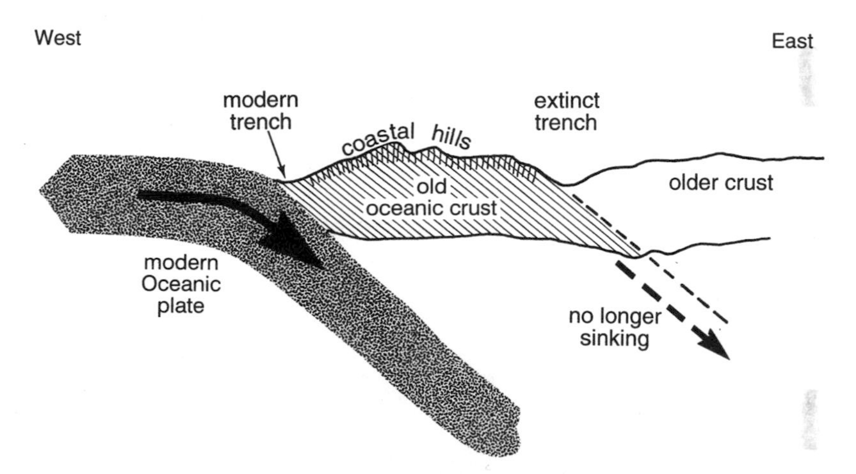

Oceanic crust stranded between an old trench and a new trench.

Southwestern Oregon contains large volumes of Eocene volcanic rocks, mostly basalt. Some geologists interpret them as the remains of a chain of volcanoes that erupted somewhere offshore and then landed as the trench gobbled up the intervening ocean floor. That interpretation explains why those rocks are there and why some of them are squashed. Although possible, it seems less likely that those formations are the southern part of a volcanic chain that erupted where we now see them. If so, they should continue northeast to connect with the Ochoco volcanoes of central Oregon, which also erupted during Eocene time. The connection, if it exists, is hopelessly buried beneath the much younger Cascade volcanoes.

The easiest way to imagine how the Crescent and Siletz terranes docked is to suppose that the oceanic trench jumped west. That would leave a broad strip of oceanic crust stranded between the old and new positions of the trench, unable to sink. It had been destined to sink through the old trench, then suddenly found itself landward of the new trench. If so, that ocean floor was in all likelihood simply attached to the western edge of the continent, not jammed into it by force. However it happened, the Crescent and Siletz terranes were still deep ocean floor when they joined the Pacific Northwest. Their rise above sea level is the subject of another chapter.

Chapter 25

CORE COMPLEXES

50 Million Years Ago

In a sense, the continental crust within the northern Rocky Mountains stretched during late Cretaceous time, as the early mountains shed their superstructure into western Montana. Quite a bit of evidence suggests that crustal stretching continued well into Tertiary time. Most of that evidence exists within the western part of the northern Rockies, the part where rocks that formed at great depth lost their cover of younger rocks. It is not clear whether the stretching was a continuing response to all the unloading of the late Cretaceous period, or an entirely new event.

Metamorphic core complexes form a belt west of the overthrust belt. —Modified from Crittenden and others, 1980

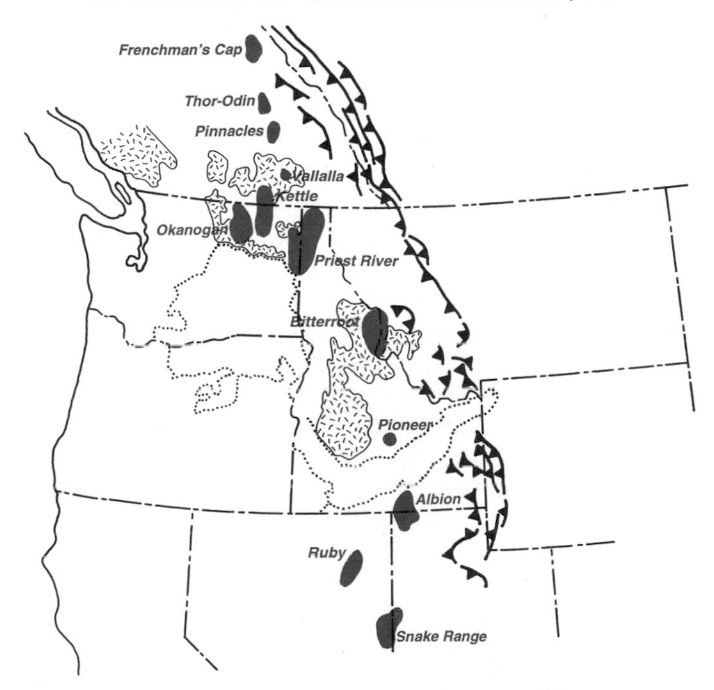

The Newport Fault, and Dead Herring Tectonics

The Newport fault is probably the best single item of evidence that the crust in the northern Rocky Mountains stretched sometime around 50 million years ago. It is a curious structure, a sort of hairpin open to the north and centered symmetrically on the border between Washington and Idaho. Enough mapping has been done to establish beyond any reasonable doubt that its northern end really is open.

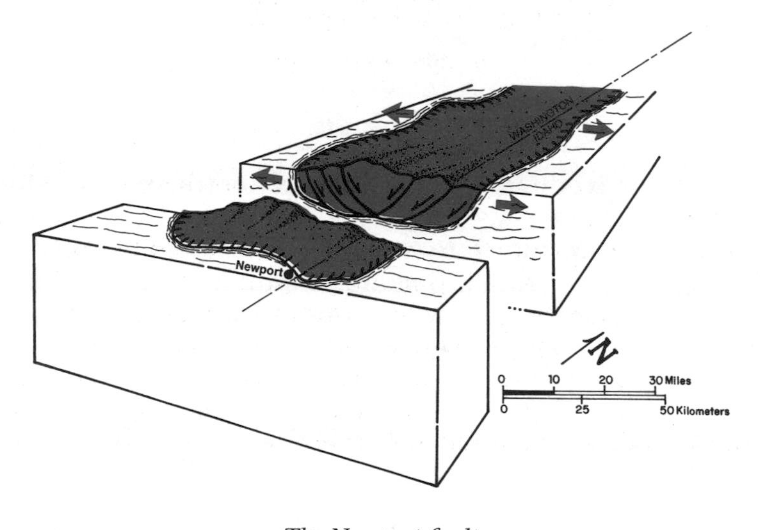

The Newport fault.

Rocks enclosed within the loop of the Newport fault all formed at shallower crustal depth than those outside; they appear to have dropped. A weakly developed shear zone along the fault contains textures of the kind that would form as the rocks beneath pulled out from under those above, a movement that would drop the rocks above the fault.

How, exactly, would the continental crust stretch? Some geologists now imagine it behaving like a sack full of dead fish. They picture gently curving and nearly horizontal faults breaking the continental crust into big pieces shaped about like fish that slip past each other as the faults move. As fault movement continues, some of the fish near the surface sink and others rise as they all pull away from each other. In that view, the rocks enclosed within the Newport fault are a sinking fish.

Where are the rising fish?

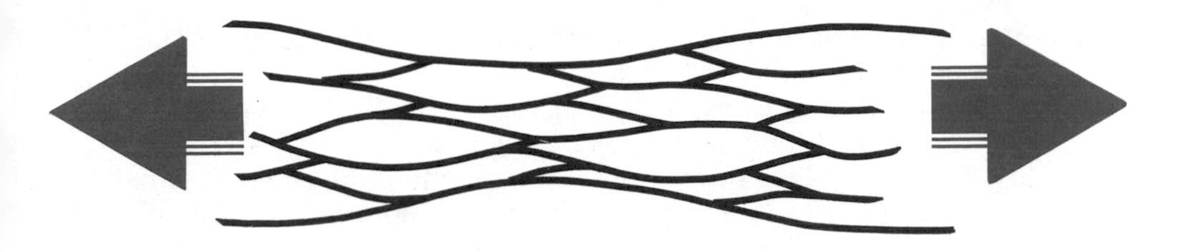

The fish model of continental stretching.

Core Complexes, the Rising Fish

A good many geologists now see the metamorphic core complexes scattered in the western part of the northern Rocky Mountains as the rising fish. A core complex is an area where rocks that had been deep within the crust rose to the surface, while the rocks that once covered them slid off to the side. Most are many miles across.

South and east of the Newport fault is the Priest River complex. Its central part, the Spokane dome, is one of the rising fish. Prominent stretch lines in the metamorphic rocks point east, the direction of movement of the cover rocks as the dome rose. The Bitterroot dome that rose as the Idaho batholith shed the Sapphire detachment is another core complex. Age dates appear to show that it

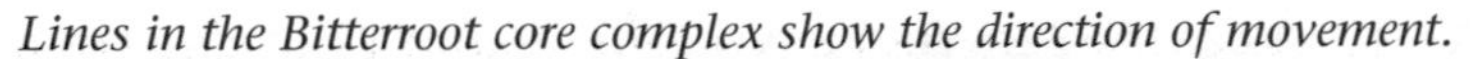

Lines in the Bitterroot core complex show the direction of movement.

began to develop during late Cretaceous time; movement either continued or renewed in Eocene time.

Other core complexes continued to rise here and there in the northern Rocky Mountains through much of the early part of Tertiary time. Many geologists now interpret the Pioneer Mountains near Stanley, Idaho, as the upper part of a core complex. The Albion Mountains of southern Idaho are another large core complex. Many others continue south in a narrow belt through Nevada and Arizona.

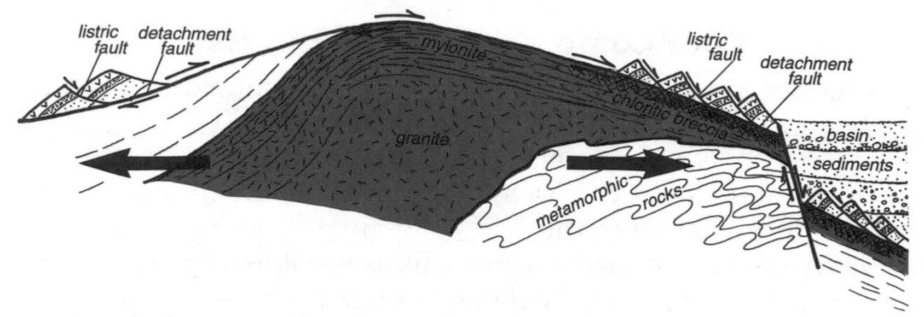

A typical core complex: deep rocks and granite in the center, younger rocks off to the sides. The deep rocks grade upward through a mylonitic fault zone to a sharp detachment fault capped by fault blocks that moved off to one side as the crust stretched. —Modified from Reynolds and Lister, 1990

Granite and metamorphic rocks that rose from deep in the continental crust are the center of a typical metamorphic core complex. The rocks that once covered them moved off a distance of some miles along a nearly horizontal zone of intensely sheared metamorphic rocks, a mylonite zone. The mylonite is commonly a half mile to a mile or more thick. Its shearing is most intense at the top and fades downward. A sharp detachment fault nearly parallel to the shearing cuts off the top of the mylonite. For several tens of feet below the detachment fault, the mylonite is chloritic breccia, a finely broken rock in which many of the original minerals are altered to pale green chlorite.

Above the detachment fault is a series of fault blocks, each sliding downslope on a fault surface that is concave upward. Imagine a boy holding a horizontal plate as he starts down a playground slide. As he slides down the curving surface, the plate tilts backward. Originally horizontal sediment layers in these fault blocks tilt backward as the blocks slide.

The deformation that produces the chloritic breccia, detachment fault, and the fault blocks above it is probably a result of crustal stretching. In our region, this deformation occurred in Eocene time, about 40 or 50 million years ago. The Kettle and Okanogan domes provide good examples.

Kettle and Okanogan Domes

Large masses of granite magma invaded the Okanogan Highlands during late Cretaceous time. Then, in latest Cretaceous or earliest Tertiary time, the Okanogan Highlands developed into a pair of core complexes: the Okanogan dome on the west, the Kettle dome on the east. The dropped fault block of the Republic graben, with its filling of volcanic rocks, separates them. Strongly developed mylonites flank both sides of the Okanogan Highlands.

As the Kettle and Okanogan domes rose, the granite and its surrounding metamorphic rocks were intensely sheared into mylonites on both sides of the Okanogan Highlands. The Kettle mylonite dips gently down the eastern side of the Okanogan Highlands toward the Columbia River, and the Okanogan mylonite dips gently down the western side of the Okanogan Highlands toward the Okanogan River. As in other metamorphic core complexes, strong lineations that point almost directly downslope record the direction of movement.

Mylonite in the Kettle dome, west of Kettle Falls, Washington.

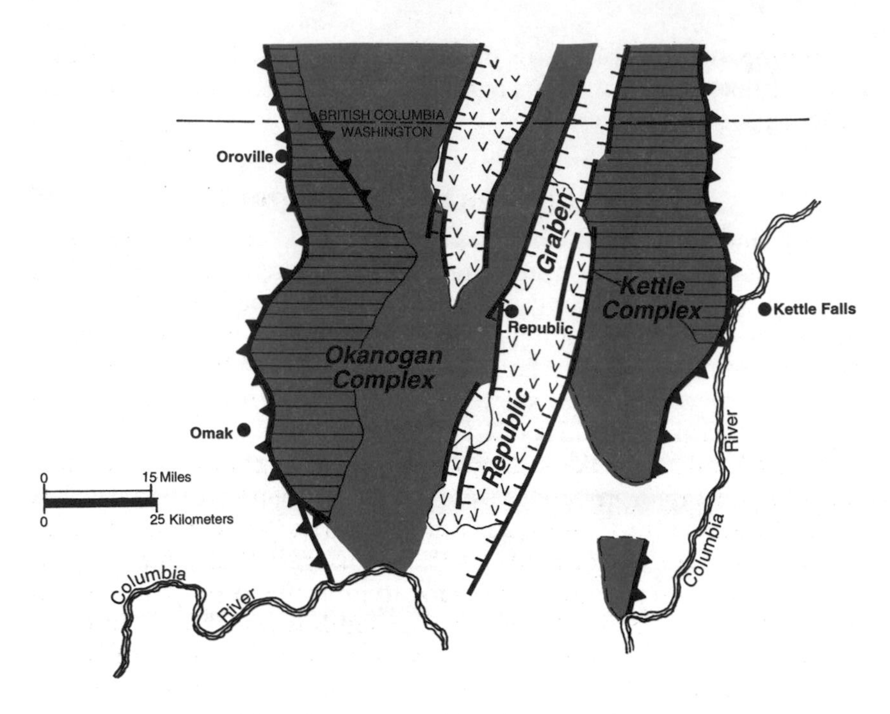

The Okanogan and Kettle domes lie on opposite sides of the Okanogan Highlands.

The final event in the drama happened during Eocene time, between about 55 and 45 million years ago. Some of the radioactive clocks that geologists use do not start running until the rock has cooled; those age dates show that the rocks cooled to surface temperature in Eocene time, so that must have been when they finally lost their roof. The continental crust of the Okanogan Highlands evidently stretched, pulling off the cover rocks and causing those below to float upward into a broad arch. Then the top of the arch dropped along faults in several slices to form grabens that trend generally north; the largest of these is the Republic graben. As usual, the rocks just below the gently dipping detachment fault are smashed into small pieces and altered to a pale greenish chloritic breccia. Meanwhile, andesite and rhyolite erupted in a chain of volcanoes that filled the grabens.

Bitterroot Dome

The Bitterroot dome, on the eastern flank of the late Cretaceous granites of the Idaho batholith, is another metamorphic core complex. The granite that crystallized deep beneath the surface was finally exposed as stretching crust pulled the last of its cover rocks off during Eocene time. Age dates on minerals in the granite indicate that it was exposed and cooled about 45 million years ago. Large masses of granite invaded the older Cretaceous granite of the Idaho batholith at about the same time. Some of those masses of magma erupted to make sheets of rhyolite.

The Okanogan mylonite forms a prominent surface sloping west into the Okanogan Valley.

Mylonite in the Priest River metamorphic core complex in eastern Washington. View is 30 inches wide.

Chapter 26

AN EOCENE VOLCANIC CHAIN

50 Million Years Ago

Eocene time brought new volcanoes to most parts of our region in a scattered and confusing pattern. Some of those volcanoes fit easily into the pattern of plate tectonic movements; most do not. We will start with those that do—the easy part.

The westernmost activity built a long chain of volcanoes parallel to the probable position of the oceanic trench, and almost certainly associated with it. The trench probably trended south somewhere west of where Puget Sound now is, then followed the western edge of the Klamath block and the present trend of the California Coast Range. It spawned a chain of volcanoes that stood somewhere between 50 and 150 miles inland, the usual distance.

Eocene Volcanic Rocks in Southwestern Oregon

Southwestern Oregon contains large volumes of Eocene volcanic rocks, mostly basalt. Some geologists interpret them as the remains of a chain of volcanoes that erupted somewhere offshore, then landed as the trench gobbled up the intervening ocean floor. That interpretation explains why those rocks are there, and why some of them are squashed. It is also possible to interpret those formations as the southern part of a volcanic chain that erupted where we now see them. If so, they should continue northeast to connect with the Ochoco volcanoes of central Oregon, which also erupted during Eocene time. But the connection, if it exists, is hopelessly buried beneath the much younger Cascade volcanoes.

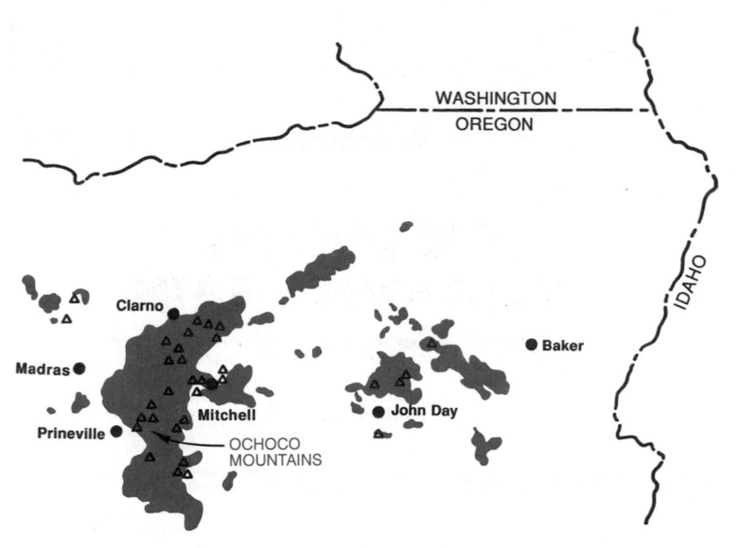

The Clarno formation. —Modified from Walker and Robinson, 1990

The Ochoco Mountains

The Ochoco Mountains of central Oregon are a volcanic pile that consists mainly of basalt and dark andesite, along with some rhyolite and an assortment of sedimentary rocks, deposited mostly during Eocene time, between 50 and 37 million years ago. Geologists call these rocks the Clarno formation. The sedimentary rocks locally include coal. Most large coal seams form in tidal swamps and marshes along the coast, so the coal in the Clarno formation strongly hints of a coastal environment. But that is only a hint because the exposed parts of the formation contain no sediments laid down in seawater. Any that may exist are buried under younger volcanic rocks.

Even so, the coal gives us some license to imagine the Ochoco volcanoes standing at the coast, looking west to the open ocean. They are probably a continuation of the chain of volcanoes of the same age that erupted in the North Cascades. And they may have continued southwest to the Eocene volcanoes in southwestern Oregon. East of those volcanoes lay an inland basin, now under the flood basalt flows of the Columbia Plateau. We do not know whether that basin was land or an inland sea.

Imagine sailing in from the Pacific Ocean through a narrow strait between towering volcanoes fringed with palm trees, then into a broad inland sea. That sea, if it existed, was rapidly filling with sediment brought in from the surrounding highlands of the Ochoco volcanoes, the North Cascades, the Intermountain terranes, central Idaho, and the Klamath block. The sediments are now deeply buried under the much younger basalt lava flows of the Columbia Plateau. They may contain oil.

Efforts to see how much sedimentary rock exists beneath the lava flows through studies of earthquake waves have revealed little. It is hard to decipher seismic records in areas where light rocks like sediments lie beneath dense rocks like basalt. A better clue comes from the simple observation that the heavy basalt lava flows of the Columbia Plateau stand high. That is possible only if they rest on a thick accumulation of much lighter rocks, such as sediments laid down in a shallow bay. The high elevation of the heavy basalt lava flows suggests that you may someday see oil fields in eastern Washington and Oregon.

Volcanic mudflows in the Clarno formation, Clarno, Oregon.

A number of the large peaks in the Ochoco Mountains look exactly like the deeply eroded remains of old volcanoes, and that is precisely what they are. After some 50 million years of weathering and erosion, those old ruins survive as perfectly recognizable elements of the modern landscape. We think they owe their survival mostly to their composition. Dark andesites tend to erupt quietly as lava flows that build a rather solid volcano. The paler varieties commonly erupt violently enough to convert most of the lava into rubble and ash, which erode rapidly. And they commonly erupt with enough violence to destroy the volcanic cone.

Although the Ochoco volcanoes trended somewhat west of north when they were active, they now trend somewhat east of north. They have rotated clockwise, as have most of the rocks in the Pacific Northwest. That rotation is probably a result of the shearing action of the Pacific plate moving north off the west coast.

Buried Eocene Volcanoes in Central Washington

The geologists who located a pair of deep wildcat wells drilled near Yakima expected them to penetrate the sedimentary rocks laid down in the inland basin that may have existed during Eocene time. As expected, they drilled through several thousand feet of basalt that erupted between 17 and 15 million years ago. Then, instead of passing into sedimentary rocks, the drills penetrated a much thicker section of volcanic rocks that erupted during Eocene time, about 50 million years ago. They never reached sediments; they never found oil. We suspect that the Eocene basalt and andesite are part of the northern extension of the Ochoco volcanic chain. The large Teanaway basalt flow in the North Cascades, west of the Yakima area, erupted during Eocene time from an unknown source somewhere to the east. Perhaps it erupted from one of those buried volcanoes.

Chapter 27

ANDESITE, RHYOLITE, AND GRANITE

55 to 45 Million Years Ago

Nearly anywhere you study them, igneous rocks appear to have formed during certain well-defined periods of activity, which in most cases reflect the changing patterns of plate movement. They form episodically, not continuously.

Few igneous rocks formed in our region during the 15 or 20 million years after the intense activity of late-Cretaceous time created the Idaho batholith and its relatives in western Montana. The late Cretaceous pattern of geologic activity ended with Cretaceous time. Then, during Eocene time, about 55 million years ago, igneous activity resumed in much of our region. Large volcanic piles erupted in the Okanogan Highlands, central Idaho, and western Montana. In all those places, andesite lavas appeared first, then rhyolite. And large masses of granite crystallized at shallow depths. Just what happened then and why remains a subject of vigorous debate.

The Great Eocene Dike Swarm

Thousands of dikes, vertical fractures that filled with magma to become sheets of igneous rock, cut through central Idaho and southwestern Montana. The individual dikes trend generally northeast, as does the swarm. In general, the dikes are largest and most numerous near the southwestern end of the swarm, and dwindle in size and number to the northeast. Almost all are andesites and rhyolites, generally granitic compositions. Age dates show that they are about 50 million years old.

If a fracture opens and fills with igneous rock, the earth's crust must unavoidably extend by the width of the fracture. It is equally unavoidable that thousands of such fractures must reflect considerable extension of the earth's crust. So the great Eocene dike swarm

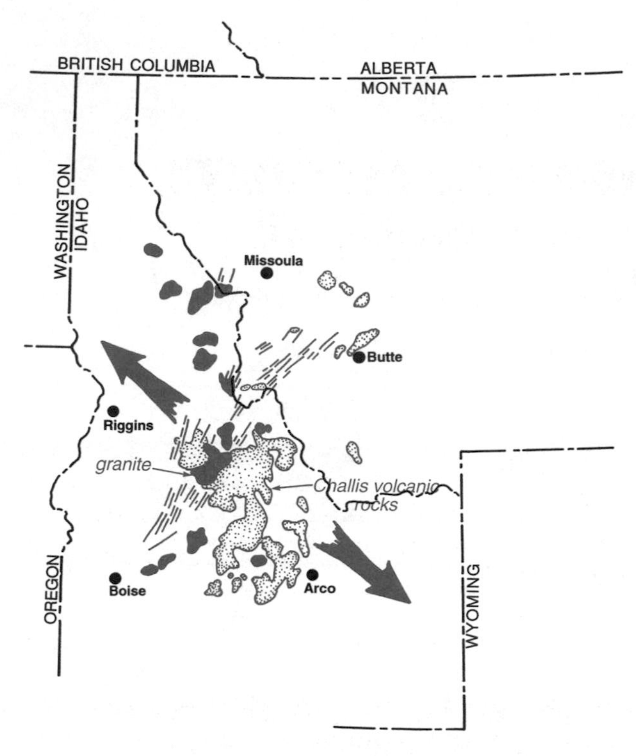

The Eocene dike swarm reaches from near Boise, Idaho, northeastward to the vicinity of Butte, Montana. The large arrows show the direction of crustal stretching.

is a sort of incipient crustal rift that did not open very wide. Some major event happened along its trend sometime within a few million years one way or the other of 50 million years ago.

Perhaps something stretched the crust from northwest to southeast, or possibly something compressed it from southwest to northeast. Either way, the mechanical effect would be the same. If some stretching or compression did happen, we would expect other faults or folds as well.

The Osburn fault zone cuts through northern Idaho and continues east into the Lewis and Clark line of western Montana, trending a bit north of west. In Eocene time, the rocks south of each fault moved west an unknown distance. If the crust in the area west of the big dike swarm pulled west to accommodate the dikes, that would at once explain the movement on that fault zone and why the dikes end at the Lewis and Clark line.

How the dike swarm relates to the pattern of plate movements in Eocene time is not clear. The generally east to west direction of crustal stretching is similar to that in the metamorphic core complexes. It is hard to imagine that movement of the Pacific Ocean floor against western North America could cause such an effect 500 miles inland, and equally hard to imagine a more likely cause.

The Sanpoil Volcanic Pile and the Republic Graben

The Sanpoil volcanic rocks erupted during Eocene time, between about 54 and 47 million years ago, along a broad zone that trends to the north approximately along the midline of the Okanogan highlands. They fill the Republic graben, a long slice of the crust that dropped between faults, and several lesser grabens.

No extinct Sanpoil volcanoes survive as landmarks in the modern landscape, but you can recognize the old volcanic chain by looking at the rocks. They consist mostly of rhyolite and the paler varieties of andesite—very little dark rock. Pale volcanic rocks commonly erupt in such great volume that the volcano collapses into the emptying magma chamber beneath, demolishing itself. That could explain the scarcity of volcanic landforms. The position of the Sanpoil volcanic rocks on top of the Okanogan Highlands explains their pale color; continental crustal rocks melt to form pale igneous rocks.

Eocene volcanoes that erupted rhyolite and andesite at about the same time as the Sanpoil volcanoes lie southwest of the Republic

Sanpoil volcanic rocks fill the Republic graben, a dropped fault slice of the Okanogan Highlands.

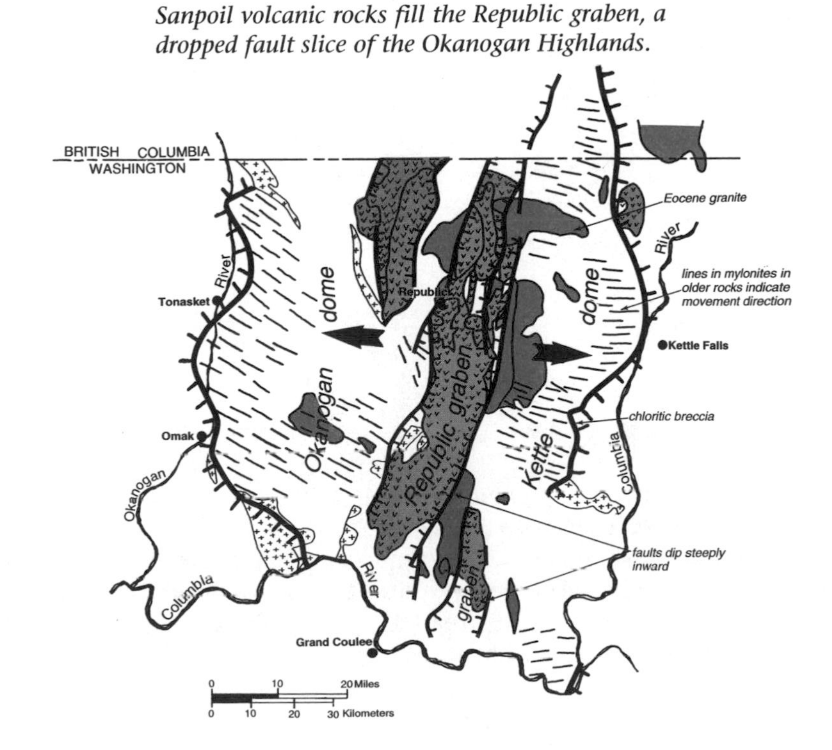

graben, through the area of Snoqualmie Pass on Interstate 90, to the area 15 to 30 miles east of Chehalis. Some geologists include them in the same chain with the Sanpoil volcanoes.

The Challis and Lowland Creek Volcanic Rocks

The Challis and Lowland Creek volcanic piles erupted during Eocene time, between 55 and 45 million years ago. The early eruptions produced large volumes of andesite. Then enormous volumes of rhyolite buried much of the andesite, raced across the landscape in great ash flows, and blew east in clouds of ash. No basalt erupted. Although some of the volcanoes were real monsters, only a few in the Challis area partly survive as distinctive peaks in the modern

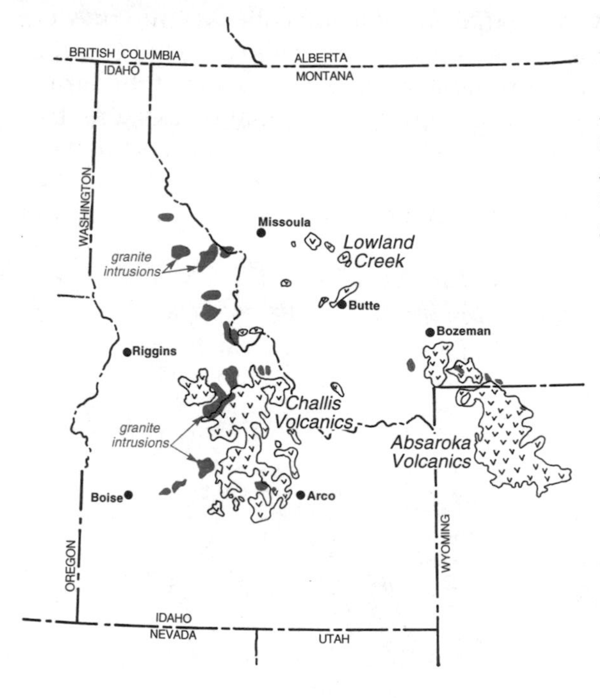

Major Eocene igneous centers in Idaho and Montana. Intrusions in solid color.

landscape, mainly because the lavas in and near the volcanic vent better resist erosion than the ash farther away. You can also see them in the rocks they produced, and in the patterns those rocks make on the geologic map.

Most of the volcanic rocks lie on or near the great swarm of Eocene dikes, have similar compositions, and are the same age. The Challis volcanic pile centers in the area where the dike swarm crosses the eastern edge of the Idaho batholith; the Lowland Creek pile is at its

Horizontal cordwood jointing in a vertical dike cutting the Challis volcanic rocks.

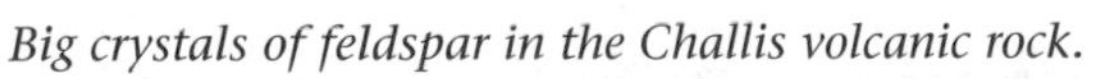

Big crystals of feldspar in the Challis volcanic rock.

northeastern end near the Boulder batholith. Those associations in time, place, and composition make it seem reasonable to suppose that the two volcanic piles may be somehow related to the dike swarm and to the big masses of late-Cretaceous granite.

In many places, rhyolite ash in the Challis and Lowland Creek volcanic piles contains large amounts of fossilized wood. In a few places, the Challis volcanic rocks contain large petrified logs, even standing stumps, petrified forests of dawn redwoods. The climate of Eocene time was evidently wet and mild.

Eocene Granite

While some of the magma of 50 million years ago erupted to become Challis or Lowland Creek volcanic rocks, the rest crystallized quietly beneath the surface to become granite. A good many of the granite intrusions lie within or near the big dike swarm; many others do not.

It is easy to distinguish between the Cretaceous and Eocene granites of the northern Rocky Mountains. The older granites consist of smaller mineral grains, are pale gray, and in some cases have a faint directional grain. The Eocene granites tend to be rather coarsely granular, distinctly pink, and perfectly massive—with no hint of a grain.

Layers of ash in the Challis volcanic pile.

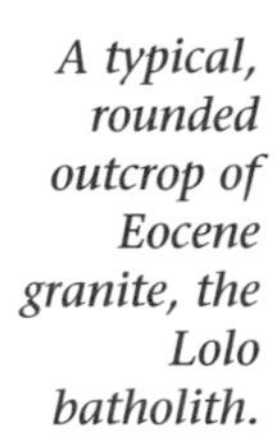

A typical, rounded outcrop of Eocene granite, the Lolo batholith.

The rising magma either exploded into rhyolite ash or crystallized below the surface to become granite. Which happened was a close call. All the Eocene granite crystallized at shallow depths, within a few thousand feet of the surface. Many of the intrusions contain masses of rhyolite, old volcanic vents that erupted part of the magma before the rest crystallized. In some places, it is hard to decide whether the rock is granite trying to look like rhyolite, or rhyolite trying to masquerade as granite. When it is hard to tell two kinds of rock apart, the distinction is probably not important.

Most of the Eocene granites are full of tiny cavities that mark places where gas bubbles formed while the magma was crystallizing. They are good evidence that it crystallized at a very shallow depth. Most are so small that you need strong light and a good magnifier to appreciate the perfectly formed little crystals that line them. Some of the Eocene granites also contain a few much larger cavities lined with beautiful crystals as much as an inch or two long. Those typically include flawless dark crystals of smoky quartz, clear and splendid enough to facet into lovely gems. The Lolo Hot Springs granite exposed along U.S. 12 at the border between Montana and Idaho contains gas bubbles and smoky quartz.

What Happened?

The Sanpoil, Challis, and Lowland Creek volcanic piles are all in areas where large masses of granite crystallized during late-Cretaceous time, as are the Eocene granites. The Sanpoil and Challis volcanic rocks erupted in places where core complexes developed, apparently as the crust stretched. The Lowland Creek volcanic pile is nowhere near a core complex.

How and why did all that magma melt when and where it did?

Some geologists contend that all the magmas of Eocene time melted above a slab of ocean floor that was then sinking beneath the continent at an extremely low angle. They point to the broad distribution of Eocene igneous rocks, all the way from the Eocene volcanoes of central Oregon and the North Cascades east through central Montana. But it is hard to imagine how a slab of ocean floor could sink at such a low angle, and, if it did, how it could avoid compressing the continent all the way. Furthermore, the rocks in the Clarno volcanic pile and in the North Cascades look like a perfectly normal volcanic chain that erupted at the usual distance from the trench. If so, the sinking slab of oceanic crust was descending steeply, not at an extremely low angle.

It is easier to imagine that the rhyolite magmas melted as crustal stretching took the roof off the big core complexes, thus reducing the pressure on the late-Cretaceous granite beneath. Ordinarily, pressure relief would melt rhyolite magma only very deep within the continental crust, where the rocks are already very hot. But in this case, the magmas of Eocene time appeared in areas where large masses of granite had crystallized during late Cretaceous time. A large mass of granite would still be very hot 20 million years after it crystallized, so the magmas of Eocene time may have melted at shallow depths. The Lowland Creek volcanic rocks are nowhere near a core complex, but they are in the area where the dike swarm could feed them from the direction of the Challis volcanic pile.

It is much harder to explain the andesite magma, which melts at a higher temperature than rhyolite. Pressure relief is not enough; heat is also needed. But it does seem reasonable to suppose that enough pressure relief to melt rhyolite at a shallow depth would also melt basalt in the upper mantle, beneath the continental crust. Molten basalt rising into the granite could easily import enough heat to melt it, creating andesite magma. This idea seems plausible in most ways, but it does not clearly explain why andesite erupted before rhyolite in the Sanpoil and Challis volcanic piles. And the absence of basalt in both areas is hard to explain.

The melting temperature of granite magma rises or falls with dropping pressure in ways that depend upon its steam content. If the magmas of Eocene time had been heavily charged with steam, they would have crystallized into granite far below the surface, as their rise into regions of lower pressure raised their melting point. The melting points of dry magmas drop with falling pressure, so they stay melted as they rise to crystallize at very shallow depth, or even erupt. The magmas of Eocene time must have been very dry when they melted.

The climate of Eocene time was wet, so plenty of water was available at the surface. As the rising rhyolite or andesite magma absorbed water, it became capable of quietly crystallizing into granite, or of violently erupting as sheets of white rhyolite ash. The masses of magma that absorbed water at depths of several thousand feet had enough room above them to rise far enough to raise the melting temperature of the magma to its actual temperature. Those masses of magma became granite. Bodies of magma that rose to shallow depths before they absorbed water were still molten when they broke the surface, erupting in great steam explosions. The margin between granite and ash must have been very close in all cases.

Beige volcanic ash of the Eocene Challis volcanic pile, eight miles north of Challis, Idaho.

Chapter 28

PECULIAR IGNEOUS ROCKS IN CENTRAL MONTANA

55 to 45 Million Years Ago

While the Challis and Lowland Creek volcanic rocks and their associated granites were forming in Idaho and western Montana, other quite different and varied igneous activity was underway across a wide area of central Montana. Almost all of the rocks are strange, in one way or another. The Absaroka volcanic rocks are perfectly normal andesites, but they are in the wrong place. Most of the other rocks are abnormally rich in either potassium or sodium, or both. We know of no theory that even begins to explain why they formed when and where they did, or why their composition is so peculiar.

The Absaroka Volcanic Pile

The Absaroka volcanic rocks erupted within a few million years, centered on about 50 million years ago, during Eocene time. They covered a roughly elliptical patch of northwestern Wyoming and nearby Montana. They are all andesites, rocks that would look perfectly at home in the Cascades.

The typical native habitat of andesite is in long volcanic chains that erupt along a line parallel to an oceanic trench and 100 or so

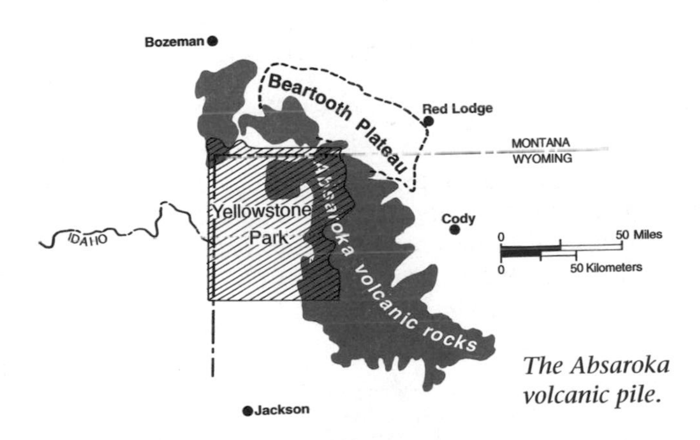

The Absaroka volcanic pile.

Petrified tree trunks near Yellowstone Park —J. R. Stacy, U.S. Geological Survey

miles away from it. The Cascade volcanoes and the Aleutian Islands are typical andesite chains. The Absaroka volcanic pile is just a pile, not in any sense a chain. And it is more than 500 miles east of the probable position of the trench. They have absolutely no theoretical business erupting where and when they did.

Visitors to the northeastern part of Yellowstone Park see the petrified forests in the Absaroka volcanic pile, its main claim to fame. They are the remains of forests that were transported and buried in great volcanic mudflows. Much of the wood is so well preserved that botanists can identify the species of the trees. It seems that the mudflows started high on the slopes where spruce and fir trees grew, then swept down into lowlands where tropical hardwood trees flourished. That happens today on the high volcanoes in Central America.

Those buried forests are another of many items of evidence that tell us of a wet and warm climate during Eocene time. That wet climate almost certainly explains the scarcity of Eocene sedimen-

Mudflow deposit in the Absaroka volcanic pile.
—J. R. Stacy, U.S. Geological Survey

tary formations deposited on land. The climate of Eocene time was seemingly wet enough to maintain streams capable of carrying eroded sediment to the ocean.

Three Big Shonkinite Volcanoes

Project the trend of the great Eocene dike swarm northeast beyond Butte, and it passes through three big volcanic piles: the Adel, Highwood, and Bearpaw Mountains of central Montana. Age dates are sparse and probably not altogether reliable, but the few that exist suggest that at least the Highwood and Bearpaw volcanoes were erupting at about the same time as the Challis and Lowland Creek volcanoes. The Adel Mountains volcanic pile may have erupted during latest Cretaceous time.

The dominant rock in those aligned volcanic centers is shonkinite, a rare and odd rock with the composition basalt would have if it

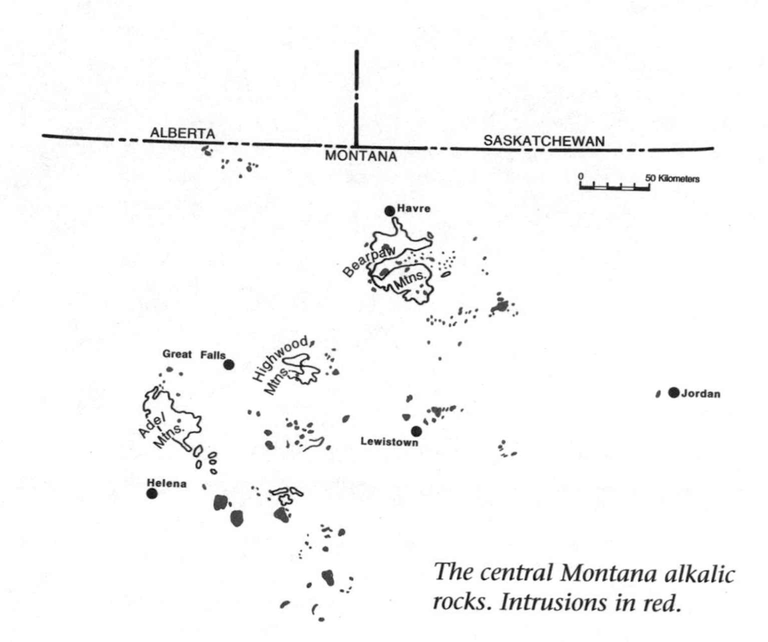

The central Montana alkalic rocks. Intrusions in red.

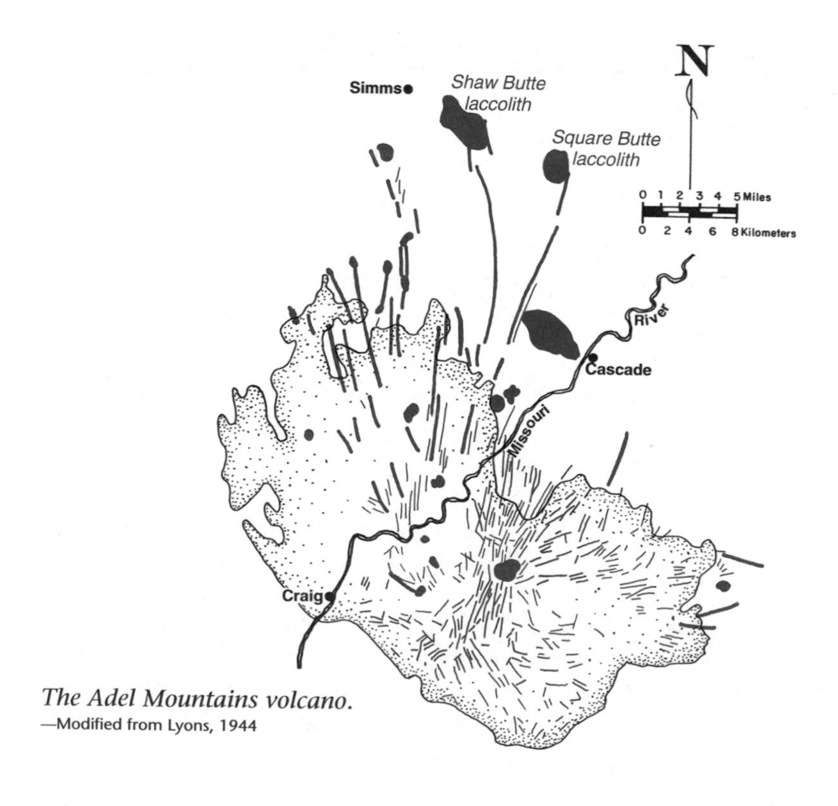

The Adel Mountains volcano.
—Modified from Lyons, 1944

were greatly enriched in potassium. A close look reveals shapely crystals of glossy black augite pyroxene set in a darkly greenish matrix of tiny crystals of augite and potassium feldspar.

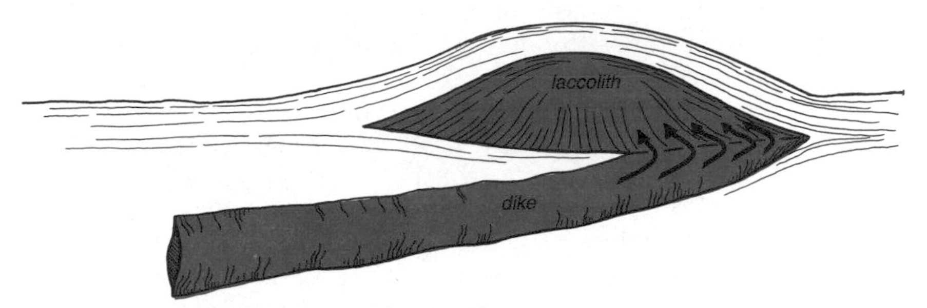

A dike with a laccolith at its end. A typical laccolith is a mile or so across and several hundred feet thick.

After most of the volcanic rocks had erupted, the last of the shonkinite magma intruded them to form swarms of dikes and great blisters of igneous rock called laccoliths. The laccoliths form at the farthest ends of the dikes, where they roll over and inject magma between layers of older rock. The Adel Mountains may be the best place in the world to see dikes connected to laccoliths that extend

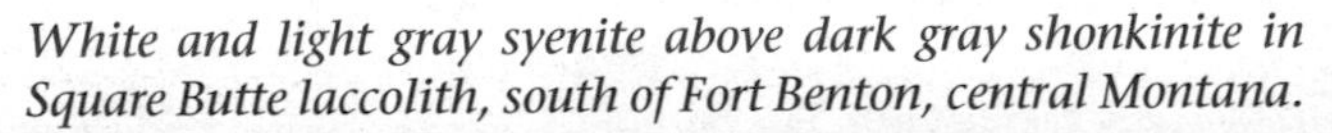

White and light gray syenite above dark gray shonkinite in Square Butte laccolith, south of Fort Benton, central Montana.

laterally from their tips, instead of standing above them like mushrooms on a stem in the way of the standard diagrams in geology textbooks.

A few of the laccoliths have a thin cap of white syenite lying on a much thicker base of dark shonkinite. The syenite consists almost entirely of white potassium feldspar, lightly peppered with crystals of black augite.

For many years, geologists believed that the syenite formed as growing crystals of feldspar floated up through the shonkinite magma to accumulate at the top of the laccolith, while the heavier crystals of black augite sank. Shonkin Sag laccolith in the Highwood Mountains became the classic example of a body of rock that separated into layers of different composition as crystals floated and sank through the molten magma. If that did happen, you would expect the biggest crystals of heavy augite to settle to the bottom of the laccolith while it was still molten, and you would expect the dark shonkinite to shade upward into the pale syenite. But the biggest augite crystals are not at the base, and the contact between shonkinite and syenite is perfectly sharp.

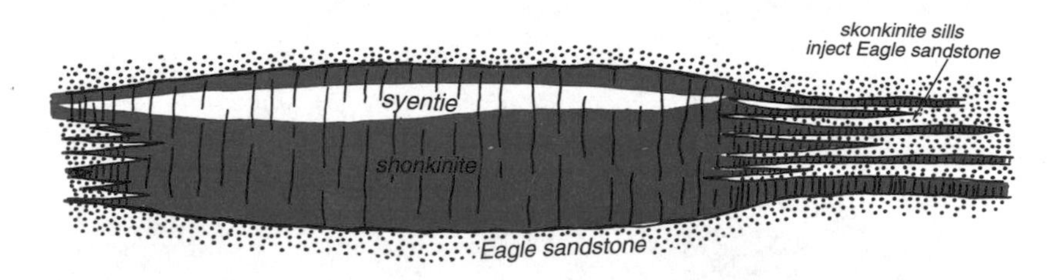

Shonkin Sag laccolith has white syenite over black shonkinite.

The dark shonkinite in several laccoliths contains round blobs of white syenite about the size of citrus fruit. Those could not form through any process of crystal floating and settling. They make it seem likely that the original magma separated into two immiscible fractions, the way a cruet of salad dressing separates into a layer of oil above and vinegar below. The lighter syenite magma floated to the top of the laccolith, leaving a few blobs suspended in the shonkinite magma below.

The Crazy Mountains

Tradition has it that the Crazy Mountains of central Montana were originally called the Crazy Woman Mountains in honor of an early settler. They could as well have been named for their weird rocks, most of which resemble shonkinite except in being enriched in sodium instead of potassium.

The southern Crazy Mountains are a high range of ragged peaks in which ice-age glaciers carved the only alpine landscape in central Montana. Their core is a large mass of igneous rock that invaded the Fort Union formation, baking it so hard that it resists erosion as well as the igneous rocks it encloses. The northern part of the Crazy Mountains contains numerous much smaller intrusions of generally similar igneous rocks. While the magmas baked the Fort Union formation, they also loaded it with sodium and iron. The result is a type of contact metamorphic rock called fenite, which is as rare and peculiar as the igneous rocks that created it. The iron silicate minerals color some of the fenites a clear shade of green that would look good in a St. Patrick's Day parade.

Alkali Syenite Intrusions

Many of the isolated buttes and clusters of buttes widely scattered across central Montana are large intrusions of another breed of peculiar igneous rock called alkali syenites. Unlike shonkinite, which is much the same everywhere, the alkali syenites vary greatly. Most are pale rocks that consist primarily of white feldspars, along with scattered dark minerals and little or no quartz. Most contain considerably more than their fair share of potassium or sodium. A few contain enough quartz to qualify as granite. Their ages range from 65 to 35 million years, with most between 55 and 45 million years.

The alkali syenite intrusions appear to have arrived one after the other while large blobs of magma, each of several cubic miles, rose into the sedimentary rocks beneath the High Plains. Had you been there then, you might have seen a low hill slowly bulge the surface over a period of a few months. All the intrusions seem to have crystallized within a few thousand feet of the surface. A few actually broke the surface and erupted. Now, 50 million or so years later, erosion has stripped the cover of sedimentary rocks off many of those intrusions. The more resistant igneous rocks stand high, in bold erosional relief, to make mountains and clusters of mountains. Sharp bulges that dome the sedimentary rocks of the plains, making hills and groups of hills, probably mark places where large igneous intrusions are still buried.

Most of the alkali syenite intrusions contain uncommonly high concentrations of gold, and a good many added gold deposits to the older rocks that enclose them. A few also brought silver. Prospectors swarm through the hills when the price of gold and silver is high. Some of the mines have produced large amounts of gold.

Diatremes

Diatremes are distinctive little igneous intrusions, dikes, or small plugs that typically cover only a few acres. Their rocks tend to weather easily, so they are hard to find. Nevertheless, several dozen are known to exist in central and western Montana. Quite a few more no doubt await discovery.

Broadly speaking, all diatremes consist of peridotite, the dense black igneous rocks of the earth's mantle. In detail, they come in varieties that differ in their proportions of augite pyroxene, garnet, and mica. Most diatremes are a chaotic mass of rounded fragments of igneous rocks mixed with rounded pieces of the older rocks they intrude. Some of the chunks of older rocks came up from below; others came down from above.

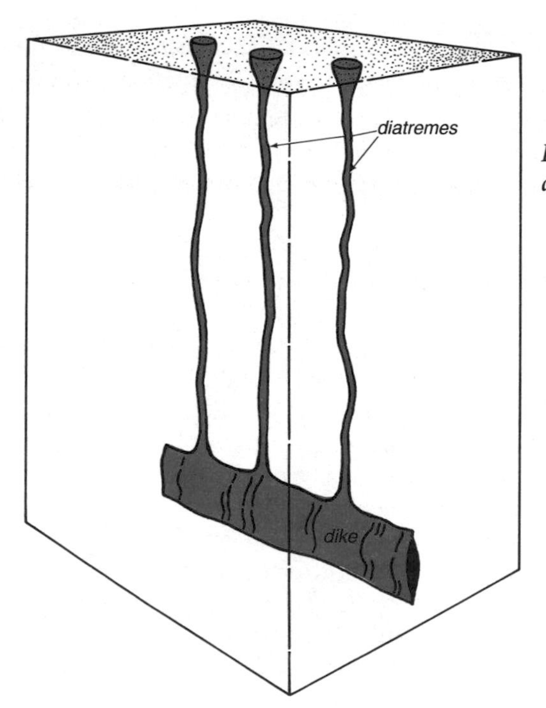

Diatremes merge downward into dikes.

The generally accepted explanation for the formation of a diatreme envisions the magma blowing off tremendous volumes of carbon dioxide as it rises through the crust. The escaping gas breaks an upward path through the older rocks. It breaks the crystallizing igneous rock and keeps all the pieces milling about in suspension long enough to round their sharp edges.

Diamonds and Sapphires

About 10 percent of known diatremes contain diamonds, including at least one in central Montana. That tells something of their origin. Diamonds crystallize under such extremely high pressure that they must form deep within the mantle, then rise into the crust with the magma. It follows that diatremes rise from great depth in the mantle, instead of from the shallower regions that probably generate most basalt magma. In 1988, a jogger picked up a magnificent yellow diamond from a path in Craig, Montana, a few miles south of Great Falls. It sold, uncut, for some $80,000. Despite that find, the diatremes seem unlikely to support a new mining industry because most industrial diamonds are now made synthetically.

Several diatremes in central and western Montana have produced large numbers of beautiful sapphires, gems quite unrelated to diamonds except that both are found in similar rocks. Diamonds consist of carbon, sapphires of aluminum oxide. Montana sapphires were originally mined mainly for use as bearings in watches and delicate instruments. Ever since the advent of inexpensive synthetic sapphires killed that market more than 70 years ago, the Montana sapphire mines have operated off and on to produce gems, always in precarious financial circumstances. Stones from Yogo Gulch in central Montana are widely celebrated for their rich color, a deep blue with no hint of the greenish tinge typical of most blue sapphires. Other Montana mines produce sapphires in a rainbow of brilliant colors. The largest of those deposits is west of Philipsburg, where the sapphires seem to occur in white syenite instead of dark peridotite.

Square Butte, a shonkinite laccolith west of Great Falls, Montana.

Chapter 29

A NEW GENERATION OF MOUNTAINS

50 to 30 Million Years Ago

As Cretaceous time ended, the new northern Rocky Mountains looked east from the overthrust belt across a broad plain that reached all the way to the shifting shores of the shallow inland sea. That sea, then in the Dakotas, was in its final retreat. The elevation of that plain must have been very near sea level. It was nearly featureless, except perhaps for a few central-Montana volcanoes marking the sites of early alkali syenite intrusions. A swarm of new mountains would rise from that plain during Eocene time.

The Cypress Hills of southern Alberta and Saskatchewan may be the northern end of those mountains east of the overthrust belt. They become progressively more impressive southward from central Montana through Wyoming, reaching a magnificent culmination in Colorado, where all the mountains stand east of the overthrust belt. Then they dwindle southward into New Mexico.

Thrust Fault Mountains

Most of the mountains east of the overthrust belt are blocks of the earth's crust that rose along faults. Many of those rising blocks broke and folded the Fort Union formation, which was deposited during Paleocene time. But they did not much disturb the Renova formation, which began to accumulate about when Eocene time ended. That brackets their rising somewhere within the 21 million or so years of Eocene time.

Geologists used explosions to make artificial earthquakes and recorded their echoes from the rocks to make deep seismic profiles across several of the mountain blocks in Montana and Wyoming. Those seismic sections clearly show faults that dive steeply into the continental crust beneath the mountains. The mountain blocks rising along those steep thrust faults carried older rocks over younger

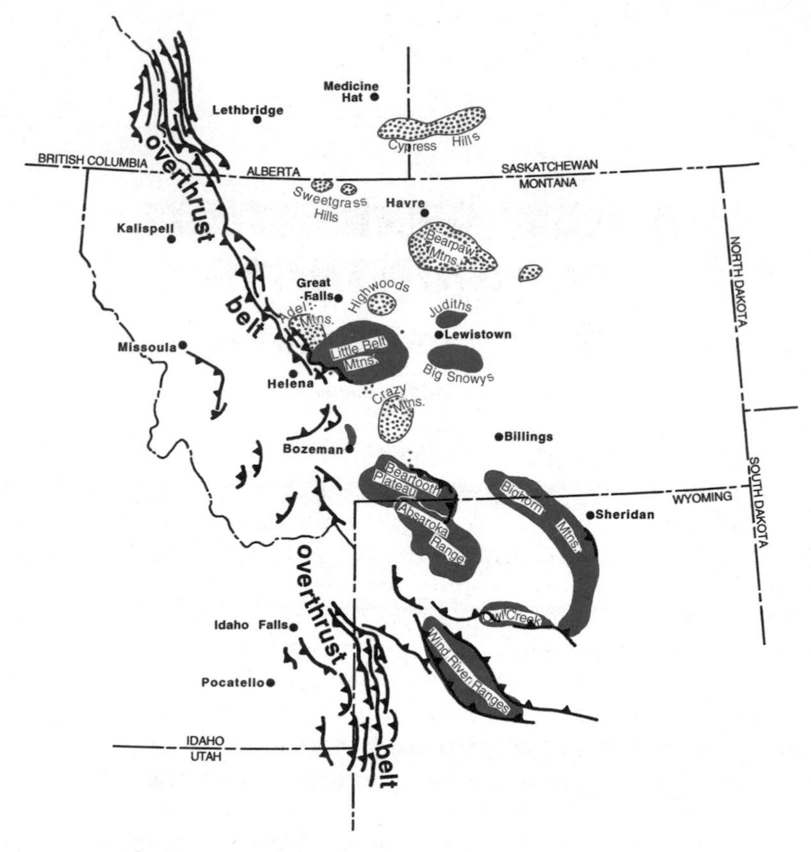

The overthrust belt, and the younger mountains that rose east of it, after the northern Rocky Mountains formed.

rocks. Some brought basement rocks to the surface in a vertical lift of more than two miles.

Movement along such thrust faults shortens the earth's crust, so these mountain blocks rose during one or more episodes of compression. Some of the ranges, such as the Beartooth Plateau along the border between Montana and Wyoming, moved generally north along a fault that trends southeast. Others, such as the Big Snowy Range of central Montana, moved south along a fault that trends east. Those ranges and their faults suggest that the region was compressed in a north to south direction. But the Bridger Range of central Montana apparently moved west along a fault that trends generally north, and the Bighorn Range of northern Wyoming moved east along a fault that trends north. Those ranges suggest compression in an east to west direction. The picture does not make sense.

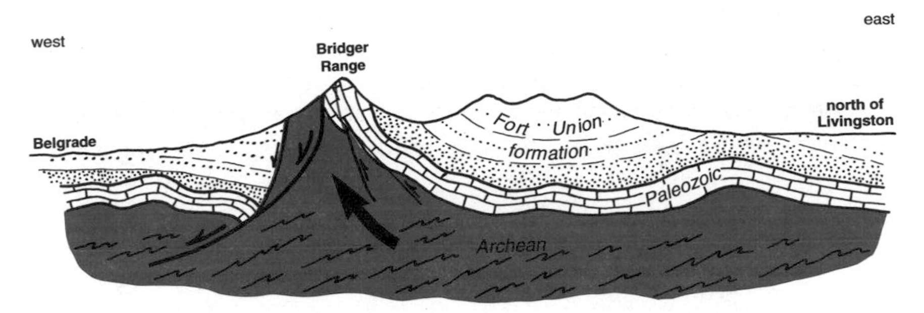

Section through the Bridger Range of central Montana.

It hardly seems possible for compressive forces to act in different directions at the same time. And how could crustal compression raise mountains east of the overthrust belt while crustal stretching was creating metamorphic core complexes within the western part of the northern Rocky Mountains? Perhaps different things happened at different times during those 21 million years of Eocene time. Or perhaps the different orientations may reflect the inhomogeneous nature of the continental crust in a region that contains ancient basement rocks broken by old and young faults, thick and thin sedimentary rocks in different places, a major overthrust belt, and igneous intrusions. More field work and better age dates may eventually clarify and simplify the picture.

Many geologists think that the rise of the mountain ranges east of the overthrust belt was the last act in the drama of the Rocky Mountains. They try to explain the younger mountains by relating them in one way or another to the great mountain-building events of late Cretaceous time. But none of those suggestions explains why the younger mountains do not follow the Canadian Rockies northward through Alberta. It is more likely that the events of Eocene time were unrelated to those that created the Rocky Mountains.

Lake Basin and Other Faults

People driving Interstate 90 a few miles east of Billings, Montana, see tilted layers of brown sandstone as they pass through the Lake Basin fault zone. It is one of several parallel and evenly spaced systems of faults that trend generally west to northwest across central Montana. They all moved horizontally, with the side north of the fault zone sliding to the northwest, and the land to the south rising. Parallel zones became large folds that also trend west to northwest. These include the Cat Creek anticline, north of the Lake Basin fault zone. It produced large amounts of oil.

Those faults break the Fort Union formation, so they must have moved sometime since it was laid down during Paleocene time, between 65 and 55 million years ago. Beyond that, it is hard to find further direct evidence of their age. We strongly suspect that they

Northwest-trending fault zones in central Montana.

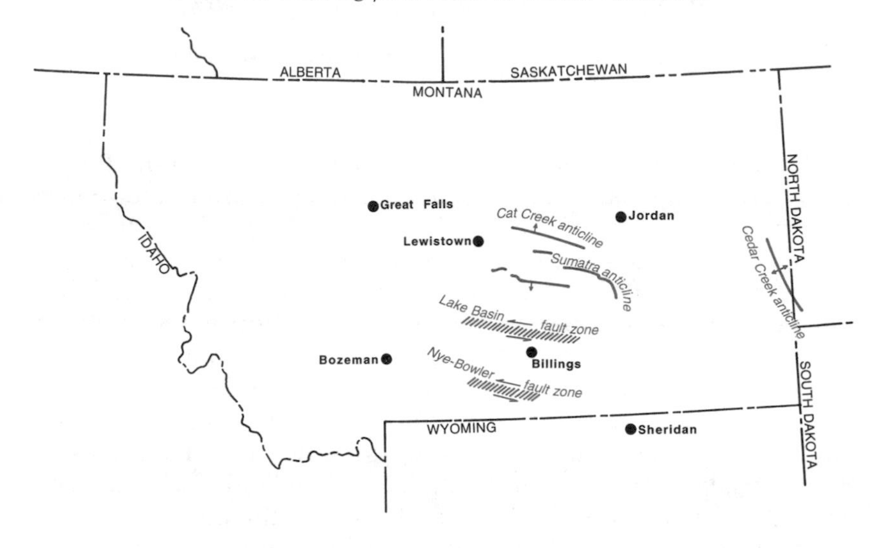

Steeply tilted layers of sandstone rise from the plains in the Cat Creek anticline.

are another aspect of the varied activity of Eocene time, but we can cite no solidly confirming evidence.

Cedar Creek and Other Anticlines

At what was probably about the same time, large folds were buckling arches into the High Plains east of the new mountains. Those folds tilt the Fort Union formation, but not the White River Oligocene beds. That brackets their formation somewhere within Eocene time.

The Cedar Creek anticline trends through eastern Montana, then southeast into South Dakota. It does very little for the landscape because the rocks along it are too soft to resist erosion. A parallel and much larger anticline brings older rocks to the surface in the Black Hills of South Dakota. The very resistant basement rocks in the center of the Black Hills and the Paleozoic formations that surround them in a series of concentric ridges make a spectacular small mountain range. Eocene igneous rocks in the Black Hills reinforce the conclusion that the range rose then.

Ranges and Basins in Western Montana

The mountain ranges and large valleys of western Montana broadly express the faults in the bedrock. Some of those ranges and valleys are clearly forming now, but most are much older, the faults that define them completely inactive. Most of the broad valleys contain a deep fill of sediments that began to accumulate during Oligocene time, so they must have formed before then. Some of those older mountains and valleys cut across sedimentary formations laid down during late Cretaceous time, so they must have formed after those were deposited. That brackets their formation to early Tertiary time, around 50 million years ago.

Many geologists argue that the older mountains and broad valleys appeared during Eocene time, during an episode of crustal stretching. Rhyolite and basalt that erupted during Eocene time reinforce their argument. One easy way to melt rhyolite and basalt magma at the same time is to stretch the crust, thus relieving the pressure on rocks deep within the continental crust and the mantle below it.

Occasional earthquakes, some of them quite large, leave no doubt that some of the mountain ranges in central Idaho and western Montana are rising now, as the valleys between them drop. Those mountains and valleys trend generally northwest, and that makes it tempting to view them as the northern end of the Basin and Range province.

The Rise of the High Plains

The shallow inland sea still flooded most of the region that was to become the High Plains during late Cretaceous time, so they were not high then. Large areas in the northern part of the region were still flooded well into Eocene time, so the region was near or below sea level that recently. Now it is high and rises westward to elevations as much as several thousand feet above sea level. That is a spectacular rise, a major geologic event by any reckoning. No one seems to know when it happened or understands why.

We know of no direct evidence that tells when the High Plains rose. Highly circumstantial evidence leads to the suspicion that they probably rose during Eocene time: That was when the last vestiges of the great inland sea that had flooded the broad middle of the continent ever since early Paleozoic time finally drained. And that was when new mountains rose east of the overthrust belt.

Many geologists believe that the rise of the High Plains was somehow associated with the formation of the Rocky Mountains, even though some time may have separated the two events. It is true that the High Plains slope down to the east, away from the front of the Rocky Mountains. And it is also true that similarly high ground also slopes down to the east from the Andes Mountains of South America. So the association of the High Plains with the Rocky Mountains is probably not a coincidence. But why should formation of a mountain belt cause the adjacent land to rise vertically?

Continents float. Areas of continents stand at certain elevations because that is how high they float. If a large area of the continental crust rises vertically, then it is floating at a higher level. That may reflect some increase in the thickness of that part of the continental crust, or perhaps a decrease in its density. We know of no direct evidence that the crust under the High Plains is either thicker or less dense now than it was 50 million years ago, nor is it easy to imagine any good reason why it should be.

Many geologists believe that the Eocene igneous rocks of central Montana formed above a slab of oceanic crust that was then sinking at a very shallow angle, moving almost horizontally beneath Montana and Wyoming. Many of them argue that the continent overrode the sinking slab, pushing it down to a gentler angle. They believe that the slab is still there, raising the continental crust to make the High Plains high. It is very hard to imagine how a coherent slab of oceanic crust could move that far under the continent—but if it did, its presence underneath would indeed float the continental crust higher.

A look at the Andes Mountains of South America makes the idea of a slab of oceanic crust under the High Plains seem slightly more palatable. Volcanoes in the high country east of the Andes are now erupting peculiar igneous rocks similar to those in central Montana, so the situation there may correspond to the one that existed in Montana some 50 million years ago. And earthquakes originating in the slab of oceanic crust now sinking beneath the Andes show that it is descending at a relatively flat angle, although not nearly as flat as would be necessary to explain the situation in central Montana.

The front of the Beartooth uplift, southwest of Billings, Montana.

Chapter 30

THE COASTAL HILLS

50 Million Years Ago, and Continuing

For humans, the deep ocean floor is the most remote part of the earth's surface, accessible only to people peering through the thick windows of submarine research vessels. It is fantastic to think of walking there, looking at abyssal sediments laid down thousands of miles from land. But you can do exactly that in the Coast Range of Oregon and the Willapa Hills of southwestern Washington, where oceanic crust is now high and dry.

Old Ocean Floor

Most of the rocks you see in the long stretch of coastal hills between the Klamath block in southern Oregon and the Olympic Peninsula in northern Washington are muddy sandstones in shades of dark gray and brown. They were deposited on oceanic basalts of the Crescent and Siletz terranes. Geologists call such sediments turbidites because they were deposited from clouds of muddy water that poured down the continental slope offshore and then billowed across the deep ocean floor. In a good exposure, you can see that the sand grains are largest at the base of each layer and become smaller upwards. That is the order in which the grains settled out of the water. You can make a similar deposit by shaking up a cloud of sandy mud in a bottle of water and then letting it settle.

The turbidites in the coastal hills were deposited somewhere offshore on the floor of the Pacific Ocean, probably under some two miles of water. They were still accumulating during Eocene time, and on into Oligocene time. The shoreline was well inland of its present location then, although the cover of younger volcanic rocks in the Cascade ranges makes it impossible to see the rocks that might tell exactly where. Sandstone, silt, shale, and coal were accumulating east of the shoreline. Those rocks include the Puget group of

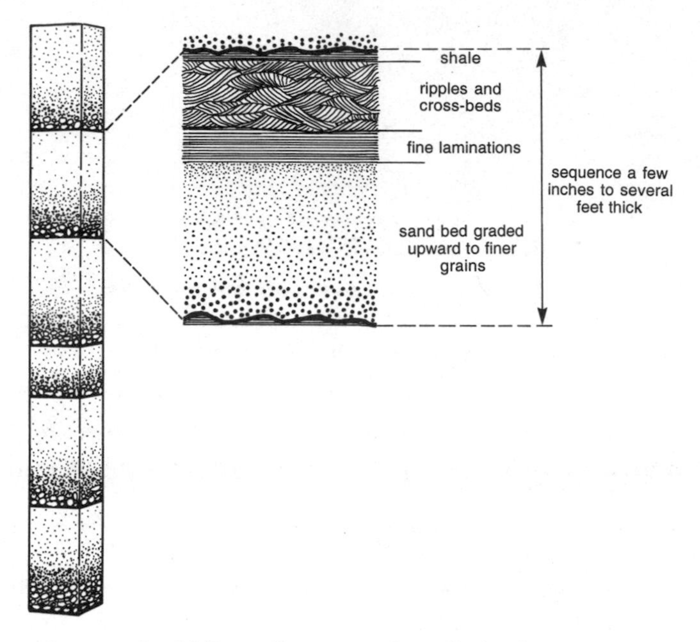

Layers of turbidite sediments make a distinctive pattern.

formations in the eastern Puget Sound area, and the Cowlitz and Skookumchuck formations. Some contain the fossilized leaves of palm trees and other tropical plants.

Here and there in the Oregon Coast Range the streams have cut through the cover of muddy sediments and into the dark basalts of the bedrock oceanic crust. The Siletz River has cut all the way through the basalt and into the layered gabbros near the base of the oceanic crust. In places the exposed oceanic crust consists of pillow basalt, actually piles of long cylinders of basalt that look like pillows only if you see them sectioned in a cliff or road cut. Anyone who has watched long cylinders of wax dribbling down a candle will recognize the process: A cold skin forms on the molten basalt as it meets the water, then the hot basalt within breaks through the skin and pours out into another long cylinder. Some submarine basalt flows do not develop into pillows.

These old basalts erupted on the ocean floor and were then buried in wet sediments; they have been in contact with water for 50 million years. The water reacted with the rock, converting the original black augite into darkly greenish chlorite, which gives the al-

Layers of turbidite sediments include pale sandstones and dark mudstones. These are on the Olympic Peninsula.

Basalt pillows broken free and lying on the floor of a quarry near Remote, Oregon. These are about the size of barrels.

tered basalt its greenish color. The pale plagioclase turned into white or glassy clear zeolite minerals, which come in many varieties. Many zeolites can swap their sodium or potassium atoms for atoms of calcium, magnesium, and various heavy metals such as lead. They are widely used in water softeners, water filters, and hazardous waste management.

High and Dry

Why does all that oceanic crust stand high and dry along the western fringe of the Pacific Northwest? Why is it not lying under two miles of water, like most of the earth's inventory of oceanic crust? What holds it up?

Building a raft out of more or less waterlogged driftwood may well be an essential part of childhood. The first big lesson comes when the raft sinks as the first kid climbs aboard. The usual solution is to shove more chunks of soggy wood under the thing until it finally becomes buoyant enough to float the child. That experience illuminates the situation along the west coast.

Most oceanic crust lies under about two miles of water because that is as high as it will float; it is much too heavy to stand above sea level. The long stretch of oceanic crust along the west coast could not stand so high unless lighter rocks beneath were floating it. It could well be riding on light sedimentary rocks that were scraped off the oceanic crust that is sinking through the offshore trench.

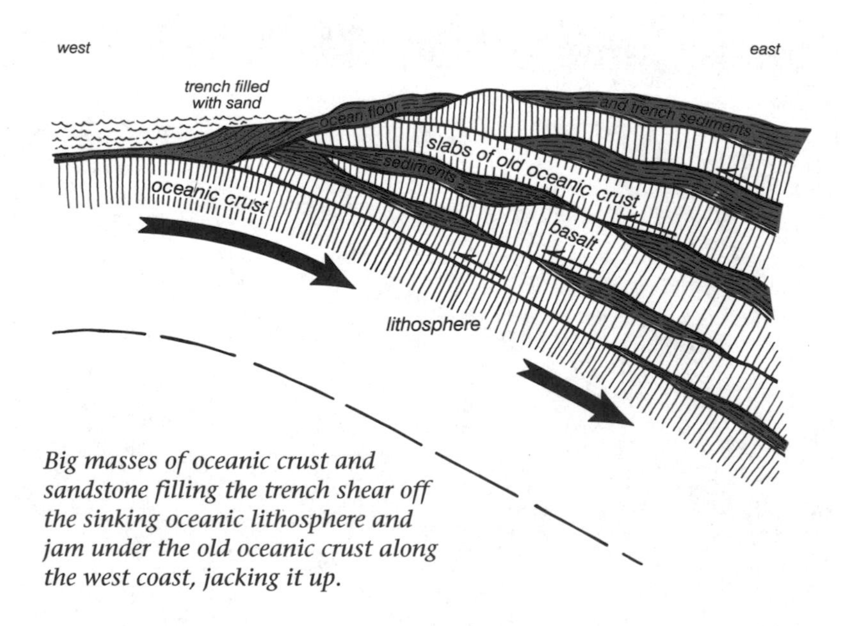

Big masses of oceanic crust and sandstone filling the trench shear off the sinking oceanic lithosphere and jam under the old oceanic crust along the west coast, jacking it up.

Most of the light oceanic sediments shear off the slab of ocean floor sinking off the west coast and stay behind in the trench. But the sinking ocean floor probably drags some big slices of those sediments under the coastal slab of oceanic crust. That stuff is lighter than the dense basalt that forms most of the oceanic crust, so it floats the heavier rocks above. Every mass of muddy sandstone and mudstone that jams under the coastal slab of oceanic crust floats the hills a little bit higher.

Most of the muddy sandstones in the coastal hills contain fossils of little animals that lived in the Pacific Ocean during Eocene time. Obviously, the slab of ocean floor was still below sea level then. Some of those rocks contain Oligocene fossils, so at least part of the slab was still below sea level 35 or so million years ago. By early Miocene time, about 24 to 15 million years ago, most of it was no longer accumulating oceanic sediments, presumably because it was above sea level. As soon as the old ocean floor rose above sea level, streams began dissecting it into the rugged erosional landscape we see today.

The Pacific Northwest in Oligocene time.

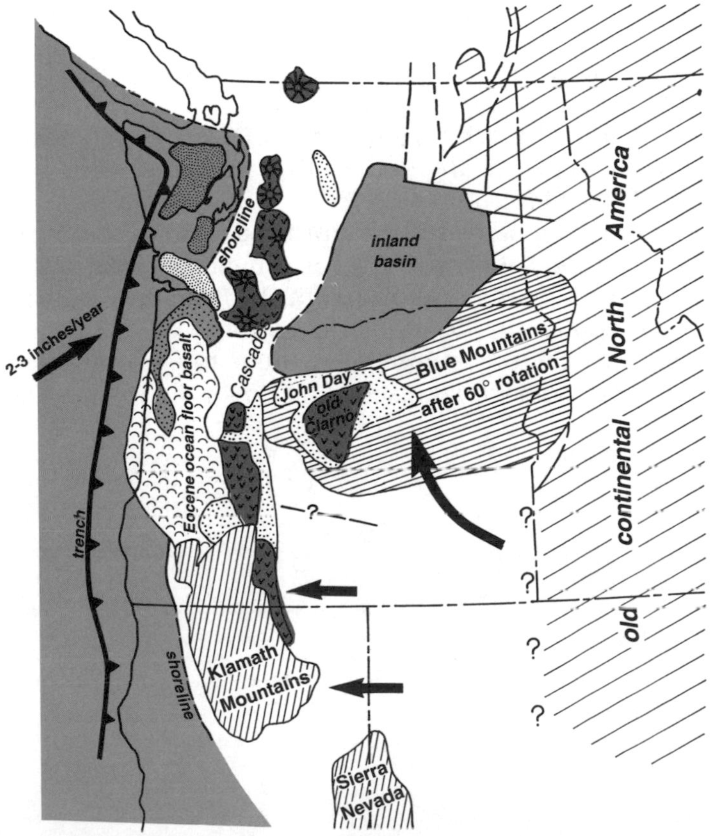

Gas

Sedimentary rocks deposited in seawater quite commonly contain petroleum, so the coastal hills are a reasonable place to expect to see oil or gas wells. Organic matter in sedimentary rocks cooks into oil within a narrow window between temperatures too low to cook anything and temperatures high enough to bake the oil into black carbon. The sedimentary rocks that cover the slab of oceanic crust in the coastal hills have never been buried deeply enough to reach a temperature high enough to cook up a batch of oil. Gas forms at much lower temperatures.

So it is no surprise that the Mist Field between Portland and Astoria in northwest Oregon produces only gas. The gas is trapped in porous Eocene sedimentary layers where they are sealed against faults under a large anticline. The future will probably bring more gas fields, but no oil fields.

Younger Rocks along the Coast

The early Miocene Astoria formation appears near the coast in southwestern Washington, in the northwestern corner of Oregon, and here and there as far south as Nehalem, Oregon. Similar rocks exist along several miles of coast south of Tillamook, and north and south of Newport. They are sandstones and shales that were deposited in shallow seawater along the coast.

The Astoria formation contains a variety of fossil snails and clams, along with a few brachiopods, corals, and barnacles. In some places people have found shark's teeth and bones of whales, seals, and turtles. Fossils are especially easy to find in the sea cliffs between Tillamook Bay and Newport.

Most of the prominent capes that protrude into the Pacific Ocean are Miocene flood basalt, the distant tongues of great lava flows that erupted in Miocene time and built the Columbia Plateau. The lavas poured down the Columbia River valley all the way to the coast and onto the soft sediments of the Astoria formation being deposited there. Some flows reached far down the coast of Oregon.

Chapter 31

THE WESTERN CASCADES

35 to 17 Million Years Ago

The Pacific Ocean floor collided with western North America in several different ways during Mesozoic time. At some times, the Pacific Ocean floor sinking under the edge of the continent generated volcanic chains like the modern High Cascades. At other times, fragments of continental crust arriving from the west or southwest either collided with the continental margin or slid along its edge.

Around the beginning of Oligocene time, about 35 or 40 million years ago, the oceanic plate sinking through the trench off the west coast began to spawn the Western Cascades, the beginning of a long volcanic career. The new volcanoes quickly grew into a formidable chain. To judge from the rocks, some of their eruptions were world-class events. Those great volcanoes spread ash across the landscape all the way from the western part of the Pacific Northwest to the northern High Plains. They erupted all through Oligocene time and the first half of Miocene time. Then they snuffed out about 17 million years ago.

The Volcanoes

You can tell the kind of filling in a pie by checking to see what bubbles up through its crust. And you can tell something about the kinds of rock in the depths by checking to see what lavas erupt from volcanoes.

In the northern part of the chain, the Western Cascades erupted mostly andesite and some rhyolite. The rhyolite suggests that those volcanoes stand on the continental crust of the North Cascades. In much of Oregon, the rocks of the Western Cascades are mainly dark andesites, probably because the slab of oceanic crust of the Oregon Coast Range, the Siletz terrane, extends to the east beneath them. Oceanic rocks tend to be dark. The Western Cascades of southern

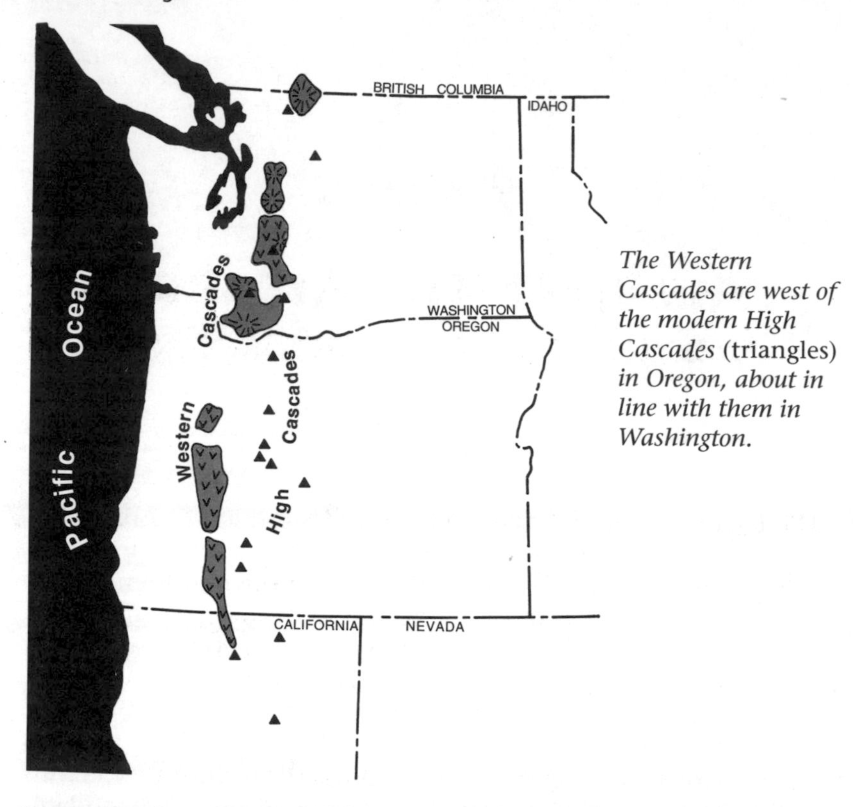

The Western Cascades are west of the modern High Cascades (triangles) *in Oregon, about in line with them in Washington.*

Oregon and northern California erupted andesite, along with enormous volumes of rhyolite. Again, the rhyolite suggests that those volcanoes stand on continental crust, in this case the Klamath block. The change from the dark volcanic rocks to the pale lavas of southern Oregon probably marks the buried northern boundary of the Klamath block.

One of the largest piles of Western Cascades volcanic rocks is the Ohanapecosh formation of western Washington, which dates from about 36 to 28 million years ago. In the area south of Mount Rainier, it is as much as 10,000 feet thick, mostly andesite and some rhyolite, mainly in volcanic ash and mudflows. Some of it doubtless erupted from Mt. Aix, the deeply eroded core of an andesite volcano about 25 miles east of Rainier. It is one of the few old volcanoes that survive as distinct peaks in the high and rugged landscape of the Western Cascades.

Most of the Western Cascades are now a perfectly ordinary erosional landscape carved from volcanic rocks; hardly any of the original volcanic landforms survive. Violent eruptions of pale andesite and rhyolite probably demolished some of the volcanoes, and then erosion eradicated the rest of the original volcanic landscape. Some of the old volcanoes survive in the much older Ochoco Mountains

Volcanic debris of the Ohanapecosh formation at Mossyrock Dam, southeast of Centralia, Washington.

of central Oregon because they specialized in dark andesite and basalt, which do not erupt violently.

The Abrupt End of the Western Cascades

After something like 20 million years of vigorous activity, the entire chain of the Western Cascades abruptly went out of business sometime between 15 and 20 million years ago. Evidently, the slab of oceanic crust that had been sinking through the offshore trench and sliding beneath the west coast was no longer feeding superhot water into the rocks above, causing them to melt.

Age dates plotted on maps show that granite intrusions emplaced below volcanic chains like the Western Cascades tend to fall on distinctly linear trends according to their age. Intrusions of one age lie along one line, those of another age along another. Presumably, the linear trends of the active volcanoes above the granite intrusions also shifted from time to time.

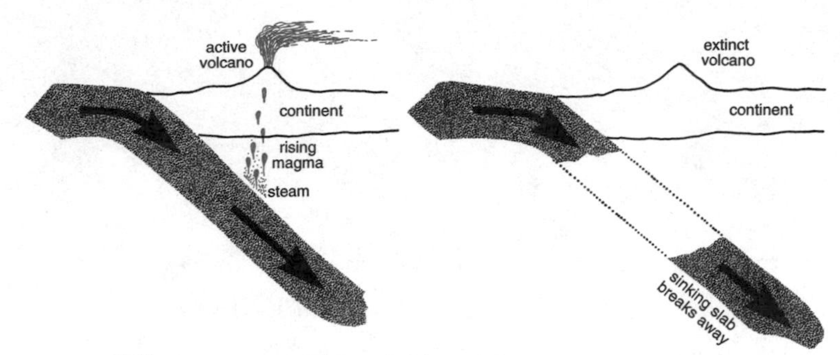

If the upper part of the sinking slab broke, the volcanoes of the Western Cascades would be put out of business.

Seismologists, who study the details of earthquake motion, can determine which way the rocks on opposite sides of the fault moved. They find that the upper parts of slabs of oceanic crust sinking through trenches break as though they are under tension, as though the weight of the lower part of the sinking slab is pulling the rest along behind. If a sinking slab of ocean floor were to break off entirely, its detached lower part would sink into the mantle faster than before because it would not be pulling anything along. Hot mantle rock would flow in to fill in behind it, but that rock can not supply the superhot water that drives volcanic chains. So the gap in the slab would cut the magma supply at its source, snuffing the volcanoes as surely as you douse a gas burner by closing its valve.

If the ocean floor continues to sink through the trench, a new volcanic chain will start to erupt some millions of years later. If the ocean floor were then sinking at a different angle than before, the

Old volcanoes drop into the gap that forms when the crest of a crustal arch collapses after the volcano dies.

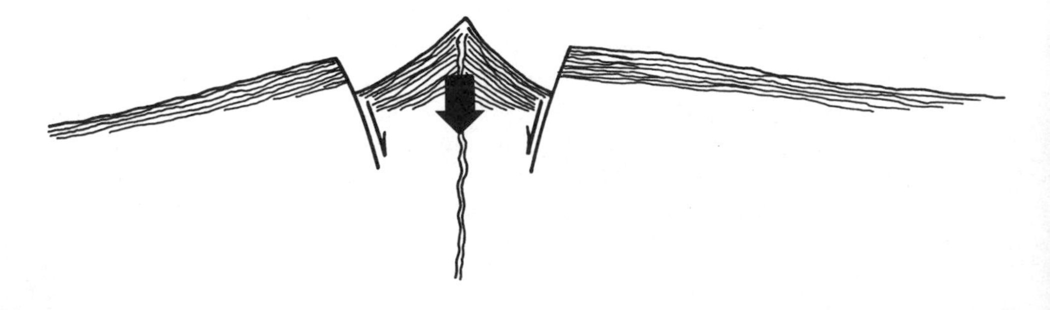

sinking slab would reach the depth at which it boils off water at a different distance from the trench. The new volcanoes would then erupt along a new trend. When the next generation of volcanoes, the High Cascades, started to erupt near the west coast, they appeared on a trend about 50 miles east of the old Western Cascades.

Active volcanic chains typically stand on broad crustal arches that rise as magma and steam heat the older rocks below the volcanoes, making them expand. After the volcanic chain dies, the rocks beneath it cool and shrink. The crest of the arch drops along faults as though it were the keystone of a sagging arch in an old building. The old Western Cascades now lie in long fault blocks that evidently dropped as the rocks at depth cooled and shrank after the supply of magma stopped.

Chapter 32

THE FIRST LONG DRY SPELL

37 to 17 Million Years Ago

Abundant petrified wood preserved in rhyolite ash that erupted in the northern Rocky Mountains during Eocene time shows that the climate was wet then. So does the general scarcity of Eocene sedimentary rocks deposited on land, which means that streams were vigorous enough then to carry eroded sediment to the ocean. All that changed as Eocene time ended.

The John Day, Renova, and White River Oligocene Beds

In the modern world, widespread deposits of sediment accumulate on land only in regions with climates too dry to maintain a connected network of flowing streams, not in humid regions where streams carry eroded sediments to the ocean. Sediments accumulate in the riverless valleys of Nevada, not in the lush valleys of the Midwest. So the very existence of widespread deposits of sediment laid down on land is good evidence that the climate was dry when they were deposited.

Sedimentary rocks began to accumulate on land throughout much of our region at the beginning of Oligocene time, or possibly in the last part of Eocene time. Their deposition continued during all of Oligocene time and the first half of Miocene time, approximately from 37 to 17 million years ago. Sediments deposited during those years include the John Day formation in Oregon, the Renova formation in the northern Rocky Mountains, and the White River Oligocene beds on the northern High Plains. They are so much alike that we think of them as parts of the same regional blanket of sediment with different names in different places. All three consist mostly of mud, sand, gravel, and volcanic ash, with the proportion of ash diminishing eastward.

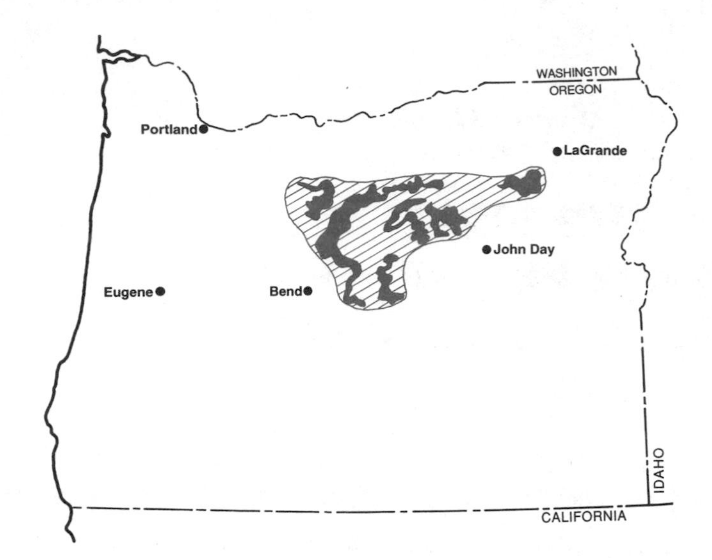

Where to look for the John Day formation. It probably extends much farther east and north beneath the younger lava flows of the Columbia Plateau.

Watch in the dry John Day country of central Oregon for the distinctive exposures of the John Day formation: picturesque pale hills with occasional bands of dark gray, brown, or pink, a sort of giant badlands. A prominent ledge of hard welded ash punctuates the barren slopes eroded in rock that is otherwise soft enough to dig. It was laid down from a cloud of molten rhyolite ash suspended in red hot steam that boiled across the countryside, then fused into solid rock as it finally settled. It is welded at least 100 miles from

Hills eroded in the John Day formation.

the nearest plausible source in the Western Cascades, an extraordinary distance. That must have been a world-class eruption.

The Renova formation fills the floors of the older broad basins within the northern Rocky Mountains, mostly in western Montana. Most of the Renova formation looks very much like the John Day formation except that it contains more gravel and sand and a smaller proportion of volcanic ash, none of it welded. Here and there, the formation contains limestone, oil shale, and coal. Most of the broad mountain valleys in western Montana contain at least several thousand feet of Renova formation, a few considerably more. Watch for road cuts in pale gray or yellowish rock, mostly silt and gravel, most of it soft enough to dig with a shovel.

Beds of gravel in the Renova formation consist entirely of pebbles eroded from the nearby mountains; none washed in from some distant source. If streams had flowed long distances when the formation was accumulating, they would have imported a few foreign pebbles. Their absence provides further evidence of a climate too dry to maintain a connected network of streams. Evidently, the valleys were then undrained desert basins like those in Nevada today.

The Renova formation fills the older valleys in the northern Rocky Mountains. —Modified from Fields and others, 1985

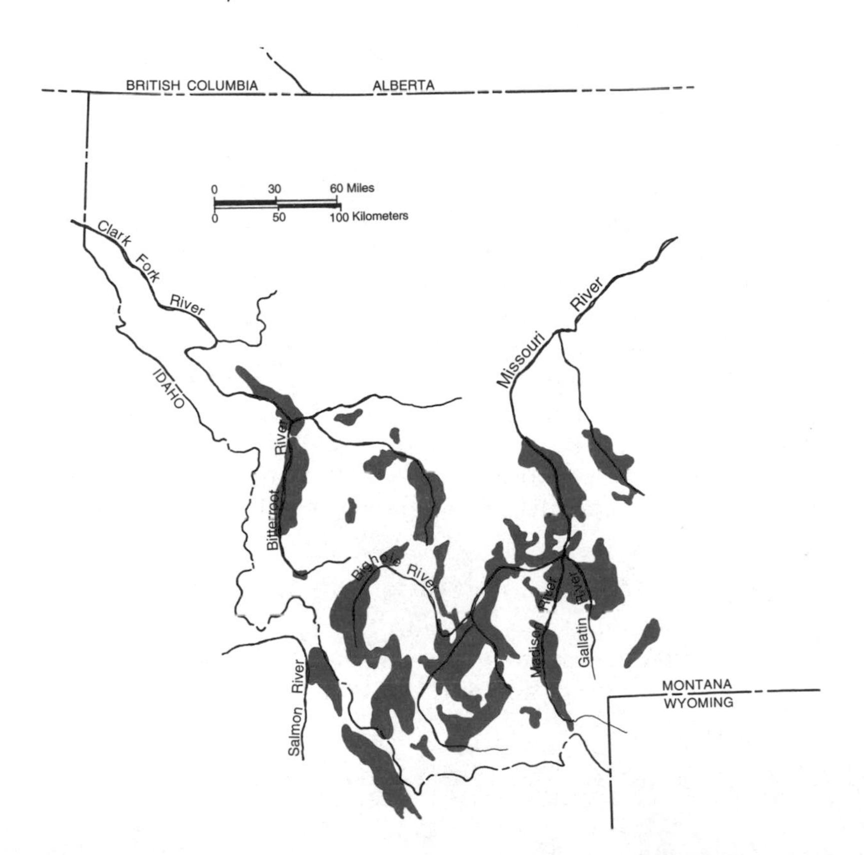

The floors of undrained desert valleys commonly contain temporary or permanent lakes, which typically contain alkaline and salty water. Limestones now forming in some of the desert lakes in the arid southwest look very much like those in the Renova formation. Although desert lakes are generally salty, the streams that flow into them may maintain areas of fresh water and marshes here and there around their edges. Many of the freshwater marshes lay down deposits of peat. If those are buried, they will become coal seams like those in the Renova formation.

Exposures of the Oligocene and Miocene sedimentary beds on the northern High Plains are scarce, so you rarely see them. When you do, they look very much like the John Day and Renova formations. More people see those rocks in the Badlands of western South Dakota than anywhere else.

Fossils

A rich abundance of bones in the John Day, Renova, and White River Oligocene beds endears those formations to vertebrate paleontologists. The bones are essentially similar in all three, which means that all were deposited during the same time. Many of the bones are the remains of early horses, camels, and rhinoceroses. Their modern descendants roam across rather barren plains, so it seems reasonable to imagine their Oligocene and Miocene ancestors grazing on the harsh plants of arid plains, gathering at desert water holes. Many other bones are remains of animals that would not look at all familiar if you were to see them in a zoo. Those include oreodonts, which looked vaguely like sheep, and entelodonts, which looked a bit like oversized swine.

Petrified wood in the Renova formation comes in small chunks that could have washed in from the mountains, but no logs or standing stumps. Some of the ash layers are full of fossil leaves that could have blown in on the wind. The fragmental character of the plant fossils suggests that trees grew on the mountains, but not in the valleys where the sediments were accumulating. Evidently the mountains caught enough rain to support trees, but not enough to maintain an integrated network of streams that drained to the ocean. If the dawn redwoods of those days liked the same kind of weather as their modern descendants, the climate was mild.

Where Did the Volcanic Ash Come From?

The pale volcanic ash in the John Day formation of central Oregon undoubtedly erupted in the Western Cascades. The similar ash in the Renova formation and White River Oligocene beds is all fine material that drifted in on the wind. It includes no welded ash, no chunks of pumice, nothing to suggest nearby eruptions. The Western Cascades were the only volcanoes in our region that erupted large volumes of volcanic ash through Oligocene time and the first half of Miocene time. They must have been the major source of the ash.

Then, about 17 million years ago, the blanket of sediments stopped accumulating and the climate became very wet and very warm. That happened as the Western Cascades volcanoes snuffed out like a spent match, and enormous floods of basalt began to erupt to the east. Our region had suddenly entered the next stage in its geologic development.

Bedding in the John Day formation at Cathedral Rock, central Oregon.

▸ Part 8 ◂

LATE TERTIARY TIME

17 to 2 Million Years Ago

Had anyone been here about 17 million years ago, they would surely have thought that the course of tectonic and volcanic events in our region was settled and destined to continue so indefinitely. Nothing foretold a radical change. Then, we believe, an asteroid struck near the southeastern corner of Oregon and set the development of our region on an entirely new course, which continues. It was a random and utterly unpredictable event.

The geologic epochs of late Tertiary time.

Millions of Years Ago	
	Recent
0.1 —(10,000 YEARS AGO)	
	Pleistocene
2	
	Pliocene
5.3	
	Miocene (late)
17	
	Miocene (early)

Astronomers can measure with great accuracy the speeds of the asteroids they see, and they have ways of estimating their sizes. Given the speed and size of an asteroid, it becomes possible to estimate its energy of motion. That leads to an estimate of how big a crater it will make, if it should strike the earth. The largest craters that exist as open holes are the Manicouagan Crater of northern Quebec and the Popigai Crater of northern Siberia. Both are about 60 miles across, and that raises a question: Some of the asteroids known to exist in orbits that cross that of earth carry enough energy to open a crater much more than 60 miles across. Where are those very large impact sites? If no craters more than about 60 miles across exist, then those larger impact sites must have some other form. It seems very likely that they develop into flood basalt provinces, like the Deccan Plateau of India and the flood basalt province of the Pacific Northwest.

Chapter 33

HORRIBLE CATASTROPHE IN SOUTHEASTERN OREGON

17 Million Years Ago

Several quite different things happened about 17 million years ago: The entire chain of big Western Cascades volcanoes, which had been erupting vigorously for at least 20 million years, abruptly became extinct. Overwhelming flood basalt lava flows began to pour across a vast region of the inland Pacific Northwest. Faults began to break the former plains of Nevada and Utah into the swarm of broad valleys and jagged mountain ranges of the Basin and Range. And the climate became very wet and warm. Was it a coincidence that all those changes came at once? It seems far more likely that they had a common cause.

The Big Bang

An observant camel placidly grazing in Oregon might have looked up to notice a speck suddenly appear in the sky one afternoon about 17 million years ago. It was a vagrant asteroid several miles in diameter, a rock the size of a mountain falling through the sky. Moments later it burned into the atmosphere, flaring to fill the sky with a light brighter than that of the sun, flooding the dark sun shadows with its glare. Then the streaking light in the sky blossomed into an enormous fireball as the asteroid struck the earth. Those who theorize about such matters estimate that such an explosion would raise the sky temperature to something like 5,000 degrees centigrade over a large area of western North America. Everything within sight of that sky died in its glare, instantly cremated. So much for the camel.

Meanwhile, a dark column of broken rock and dust shot high into the stratosphere, where it spread into a mushroom cloud that within minutes darkened the bright afternoon into night. Earthquakes of extraordinary violence shook the entire planet, perhaps continuing for hours, raising great tidal waves, tsunamis, on coasts all around the world. The ground has not shaken so violently or so long since then, never within the experience of our species.

The inevitable had finally happened, as it must from time to time. An asteroid or perhaps a comet several miles across had collided with the earth. In the moment of impact, most of the asteroid's enormous energy of motion became a sudden flash of heat so intense that the object vaporized itself in an enormous fireball. Within a few seconds, part of the energy blasted a hole that was probably something more than one hundred miles in diameter. For reasons that will emerge, we think the asteroid struck near the southeastern corner of Oregon.

Although quite a few animals disappeared then, the events of middle Miocene time did not cause a massive extinction even remotely comparable to the one at the end of the Cretaceous period. That is probably because the impact of 17 million years ago was considerably smaller than the one that created the much larger Deccan flood basalt province of India. Nor has any middle Miocene deposit comparable to the terminal Cretaceous boundary clay been found, perhaps because no one has searched for one. It will be hard to find.

The Lava Lake Volcano

Imagine yourself standing on the moon with your binoculars that terrible day to watch the spectacle—the moon was a safe place to stand. You could have seen the momentary flash as the asteroid passed through the earth's atmosphere, followed by the even brighter and more prolonged glare of its explosion. Then an enormous cloud of dust would obscure the view, covering the beautiful blue earth with its dark smudge as it spread before the high-elevation winds.

If you could have seen through the dust and glare, you might have watched the impact explosion open a crater large enough to bite deeply into the glowing hot rocks of the lower continental crust, perhaps even the upper mantle. But those rocks are too weak to support a deep crater—in the same way that a soft chocolate pudding is too weak to support a deep spoon hole. The bright glow of the deep continental rocks would quickly fade as cold rocks from around the crater slumped across them. The deep initial crater was collapsing into a much broader and much shallower crater basin.

The final crater basin was still deep enough to relieve the pressure on the already partially molten rocks in the upper mantle, the asthenosphere, permitting them to begin expanding into molten basalt. The rocks beneath the lithosphere are always ready to melt

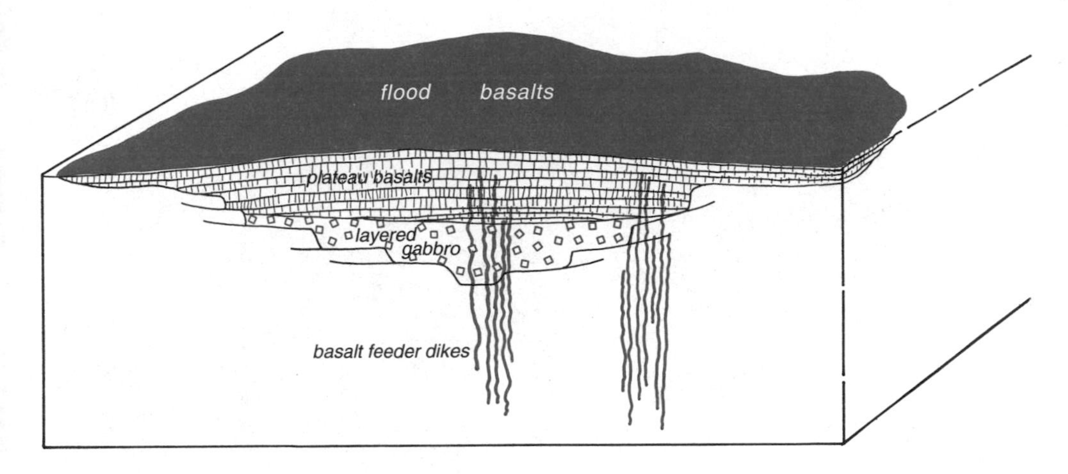

The lava lake volcano.

the moment something eases the heavy burden of the rocks above, and the impact explosion had undoubtedly raised their temperature considerably. Still watching from the moon, you would see glowing basalt lava rise into the crater basin, probably within a few days. The rising lava soon filled the crater basin, converting it into an enormous lava lake, a volcanic monster. If clouds survived its heat and passed over the lava lake at night, you could have watched the lantern glow of the lava lighting them from beneath.

As the years passed, you could have watched the rising tide of basalt magma stretch the crust north of the crater rim and inject the fractures with basalt. Pressure from the lava lake would push the magma north through fissures to spill glowing floods of lava across most of the inland Pacific Northwest, some down the Columbia River Valley all the way to the coast. Each glowing sea of lava darkened as the molten rock cooled into a black expanse of solid basalt. The lava lake in southeastern Oregon also darkened as its surface eventually cooled and solidified. But occasional eruptions broke through its hard crust from the still hot depths of the lava lake below to spread flood basalt lava flows and rhyolite ash across its solid surface and the surrounding lowlands.

Most of the eruptions came within the next million and a half years. During all that time, every overflowing flood of basalt lava again relieved the pressure on the hot mantle rocks below, permitting more melting, refilling the lava lake. Meanwhile, a column of hot rock began to rise through the mantle to replace all the rock erupting on the surface. The flood basalt eruptions finally ceased when the moving North American plate carried the crater basin

Stacked basalt lava flows near Lewiston, Idaho. —F. O. Jones, U.S. Geological Survey

lava lake beyond the column of hot rock rising through the asthenosphere. The lava lake then lost its continuing supply of basalt magma and finally cooled. But the column of hot rock rising through the mantle beneath the continent would spawn a series of new volcanoes, each one a souvenir of the event of 17 million years ago.

Many geologists dislike the idea of catastrophic events such as asteroid impacts, preferring instead to think of slow processes and weak forces working great results over long periods of time. It is certainly true that the great length of geologic time makes it possible for slow processes and weak forces to create large results. It is equally true, if less often remarked, that the great length of geologic time makes it inevitable that inherently unlikely events will happen, sooner or later.

Less catastrophic ideas have been proposed for formation of the flood basalts, each of which has major problems. Many geologists imagine plumes of hot rock rising from the base of the earth's mantle to melt the upper mantle, generating huge volumes of basalt magma. We agree that mantle plumes exist, but do not suppose that they rise all the way from the earth's core. The melting that would certainly result from excavation of an enormous crater by asteroid impact seems a sufficient cause.

Many geologists contend that the flood basalts form a zone of crustal stretching behind the Cascade volcanoes. That sort of thing does happen, most notoriously in the Sea of Japan, in the region behind the great Japanese volcanoes. The main problem with this idea is that the flood basalts of the Pacific Northwest are the only ones that erupted anywhere near an active volcanic chain, or collision plate boundary. If crustal stretching above a sinking plate can create one flood basalt province, then the same mechanism should also have caused others. Furthermore, no flood basalts erupted during the 35 million years while the Western Cascades were active volcanoes, and none have erupted while the modern High Cascades have been active.

Miocene Columbia River Plateau basalts exposed by the Columbia River near Vantage in central Washington.

Chapter 34

FLOODS OF BASALT

17 to 15 Million Years Ago

We ordinarily think of a volcano as a centralized volcanic pile erupted from a central crater or vent. A flood basalt volcano does not stand high; it fills a depression. And the molten basalt lava is very fluid, so it spreads out to make a nearly flat surface.

Like other large flood basalt volcanoes, the one that erupted in the Pacific Northwest produced a series of immense lava flows—within a brief time, for a volcano. Within less than two million years, it spread its flows from northeastern California, the northwestern corner of Nevada, across much of Oregon, most of eastern Washington, and the western edge of Idaho. A few flows poured down the Columbia River, across the Willamette Valley, and into the Pacific Ocean.

The flood basalt province.

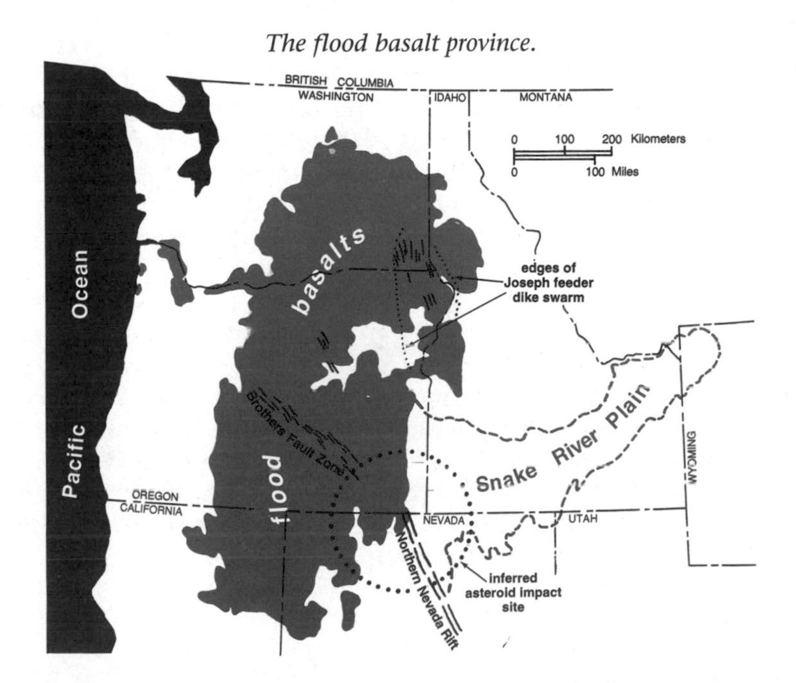

Flood basalt flows are absolutely outrageous, even by the outlandish standards of volcanoes. Several of the larger flows in eastern Washington contain more than 200 cubic miles of lava spread over an area of tens of thousands of square miles. To judge from the behavior of ordinary basalt flows, they probably covered all that ground within a few days. Imagine a single molten basalt flow pouring across an area comparable in size to the state of Maine within less than a week. The raw numbers hardly convey a feeling for such a catastrophe. It is well that no such eruptions have happened since human beings walked the earth.

Flood basalt flows raise all sorts of troubling questions: Why did all that basalt magma melt? Where within the earth did it come from? How did such enormous volumes of basalt reach the surface so fast? Where was it stored before it erupted? Did all the eruptions produce the same kind of rock? How long did the eruptions last? Why did they finally end?

The theory of an asteroid impact answers many of those questions: Pressure relief melting within the asthenosphere would produce enormous volumes of magma, and thorough fracturing of the lithosphere would let it rise to the surface. A lava lake filling a crater basin could store an enormous volume of magma, which could leak out through large fractures and erupt. The eruptions would end when the lava lake finally crystallized, after the moving lithosphere carried it away from the rising column of hot rock in the asthenosphere.

Basalt and Basalt

The Duluth complex of northeastern Minnesota lies directly beneath the Keweenawan flood basalts and is the same age, about 1.1 billion years. Most of it is an enormous mass of gabbro, a rock similar to basalt except that it consists of larger mineral grains. It seems obvious that the Keweenawan flood basalts erupted from the lava that finally crystallized into the Duluth complex. To put that in the professional jargon: The Duluth complex is the magma chamber for the Keweenawan flood basalts. We regard it as a lava lake.

The gabbro in the lower part of the Duluth complex is very dark and full of mineral grains of black pyroxene and glassy green olivine. Higher up, the olivine and pyroxene become less abundant, the white feldspar more abundant. The upper part of the Duluth gabbro contains very little olivine and pyroxene, but enough feldspar to make the rock quite pale. It appears that the heavy crystals of olivine and pyroxene tended to sink through the molten magma, while the lighter plagioclase crystals tended to float, slowly differentiating the lava lake into different compositions.

A similarly differentiating lava lake in southeastern Oregon probably fed the basalt flows that flooded much of the Pacific Northwest during middle Miocene time. That would certainly explain why the flood basalts vary through quite a range of compositions.

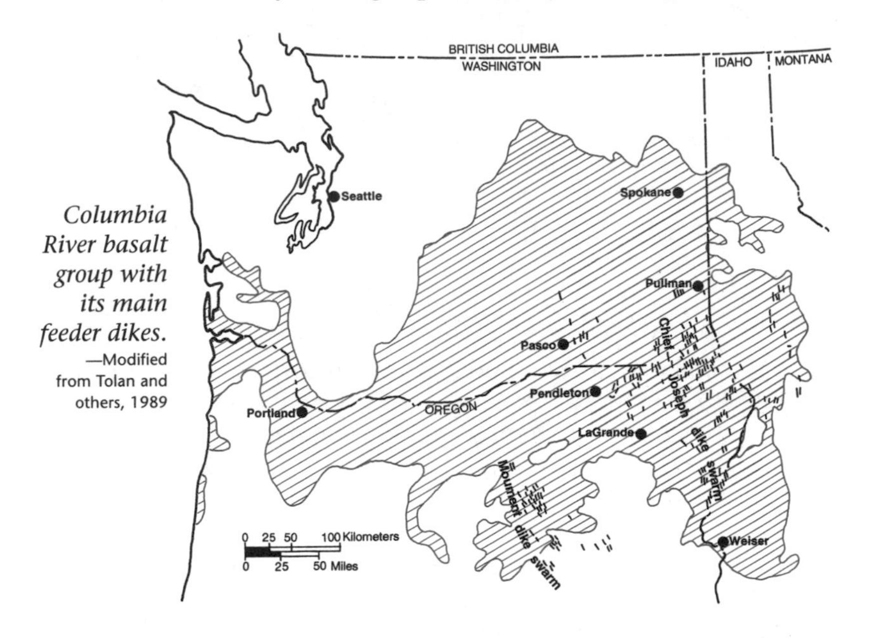

Columbia River basalt group with its main feeder dikes. —Modified from Tolan and others, 1989

All basalt looks very much alike—black and fine grained. But it comes in several varieties at least as distinctive as the various breeds of house cats. With a lot of practice, you can develop an eye to distinguish some of the more distinctive varieties of basalt. Precise chemical analysis even makes it possible to identify and trace individual flood basalt flows through their distinctive compositions, especially of minor elements. Geologists have separately identified many of the flood basalt flows in Washington and western Idaho and plotted them individually on maps.

First came the Imnaha basalts, which erupted between about 17.4 and 17.0 million years ago. The older date indicates that the big impact actually happened some time before our rounded-off time of 17 million years ago.

Flows of Imnaha basalt poured across northeastern Oregon and flooded many of the mountain valleys of western Idaho to depths as great as several hundred feet. By some estimates, they amount to 5 to 10 percent of the northern part of the province, but no one knows how much may be buried under younger lava flows. Imnaha basalt contains easily visible grains of green olivine and greenish

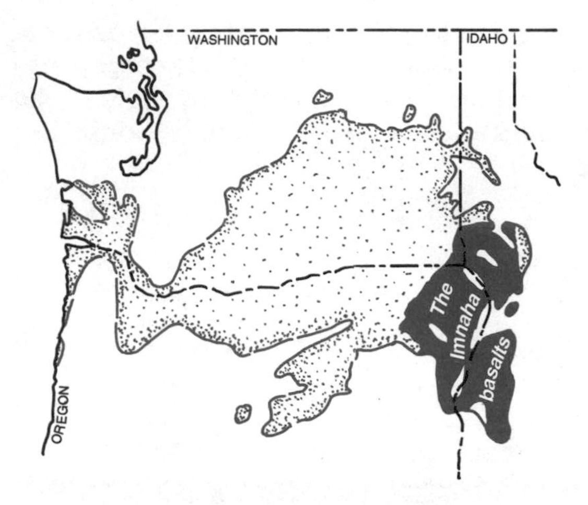

Exposed area of the Imnaha basalts. —Modified from Anderson and others, 1987

white plagioclase feldspar. Look at broken surfaces through a good magnifier to see the little glassy green crystals of olivine, but choose fresh rock because weathering quickly reduces olivine crystals to rusty specks. Like most olivine basalts, the Imnaha flows tend to weather into rounded boulders and smooth slopes.

Flows of Picture Gorge basalt covered part of Oregon south of the Blue Mountains. They also erupted early and disappeared under younger flows, so it is impossible to know how much area they may cover. Picture Gorge basalt also contains olivine and tends to weather into rounded boulders and smooth slopes. Those flows erupted from a swarm of fissures that filled with basalt to become the large dikes of the Monument dike swarm. If you project the trend of those dikes to the southeast, it passes through southeastern Oregon, which was probably the site of the lava lake.

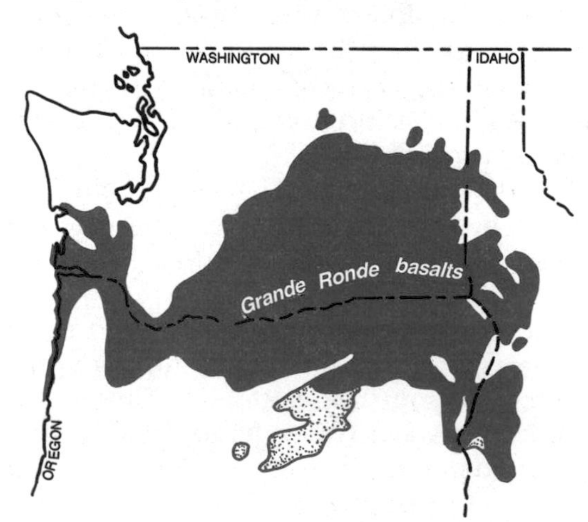

The Grande Ronde basalts. They probably cover large areas of Imnaha and Picture Gorge basalt. —Modified from Anderson and others, 1987

The long series of Grande Ronde basalt flows followed between 16.5 and 14.5 million years ago, burying most of the older flows. Some geologists estimate that they amount to about 85 percent of the volume of flood basalt flows of Washington and northern Oregon. Grande Ronde basalt contains no olivine. In fact, it contains very few visible crystals of any kind; the rock looks as smooth as black ice cream. Flows of Grande Ronde basalt make strong ledges and prominent cliffs in the slopes. They tend to weather into long palisades of columns with distinctively crisp edges.

The floods of Grande Ronde basalt erupted from fissures in southeastern Washington and northeastern Oregon and left them filled with basalt to make dikes, the Joseph dike swarm. Project the trend of those dikes to the south and they intersect the projected trend of the Monument dikes in southeastern Oregon.

One careful and thoughtful geologist who studied the Joseph dike swarm in some detail estimated that it contains more than 20,000

Columbia River Plateau basalt flows exposed in the canyon of the Palouse River. —F. O. Jones, U.S. Geological Survey

A dike of the Joseph dike swarm stands like a ruined wall in Grande Ronde Canyon near the border between Washington and Oregon. The crosswise cordwood pattern is typical of basalt dikes.

dikes. Incredible, but probably close to the truth. The fractures those dikes fill rather thoroughly trashed the eastern Blue Mountains. The magma probably moved north through those fissures from the lava lake in southeastern Oregon on its way to erupt across northern Oregon, eastern Washington, and western Idaho. Most of those dikes are more than 25 feet thick—wide conduits capable of carrying a lot of magma. They were probably even wider when the magma, under pressure, was pushing the fractures open.

When the Grande Ronde eruptions finally ended, the northern part of the flood basalt province, the Columbia Plateau, was essentially complete. Later eruptions of Wanapum and Saddle Mountains basalt covered large areas in the central part of the plateau but added very little volume.

The earth's crust sank under the weight of the Grande Ronde flows to form a broad basin in eastern Washington. The Wanapum basalts erupted into the center of that basin between 14.5 and 13 million years ago. They cover most of the central part of the Grande

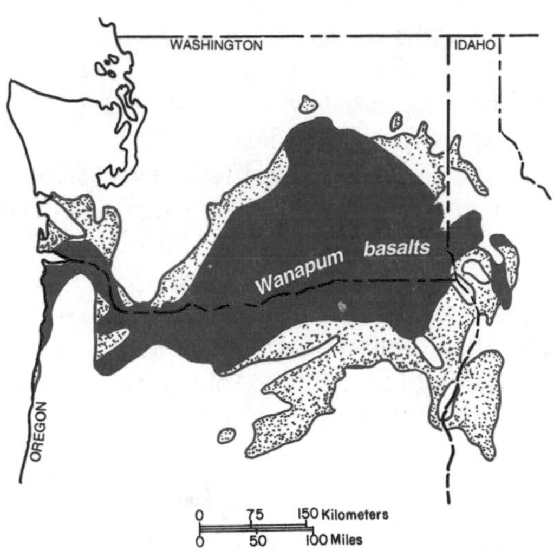

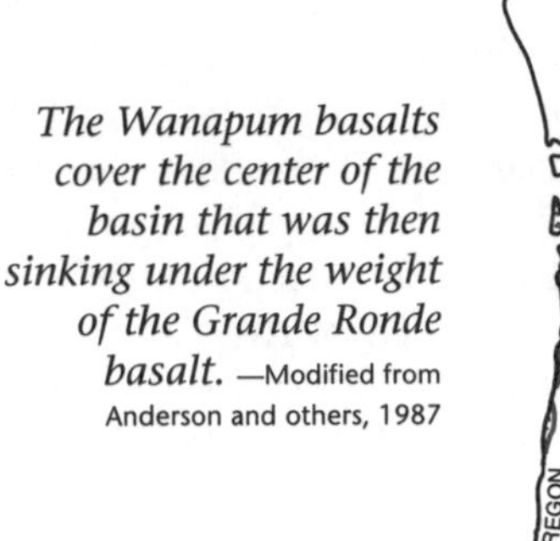
The Wanapum basalts cover the center of the basin that was then sinking under the weight of the Grande Ronde basalt. —Modified from Anderson and others, 1987

Ronde basalts and may amount to something between 5 and 10 percent of the known volume of basalt north of the Blue Mountains. Wanapum basalts vary in appearance. Some contain big flakes of pale greenish plagioclase feldspar and crystals of glassy green olivine; others contain the plagioclase but no visible olivine.

The Saddle Mountains basalts were the last in the northern part of the flood basalt province. They flooded the central part of the basin that was still subsiding, probably under the weight of the Grande Ronde and Wanapum flows, and filled old river canyons. Although they erupted over a long time and cover a large area, the Saddle Mountains basalt flows amount to a minuscule portion of the total volume of flood basalt.

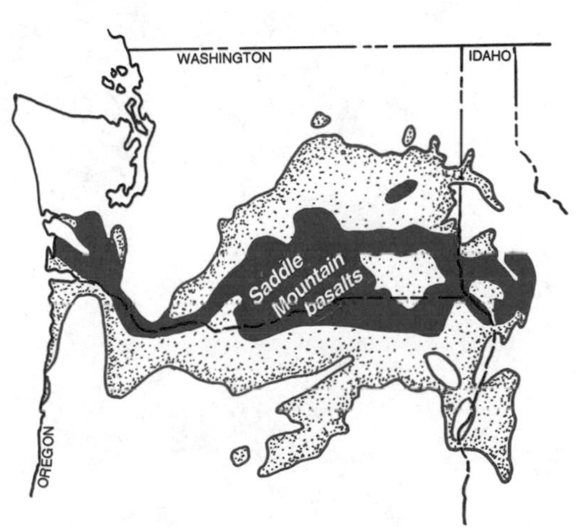

Where to look for Saddle Mountains basalt. —Redrawn from Anderson and others, 1987

The basalts south of the Blue Mountains have received very little attention, so their story is poorly known. The Steens basalts are certainly the most widespread. They contain white and greenish white flakes of plagioclase feldspar, some the size of postage stamps and a few as large as business cards. Those large plagioclase crystals reflect the abnormally high aluminum content of the rock. Steens basalts spilled across much of southern Oregon, the Modoc Plateau of northeastern California, and a bit of nearby Nevada. They erupted at the same time as the other flood basalts were pouring across the Columbia Plateau north of the Blue Mountains. All the basalt, north and south of the Blue Mountains, probably came from the same source in southeastern Oregon.

Basalt Columns

Basalt is one of those rocks that anyone can learn to recognize, even from a distance. If you see a road cut or cliff in dark rock that breaks into a palisade of more or less vertical columns, that is basalt, or possibly dark andesite. Those rows of columns normally make fairly strong ledges in the hillsides. Resist the temptation to translate a count of ledges into a count of lava flows.

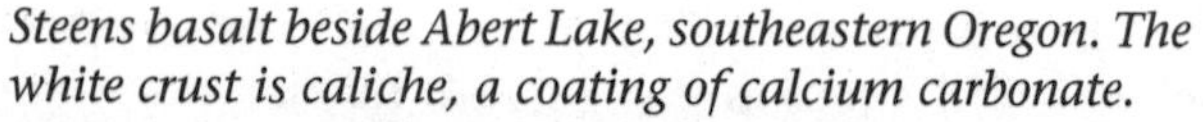

Steens basalt beside Abert Lake, southeastern Oregon. The white crust is caliche, a coating of calcium carbonate.

A basalt flow about 50 feet thick near Vantage, Washington, shows palisades of columns above and below a hackly entablature.

A view of a valley wall eroded through a thick section of flood basalt flows shows many ledges, many flows. Thin basalt flows generally break into a single palisade of columns and appear as a single ledge. Thick basalt flows typically include two palisades of columns with a zone of irregularly fractured rock between them. Each set of columns makes a separate ledge, so the number of thick flows is generally about half the number of ledges. Despite the potential for confusion, many counts in many places leave no doubt that hundreds of basalt flows erupted during the few million years of volcanic activity in the flood basalt province.

Basalt columns form because the lava shrinks between 5 and 10 percent as it crystallizes; mineral grains occupy less volume than the melt from which they crystallized. The lava flow cools first at the top and bottom, cracking like drying mud into vertical polygons, most of which have five or six sides. As the flow continues to crystallize, the cracks penetrate down from the top and up from the bottom. The two sets of cracks do not match, so the part of the flow between them breaks irregularly, making a hackly zone called an entablature. The flow cools more rapidly from the top than from the warm ground beneath, so the upper colonnade of columns is generally taller than the bottom set.

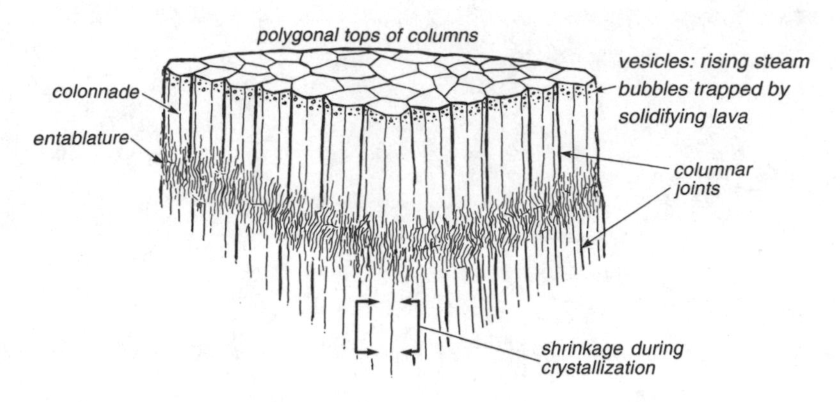

Section through a basalt lava flow. Columnar joints form because the flow shrinks as it crystallizes.

In some places, flows poured into lakes, where the lava flowed within a chilled crust to make bulbous cylinders. They look like black pillows where you see them exposed in section in a cliff or road cut. The pillows may grade upward into colonnades of basalt columns that formed when the basalt solidified above the level of the water. Basalt pouring into a lake may also break into chunks as it boils the water into steam and reacts with it to make a yellowish clay called palagonite. Watch for road cuts that expose broken chunks of black basalt embedded in the yellowish clay.

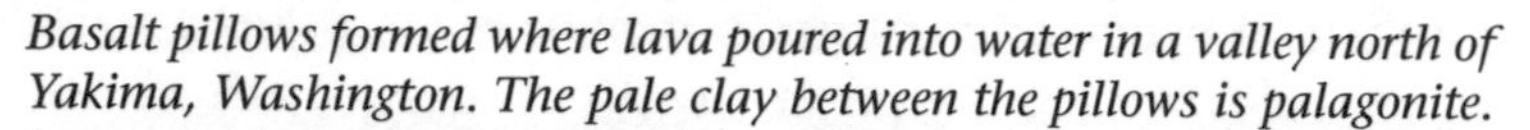

Basalt pillows formed where lava poured into water in a valley north of Yakima, Washington. The pale clay between the pillows is palagonite.

Ocean Capes

Some of the floods of basalt covered the northern part of the Willamette Valley of western Oregon. A few reached the Pacific Ocean. In fact, you can see the same variety of basalt types along the coast of Oregon that you see within the flood basalt province. The flows resist erosion well enough to make prominent capes along the coast, such as Tillamook Head. The long finger of Yaquina Head is an old stream valley filled with basalt.

Where large volumes of lava poured into the ocean, they raised enough steam to drive secondary eruptions. Those were bizarre volcanoes that grew at the outermost ends of lava flows, doubtless beneath towering clouds of steam and amid tremendous commotion. Their eruptive careers could not have lasted any longer than the flow continued to pour lava into the ocean, probably just a few days. We know of no other volcanoes like them. The easiest place to see one is at Cape Foulweather on the Oregon coast. U. S. 101 passes confusing road cuts in basalt that reacted with seawater to make messy volcanic rocks and a variety of volcanic sediments.

Rhyolite

Great quantities of rhyolite erupted in southeastern Oregon and nearby parts of southwestern Idaho and northern Nevada during middle Miocene time. In fact, rhyolite is so abundant there that in many areas it is more conspicuous than basalt. If this is a flood basalt province, then why do we see so many white rocks?

At least some of that rhyolite is a peculiar variety called granophyre. It differs from ordinary rhyolite in containing a bit more iron, and it shows a distinctive geometric texture under the microscope. Laboratory experiments show that molten granophyre separates from molten basalt in the same way that oil separates from water. That unmixing generally produces minute specks of granophyre within basalt and small patches of granophyre within gabbro. The same unmixing within the lava lake was probably at least partly responsible for the enormous amounts of rhyolite that cover so much of southeastern Oregon and nearby Idaho and Nevada.

Ordinary rhyolite magma can form by melting ordinary continental basement rocks, granite, gneiss, and schist. It is easy to imagine that large volumes of such rocks might melt around a very large impact site. If we knew how much of the rhyolite in southeastern Oregon is ordinary and how much is granophyric, it might be possible to figure out how much formed from melted continental basement rocks and how much separated from basalt magma.

Most continental basement rocks contain a minute amount of the rare element rubidium, a small part of which is the weakly radioactive isotope rubidium 87. It decays into another isotope, strontium 87, which is not radioactive. Even though the amount of rubidium 87 is very small and it decays very slowly, continental basement rocks are so very old that they contain an abnormally large amount of strontium 87. If they melt, they form rhyolite magma with the same large content of strontium 87. Basalt, on the other hand, contains very little rubidium and therefore very little strontium 87. If the rhyolite in southeastern Oregon separated from basalt magma, it should contain very little strontium 87. Geologists have measured the strontium 87 content of a few samples of that rhyolite, and its abundance is indeed very low. More such research will eventually show whether it is correct to conclude that a very large proportion of the rhyolite in the southern part of the flood basalt province separated from basalt magma.

All flood basalt provinces contain a relatively small area in which substantial amounts of rhyolite erupted; none are complete without their patch of rhyolite, and none have more than one. The area where rhyolite erupted is an essential part of the Pacific Northwest flood basalt province, not in any sense a peculiarity.

Superhot Rhyolite

The minerals in at least some of the rhyolite flows in southwestern Idaho and northern Nevada crystallized at temperatures above 1,100 degrees centigrade. The lava was evidently at least that hot when it erupted. Rhyolite normally erupts at a temperature around 750 or 800 degrees centigrade, basalt at a temperature of about 1,200 degrees. Why should these rhyolites erupt at temperatures closer to that of molten basalt than to that of normal rhyolite?

That question brings us back to the Duluth complex of northeastern Minnesota. It lies directly beneath the Keweenawan flood basalt province, which also contains lava flows of granophyric rhyolite. The two kinds of lava flows are interlayered, red granophyre sandwiched within black basalt, so both kinds of magma were molten at the same time. Most of the Duluth complex is gabbro, but the upper several thousand feet are red granophyre, which apparently separated from the molten basalt magma and then floated on it the way oil floats on water. The granophyre magma must have existed at the same temperature as the basalt magma. The flood basalts evidently erupted from the reservoir of basalt magma that became the gabbro, while the granophyric rhyolites erupted from the reservoir of granophyre magma.

All flood basalt provinces resemble each other. If the Keweenawan flood basalts erupted from a layered lava lake, then so must have the Pacific Northwest flood basalts. A layered lava lake similar to the Duluth complex beneath southeastern Oregon would explain why the granophyric rhyolite erupted in that area, and why it erupted at such a high temperature.

Any variety of rhyolite lava is much more viscous than basalt lava, much too viscous to run out into thin flows that cover large areas. It tends to stay close to home. If layered basalt and granophyre lava lakes form in very large impact craters, then the patch of rhyolite in every flood basalt province probably marks the site of the asteroid impact that started all the commotion.

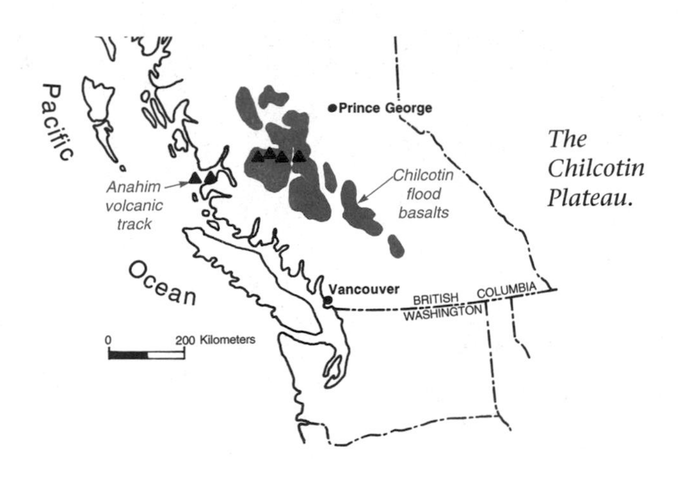

The Chilcotin Plateau.

The Chilcotin Plateau

The Chilcotin Plateau of central British Columbia is another pile of large basalt flows, apparently another flood basalt province. Like those in the Pacific Northwest, the Chilcotin basalts erupted during middle Miocene time. They may be the same age as the flood basalt flows in Oregon and Washington. Perhaps two asteroids struck our region one terrible afternoon 17 million years ago, a big one with a smaller companion. Radar imagery showed that Asteroid Toutatis, which passed close to the earth in December of 1992, consisted of two objects just barely in contact.

Miocene flood basalts in northeastern Oregon.

Chapter 35

THE CLIMATIC SIDE EFFECTS

17 to 15 Million Years Ago

During most of the last 35 or 40 million years, our region has been very dry. It enjoyed an uncharacteristically wet climate precisely while the flood basalt flows erupted. Then the climate became dry again and remained so until the great ice ages began, about two million years ago. The flood basalt volcano in southeastern Oregon probably caused the interlude of wet climate, which was also very warm.

Brown Basalt and Red Soil

It is possible to distinguish Miocene basalt from Pliocene and Pleistocene basalt by their colors: The Miocene basalt is brownish black, the younger rock black. The Miocene flood basalts owe their brownish color to a thin coating of rusty iron oxide on their surfaces. The brownish color is due entirely to weathering, not to any peculiar quality of the rock. If basalt weathered differently during middle Miocene time, the climate was different then.

Watch nearly anywhere in the flood basalt province for stripes of red sandwiched between lava flows. Most are narrow and simply oxidized rock in the top of the lower flow. Basalt quite commonly turns red as weathering breaks down its iron minerals, releasing red iron oxide. But red flow tops are brighter and far more common in the flood basalt province than in most other piles of basalt.

Some of the broader red stripes between flows are fully developed soils. Their color and their clay minerals show that they are laterites, the typically red soil of warm and wet regions such as the southeastern United States and Central America. Laterite soils do not form in cool or dry regions. The laterites sandwiched between lava flows in the flood basalt province are convincing evidence that our region was warm and very wet when all that basalt erupted.

A Columbia River basalt flow covers a dark red soil near Dale, northeastern Oregon.

In some areas west of the Cascades, the red soils become bauxite, aluminum ore, the most extreme form of laterite development. So far, no northwestern bauxite has been mined, although several deposits were sampled during the Second World War. Modern bauxites form only in truly tropical regions with heavy rainfall.

White Lake Beds

Watch the road cuts, especially in the eastern part of the flood basalt province, for broad stripes of pale yellow and white material sandwiched between the lava flows. They range from a few feet to a few tens of feet in thickness. Those pale stripes are lake and stream sediments, the Latah formation. They tell even more emphatically of a very warm and wet climate during middle Miocene time.

Streams flowed west out of the northern Rocky Mountains during middle Miocene time just as they do now, through the same valleys. The great flood basalt flows dammed those streams to impound lakes, most of which were in eastern Washington and along the western edge of Idaho. Those lakes must have been about as

A massive basalt flow caps pale beds of the Latah formation near Clarkia, Idaho.

deep as the thickness of the impounding flow, generally between 50 and 200 feet. Some were quite large. Floods washed gravel, sand, and clay from the northern Rocky Mountains into them, while the wind blew clouds of rhyolite ash in from southeastern Oregon. Eventually, another flood basalt flow would fill the lake, burying the pale sediments that had accumulated on its floor.

People who split layers of the Latah formation often find impressions of leaves so clearly etched on the white bedding surface that they look almost like delicate ink drawings. In some cases, the leaves are still green when you first expose them, but contact with the air immediately turns them black. The preservation is so perfect that DNA has been found in fossil magnolia leaves near Clarkia, Idaho. Here and there, a lava flow or a fall of rhyolite ash covered sunken snags too waterlogged to burn, and those eventually petrified. The best place to see petrified logs is in Gingko State Park, on the north side of Interstate 90 near Vantage, Washington. The fossils there include many kinds of wood, very little of it gingko wood.

Botanists identify the fossil leaves and wood as essentially the same assemblage of trees and bushes that now thrives in the southeastern United States and the warmer parts of southern Japan and China. Subtropical hardwood forests flourished in the Pacific Northwest during middle Miocene time. Imagine the fires the flood basalt flows ignited as they seared their way through those forests.

Fossil bones are scarce in the Latah formation: a few fish, very little else. It is most disappointing to realize that in all likelihood we will never know what exotic birds sang in the trees around those lakes, what colorful butterflies flitted through the leaves, what animals prowled the shores.

Acid Waters

The very pale color of the Latah formation sediments, the leaf fossils they contain, and the bones they do not contain all suggest that they accumulated in distinctly acidic water. The pale color is especially surprising. Sediment that accumulated in the lakes must have eroded from a countryside covered with red laterite soil. So why are the lake sediments white, not red?

Laterite consists essentially of three minerals: the clay mineral kaolinite, the aluminum oxide mineral gibbsite, and red iron oxide, hematite. Pure kaolinite and gibbsite are snow white. Laterites owe their red color to the iron oxide, which is also familiar as the pigment in red barn paint, red primer, and common rust. Decaying vegetation tends to make water acidic. If the middle Miocene lakes were acidic enough to dissolve iron oxide, that would explain the white color of the Latah formation. Decaying vegetation makes the water in most of the lakes and swamps in central Florida acidic. Those lakes commonly have snowy white sand beaches, even though the soil on the nearby hills is red laterite.

Kaolinite clay is one of the major ingredients in porcelain. If the porcelain is to be snowy white, it must be absolutely free of iron oxide. The Latah formation has for many years provided kaolinite clay to the ceramics industry—pure white kaolinite for porcelain, less pure for firebricks.

The mineral matter in bones is calcium phosphate, which easily dissolves in weak acids; that explains why herring pickled in vinegar are boneless. An acidic environment of deposition would at once explain the extreme rarity of bones in the Latah formation, the excellent preservation of leaves, and some fine chinaware.

An Episode of Erosion

While the flood basalts were erupting and the Latah formation was accumulating, streams were eroding a ruggedly hilly landscape into the pale gray sediments of the Renova and White River Oligocene formations. It is mostly buried now, and red laterite soil still covers it in many areas. That erosion surface could not have developed unless streams were exporting sediment to the ocean. A climate wet enough to develop laterite soils was surely wet enough to maintain a connected network of constantly flowing streams. The only sediments deposited on land in our region during the period of flood basalt eruption are those in the Latah formation.

What Changed the Climate?

Our earth has seen many climatic changes that have no apparent association with asteroid impacts, volcanic eruptions, or any other geologic event. Nevertheless, the coincidence in the timing of the flood basalt province and the period of wet and warm climate is so very close that the possibility of a cause-and-effect relationship seems worth considering.

Hot air rising above a red-hot flood basalt flow would create a strong low-pressure area that would pull moist air in from the Pacific. That might cause very strong storms, perhaps torrential rains, but only during the several weeks while the lava cooled. If you divide the probable number of those enormous flood basalt flows into the total length of time in which they erupted, the average time between them turns out to be a few thousand years. Whatever their immediate effect on the weather, it is hard to imagine that the flood basalt flows could establish a climate capable of creating laterite soils.

Atmospheric carbon dioxide or methane cause a greenhouse effect: They warm the climate by transmitting the energy of incoming visible light while absorbing outgoing heat radiation. During the last century, the carbon dioxide content of the atmosphere has risen from about 280 to about 320 parts per million. Many people believe that the small increase may now be warming the earth's climate and changing its rainfall patterns. An even smaller increase in the methane content of the atmosphere would have a similar effect.

Volcanoes erupt large amounts of both carbon dioxide and methane. If the crater basin volcano in southeastern Oregon erupted as much carbon dioxide and methane in proportion to lava as do many modern basalt volcanoes, it could easily have created a greenhouse effect worldwide. If the climatic changes that coincided with eruption of the flood basalt province involved the entire earth, as much evidence suggests, that would be good evidence of a global greenhouse effect.

Many volcanic eruptions also produce enormous tonnages of sulfur dioxide, which reacts with water vapor in the atmosphere and comes down as strongly acid rain. If the flood basalt flows emitted comparably large amounts of sulfur dioxide, they would have caused acid rain on a grand scale. That might help explain the whiteness of the kaolinite clay in the Latah formation.

Chapter 36

THE BASIN AND RANGE

17 to 15 Million Years Ago

An early government geologist compared the isolated mountain ranges of the Basin and Range to an army of giant caterpillars crawling north out of Mexico. It was an appropriate turn of phrase. The Basin and Range is a vast region of scattered mountain ranges that stand between broad valleys, all trending north. It extends south from southern Oregon, in the region between the Sierra Nevada of California and the Wasatch Front of Utah. We know of no other geologic province quite like it.

The Basin and Range as seen from space and drawn by computer. —G. P. Thelin and R. J. Pike, U.S. Geological Survey

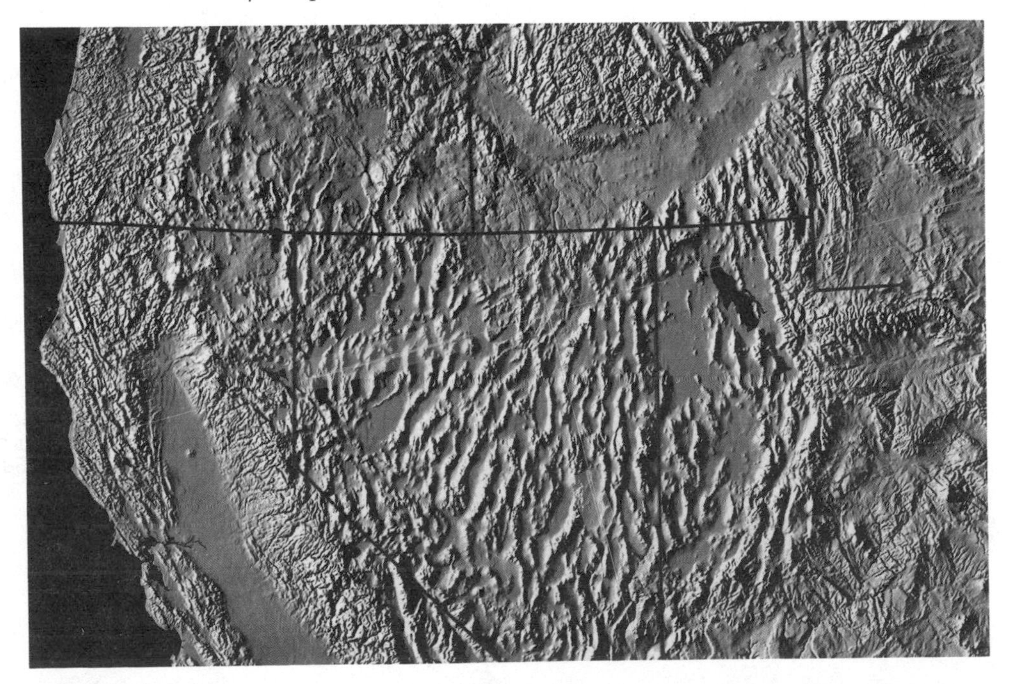

Structure of the Basin and Range

Those mountains and valleys of the Basin and Range formed as the earth's crust broke into great blocks that dropped along curving faults, concave upward. Each slice rotated as it slid down the curve of the fault, turning one edge down to make a basin, the other edge up to make a range. The slices in the western part of the Basin and Range moved east, those in the eastern part, west. The two sets of facing fault slices stare at each other across the northern Nevada rift, a line that extends south from southeastern Oregon.

The steep eastern slope of the Sierra Nevada is a high fault scarp that looks east across the broad expanses of the Basin and Range. It is the western boundary of the province. Movement on the Sierra Nevada fault raises the range as though it were a cellar door hinged at its western edge. The Sierra Nevada region was a lowland before it began to move. The steep scarp of the Wasatch Front in Utah is the eastern counterpart of the Sierra Nevada fault. It looks west across the Basin and Range. Follow the eastern margin of the Basin and Range north and you find that it passes right through the Yellowstone volcano.

If you were to slide all the displaced slices of crust back where they came from, the Basin and Range would become much narrower than it now is, perhaps half as wide. Movement on the Basin and Range faults more or less doubled the distance between Salt Lake City and Reno, moving California that much farther west.

Structure of the Basin and Range, schematically shown.

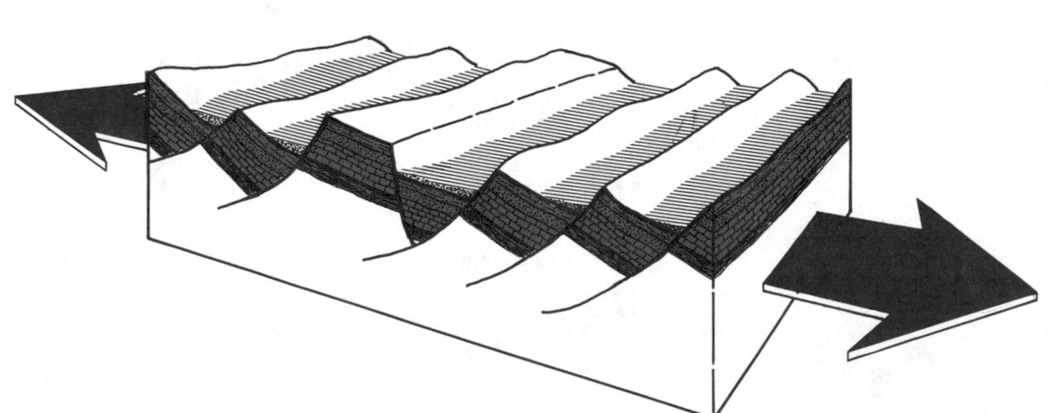

Abert Rim, a fault scarp that faces west in southeastern Oregon. Abert Lake is on the dropped block and lies against the scarp because the dropping block rotated as it moved. —K. M. Phillips, U.S. Geological Survey

When Did it Start?

As recently as 20 million years ago, big eruptions in the Western Cascades spread sheets of volcanic ash across the region that is now the northern Basin and Range. Those ash sheets look about the same from one Nevada mountain range to the next. Their monotonous sameness shows that mountains and valleys did not influence the spread of ash flows, that faulting had not yet begun in the northern Basin and Range. The region was still a plain.

For various reasons, most geologists now place the start of fault movements in the northern Basin and Range at about 17 million years ago. The most convincing evidence is in the northern Nevada rift. It is most obviously visible as a long and narrow band on magnetic maps, which depict variations in the strength of the earth's magnetic field. Age dates on dikes that trend north to northeast within the northern Nevada rift fall between 17 and 14 million years, a range that includes our magic date of 17 million years ago.

Most of the fault movement in the Basin and Range appears to have happened some time ago. Actively moving faults are now concentrated at the eastern and western edges of the province, where they cause frequent earthquakes. The central part produces very few earthquakes, and the faults there do not break the smooth surfaces of the valley fill sediments.

What Caused the Basin and Range?

The Basin and Range started to spread at about the same time that the San Andreas system of faults began to carry the western part of California north. The San Andreas fault system was created when the last remnants of the Farallon plate offshore sank beneath the continent and California came into contact with the Pacific plate, which moves northwest. The San Andreas fault is now the plate boundary between North America and the Pacific plate.

Since the early 1970s, many geologists have argued that the northward movement of the Pacific plate drags the rocks that lie west of the San Andreas fault north, thus causing oblique stretching of western North America farther inland. They infer that large amounts of hot mantle rock welled up behind the trailing edge of the vanished Farallon plate as it descended beneath North America. In one way or another, they envision that rise of hot mantle as the cause of the Basin and Range. Unfortunately for that argument, the Queen Charlotte fault off the coast of British Columbia is similar to the San Andreas fault and appears to have formed in the same way, but nothing even remotely resembling the Basin and Range exists inland from it. If northward movement of the Pacific plate caused both the San Andreas fault and the Basin and Range, then similar movement that formed the Queen Charlotte fault should have caused another Basin and Range in British Columbia. Evidently we need to look for some other explanation.

It seems more likely that the northern Basin and Range is related to the flood basalt province. The time of its starting appears to coincide with eruption of the flood basalts, and the two provinces slightly overlap. Furthermore, relating the Basin and Range to the flood basalt province fits a lot of things that might otherwise seem unrelated into a coherent general picture.

The eruption of most flood basalt provinces coincides with the opening of a new ocean basin: The Red Sea, for example, started to open with eruption of the Arabian and Ethiopian flood basalts; the Arabian Sea started to open with eruption of the Deccan basalts of India. The South Atlantic Ocean started to open with eruption of the Parana and Entedeke flood basalts of South America and West

The Wasatch Front south of Salt Lake City.

Africa, and the North Atlantic Ocean started to open with eruption of the flood basalts in East Greenland and the northern British Isles. Again and again, a continent ruptured and a new ocean basin started to grow as flood basalts erupted. The Basin and Range may be a failed attempt at such a continental rift.

Active Faults

Frequent earthquakes along the western and eastern edges of the Basin and Range show that the faults in those areas are still moving. The zone of earthquake activity along the eastern margin extends north from the Wasatch Front at Salt Lake City through the Grand Tetons to the Yellowstone volcano. From there, the zone continues north through Montana, following the west front of the Madison Range, passing through Butte and Helena, then northwest toward Kalispell. Whether that more northern extension is really related to the Basin and Range is open to debate.

Another zone of earthquake activity trends north from the Snake River Plain into central Idaho. The big Borah Peak earthquake of October 1983, magnitude 7.1, raised the Lost River Range and dropped the adjacent valley. Yet another zone of earthquake activity extends south of the Snake River Plain. It is not clear whether the movements are related to the Basin and Range, to cooling and contraction of hot rocks near the Snake River Plain, or perhaps to some other cause.

Fault scarp of the 1983 Borah Peak earthquake near Mackay, Idaho. —H. E. Malde, U.S. Geological Survey

We tend to think of the mountains rising when a large earthquake produces a fault scarp like that at Borah Peak. Actually, precise surveys show that the valley floor drops two or three times farther than the mountains rise. That happens because the earth's crust is stretching and thinning. In many places, the valley floor is lowest next to the fault and the highest mountains. Any lakes in the valley floor tend to huddle near the fault, against the foot of the high mountains. Abert Lake in Oregon is an excellent example.

Meanwhile, the faults along the western margin of the Basin and Range have provided California with some of its greatest historic earthquakes, with no help from the famous San Andreas fault. The most famous of those was the Owens Valley earthquake of 1872, which may have been the biggest yet in California.

Chapter 37

THE SNAKE RIVER PLAIN

13 Million Years Ago, and Continuing

At first sight, the Snake River Plain looks very much like the flood basalt province to the west, another vast expanse of basalt lava flows. In fact, the two are utterly different. The Snake River Plain consists mostly of rhyolite with a thin veneer of basalt poured across its surface. It is a rhyolite province with a crust of basalt, a white cake with chocolate icing. The lava flows that cover its surface are not flood basalts; they are perfectly ordinary flows erupted from perfectly commonplace volcanoes. You see their low rises from the road.

A digital view of the Snake River Plain as it appears from space. —G. P. Thelin and R. J. Pike, U.S. Geological Survey

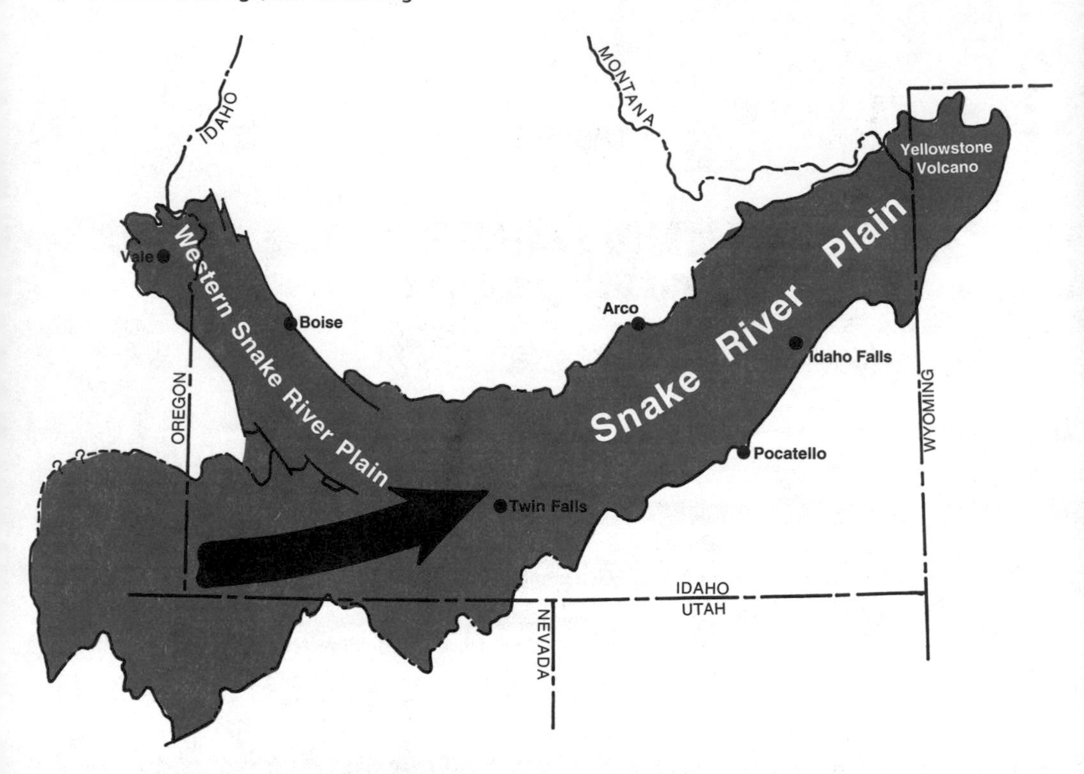

The main and western parts of the Snake River Plain.

The main part of the Snake River Plain sweeps a broad and nearly level swath through the mountains from southwestern Idaho to the Yellowstone volcano at its northeastern tip. The Western Snake River Plain projects sharply northwest from that trend. Although they look much alike on the surface, the two parts have little in common.

A Continental Hotspot Track

Most of the world's volcanoes erupt on or near a plate boundary. A hotspot volcano erupts in a place remote from any plate boundary and is not explainable in terms of plate boundaries. Most of them stand at one end of a long row of extinct volcanoes that become progressively older with increasing distance from the hotspot volcano. Geologists now believe that a hotspot volcano erupts above a stationary plume of hot rock rising within the mantle beneath the lithosphere. As the lithosphere moves across the hot mantle plume, the crust melts. A hotspot volcano erupts directly above the mantle plume, then goes out of business as the moving lithospheric plate carries it beyond the plume. Meanwhile, a new hotspot volcano begins to erupt above the mantle plume. The result is a hotspot track, a long row of extinct volcanoes, with an active hotspot volcano at the end currently above the plume of hot rock rising within the mantle.

Most flood basalt provinces are the source of a hotspot track, which invariably emerges from the part of the province where rhyolite erupted. Evidently, the plume of hot rock that begins to rise within the mantle as the flood basalt flows erupt at the surface continues to rise after the great flood basalt volcano becomes extinct. It melts the crustal rocks above, generating a row of hotspot volcanoes on the surface.

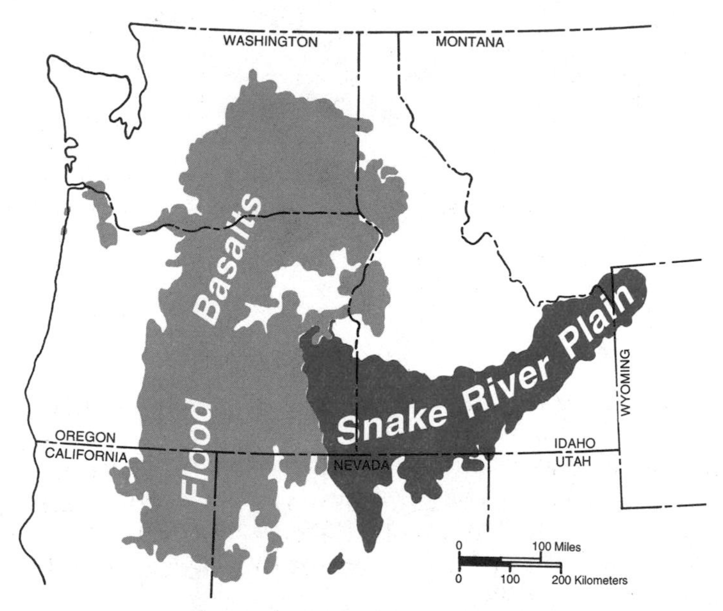

The Snake River Plain emerges full grown from the eastern margin of the flood basalt province.

Geologic maps show the Snake River Plain starting in the southwestern corner of Idaho and emerging full grown from the eastern margin of the flood basalt province, precisely from the area where rhyolite erupted along with the flood basalts. As the moving plate carried uncratered continental crust across the rising plume of hot rock in the mantle, the basement rocks in the lower crust melted into enormous volumes of granitic magma. It began to erupt about 14 million years ago from the Owyhee-Humboldt volcano in southwestern Idaho. Then the enormous Bruneau-Jarbidge volcano began to erupt just northeast of the Owyhee-Humboldt volcano about 12 to 13 million years ago. More giant rhyolite volcanoes formed one after the other, each northeast of the last, leaving the main trend of the Snake River Plain in their wake. The Yellowstone volcano is the latest in the series, the one now active. It is at the northeastern tip of the Snake River Plain, directly above the part of the mantle that was beneath southeastern Oregon 17 million years ago.

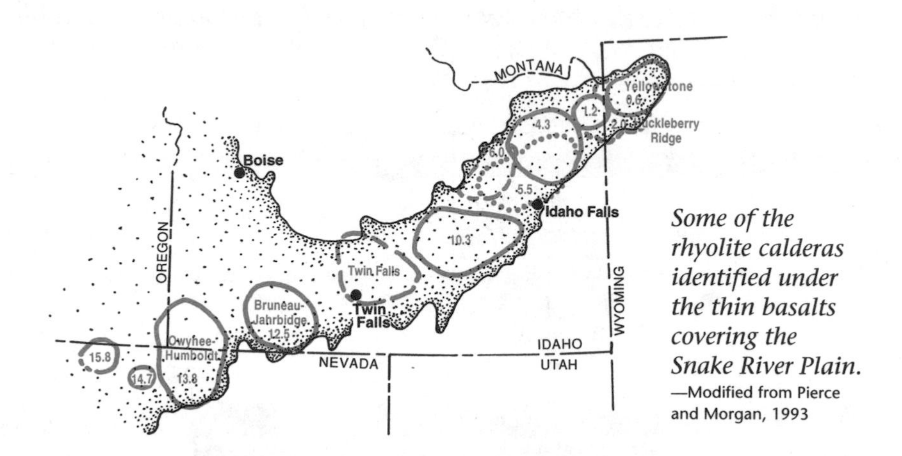

Some of the rhyolite calderas identified under the thin basalts covering the Snake River Plain.
—Modified from Pierce and Morgan, 1993

Some geologists maintain that the hotspot started somewhere remote in the Pacific Ocean; others maintain that it started at the Mendocino triple junction off the coast of California. In either case, it should have left a track between the coast and southwestern Idaho, where we first see it. The flood basalts could cover part of such a track, but they do not cover the area farther west, and no hint of a hotspot track exists there.

The Younger Volcanic Surface

Every rock you see as you drive across most parts of the Snake River Plain is basalt, a veritable sea of black rock. But here and there, if you look for it, you can see rhyolite beneath the basalt. It appears in deep canyons and where faults bring it to the surface. Seemingly the rhyolite erupted first, then a thin cover of basalt flows covered it.

Many geologists have interpreted that thin crust of basalt as evidence that each version of the hotspot volcano began its career with a series of violent rhyolite eruptions, then finished by spreading a decent cover of basalt over the wreckage. Unfortunately for that

A thin layer of black basalt caps a thick section of white rhyolite in the wall of this valley near Glenns Ferry, Idaho.

idea, the basalt flows do not become progressively younger to the northeast, as do the big rhyolite volcanoes. In fact, the youngest basalt flows lie at Craters of the Moon, near the middle of the Snake River Plain. Many of those flows are only about 2,000 years old; the rhyolite in that area erupted some five or six million years ago.

So why all the basalt? It seems to fit into the picture of the Basin and Range much more comfortably than into that of the Yellowstone hotspot. The basalts of Craters of the Moon erupted from rifts that

A thin flow of black basalt caps pale rhyolite along the southern fringe of the Snake River Plain south of Twin Falls, Idaho. Settling of fine ash produced thin layers in the upper rhyolite.

trend to the northwest, nearly parallel to the nearby faults of the Basin and Range. Small basalt volcanoes exist here and there throughout the Basin and Range, but far more abundantly in the Snake River Plain than elsewhere—probably because passage of the Yellowstone hotspot left the lithosphere hotter there.

Watch as you drive across the Snake River Plain for subtle rises in the land as much as a few hundred feet high and a mile across. Those are small basalt shield volcanoes. In places, you see steep basalt cinder cones a few hundred feet high. No such volcanoes exist in the flood basalt province to the west. A few of the volcanoes on the Snake River Plain erupted rhyolite. Big Southern Butte, on the horizon south of Craters of the Moon, is the largest of several rhyolite domes that rise like islands above the level sea of basalt flows. They erupted as masses of extremely viscous rhyolite magma that bulged up through the basalt veneer long after the hotspot passed.

Volcanoes Along the Great Rift

A few dirt roads offer the only easy access to the open fractures of the Great Rift between Craters of the Moon lava field near Arco and the Wapi and Kings Bowl lava fields near American Falls. It is awe-inspiring to stand at the edge of one of those broad gashes, trying to peer into its shadowed depths, waiting to hear a tossed rock hit bottom. What could open such a ghastly fissure?

The Great Rift in the Snake River Plain is part of Basin and Range faulting.

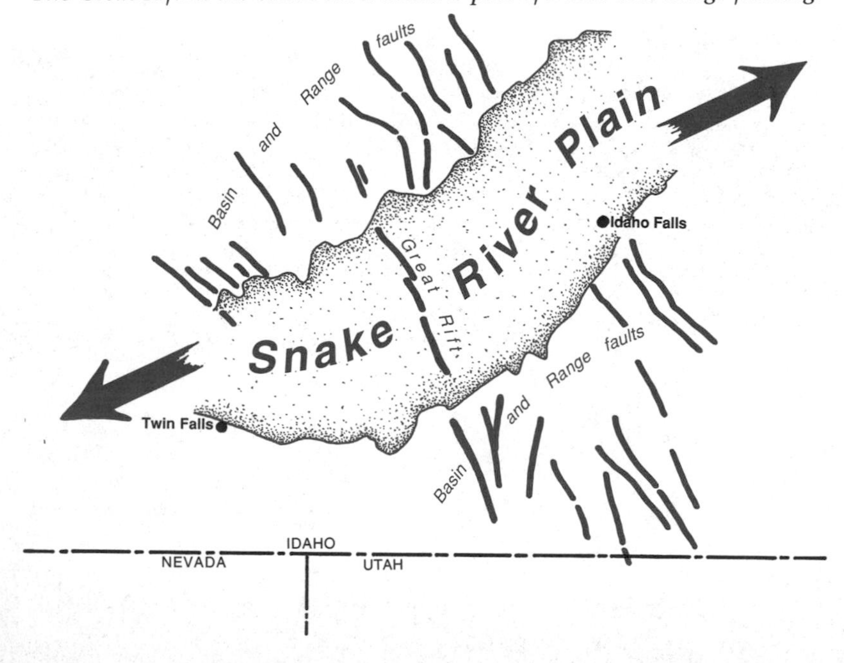

Those open fractures trend a bit west of north, parallel to the Basin and Range faults that are now slicing the Rocky Mountains both north and south of the Snake River Plain. The fractures of the Great Rift probably mark the line of one of those faults. In any case, the open fissures certainly express active crustal stretching in an approximately east to west direction. That argues against the view of the Snake River Plain as a great continental rift that is slowly extending northeast, filling itself with volcanic rocks as it opens. A rift trending northeast would require crustal stretching from northwest to southeast, opposite the direction that the Great Rift requires.

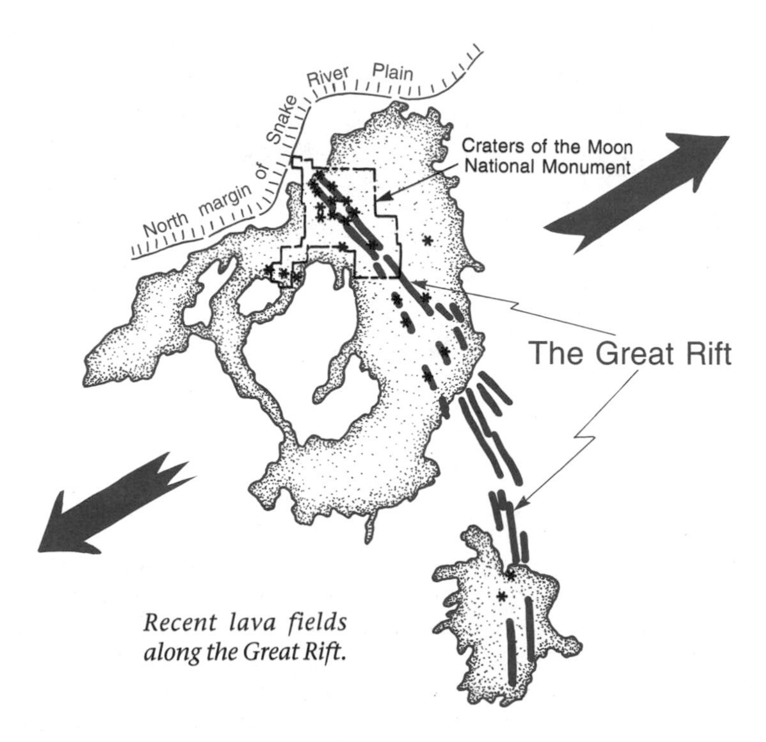

Recent lava fields along the Great Rift.

The eruptions of about 2,000 years ago along the Great Rift were among the most recent in our region. In both the Craters of the Moon lava field at the northern edge of the Snake River Plain and in the Wapi and Kings Bowl lava fields near its southern edge, basalt lava welled up from the Great Rift. It built rows of little cinder-cone volcanoes aligned along the fissures and poured across the ground in lava flows. Between the lava fields, the Great Rift is a swarm of open fissures.

Cinder cones typically form during the early stages of a basalt eruption, while the magma is blowing off the steam it acquired on its way to the surface. Escaping steam coughs blobs of molten basalt out of the vent; they land nearby to make a black pile of basalt cinders a few hundred feet high with a crater in its summit. Meanwhile, the escaping steam blows smaller shreds of lava high into the air, where they drift downwind as an ominous plume of dark ash. The glowing fountains of red hot steam and molten basalt make spectacular fireworks, but such eruptions never progress into extreme volcanic violence; basalt lava is not viscous enough to create a truly violent explosion as steam expands within it.

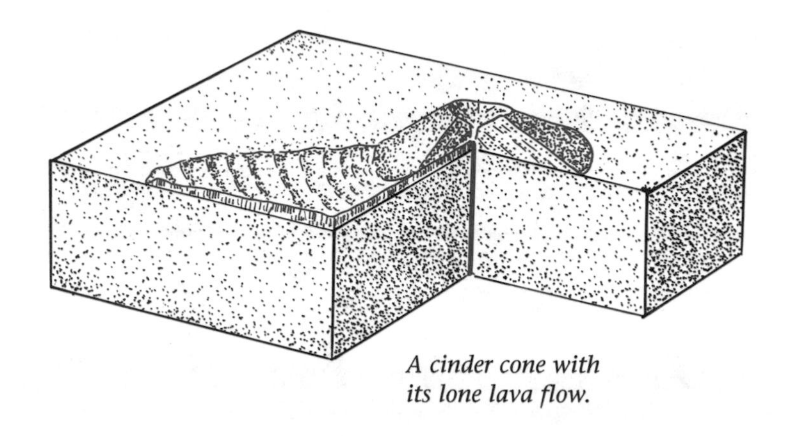

A cinder cone with its lone lava flow.

After most of the steam is gone, basalt lava bursts through the base of the cinder cone and pours fluidly across the countryside as a flow that may cover an area of several square miles to a depth of 50 or more feet. Cinder cones are piles of loose volcanic rubble much too weak to hold a column of heavy lava within them.

Pahoehoe lava surfaces are fairly smooth with a variety of wrinkled and ropy forms; they form on flows that contain enough steam to make the lava very fluid. Aa lava surfaces are rubbly and angular; they form on basalt lava that is viscous because it contains very little steam. The lava develops a bubbly crust that breaks into sharply angular fragments as the flow moves. Some flows have a pahoehoe surface near the vent where the lava still contained some steam, then develop an aa surface farther from the vent, where most of the steam has escaped.

Basalt flows cool from the top down, so they pass through a fascinating stage in which the flow surface is fairly solid while the interior is still molten. In many cases, the moving lava beneath

Ropy surface of a pahoehoe flow, Craters of the Moon, Idaho.

rafts the solid surface along in great blocks that jam together to make pressure ridges very much like those in Arctic pack ice. And the fluid lava within may drain out from beneath the solid surface, leaving the flow hollow. Many basalt flows contain long and winding caves, lava tubes that eventually collapse into rows of sink holes. Drips and dribbles of basalt make bizarre decorations on the cave walls and ceiling, lavacicles.

What the Future Will Probably Bring

If present trends continue, the Snake River Plain will extend itself far to the northeast. Meanwhile, erosion will strip the bland volcanic surface, exposing the rocks at depth. Rhyolite forms when granite magma erupts. Large volumes of the granite that rose into the great volcanoes of the Snake River Plain probably did not erupt, but instead crystallized at depth to become masses of granite. If you could come back 50 million years from now to see how things turned out, you might see a long row of granite intrusions exposed where erosion has stripped the volcanic cover off the Snake River Plain. They will become younger to the northeast, through the Northern Rocky Mountains and beyond. Swarms of black basalt dikes, the plumbing for the lava flows that now cover the Snake River Plain, will surely cut across the granite, trending slightly west of north, parallel to the open fractures of the Great Rift.

Chapter 38

THE YELLOWSTONE VOLCANO

1.8 Million Years Ago to the Present

Yellowstone Park is best known for its geysers and wildlife. It would be better known if more people realized that most of the park is the top of a giant volcano that may well erupt. It is one of the largest volcanoes in the world and one of the most dangerous. It is an uncommon type of volcano called a resurgent caldera, the latest of a long line of them that built the main trend of the Snake River Plain. It may not resemble Rainier, but the Yellowstone volcano does look exactly like a large resurgent caldera.

Volcanic Monsters

No resurgent caldera has erupted during recorded human history, so our knowledge of their habits comes entirely from study of the rocks that erupt from them, and from observation of lesser volcanoes that erupt rhyolite. According to their rocks, they are an especially ghastly species of monster.

Enormous sheets of solidly welded rhyolite ash tell of eruptions that begin as steaming rhyolite magma boils out of the ground into a dense cloud of small shreds of molten magma suspended in red-hot steam—an ash flow. Rhyolite ash flows are so hot that they ignite trees on contact. When they finally settle, they are still so hot that the particles of ash in parts of the flow are molten; they weld themselves into solid rock. Some ash flows on the eastern Snake River Plain are welded as far as 50 miles from their source, an extraordinary distance. Unwelded parts of the same ash flows went much farther.

Much larger sheets of soft and unwelded rhyolite ash tell of dark plumes of ash blown high into the sky, where they drift downwind,

Yellowstone geyser. —I. J. Witkind, U.S. Geological Survey

dropping ash like snow across hundreds of thousands of square miles. You can estimate the volume of an ash flow by multiplying the area it covers by its average thickness. Some of those from the resurgent calderas of the Snake River Plain contain about a 100 cubic miles.

As the eruption belches those enormous volumes of rhyolite lava, the ground surface collapses into the emptying magma chamber below to form a broad basin called a caldera, which then fills with rhyolite during the last stages of the eruption. Most calderas are so

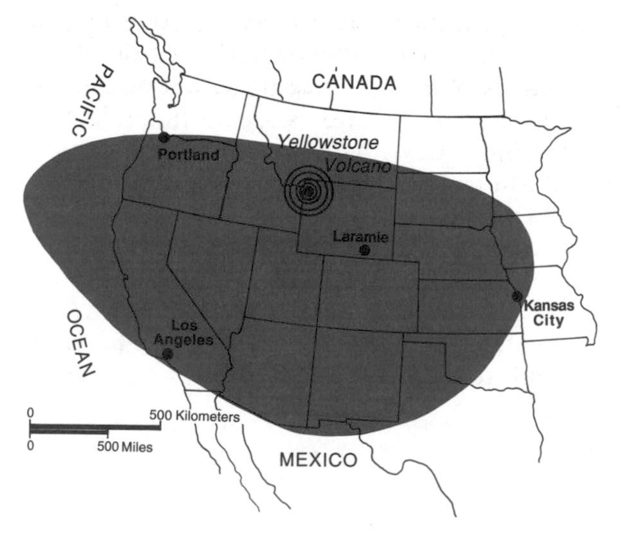

Distribution of the Huckleberry Ridge ash, which erupted from the Yellowstone volcano about 1.8 million years ago. —Redrawn from Sarna-Wojcicki and Davis, 1991

completely filled that it is hard to see them just by looking at the landscape. You need a geologic map detailed enough to distinguish between generations of rhyolite erupted at different times. No one clearly saw the big caldera of the Yellowstone volcano until it appeared on a geologic map. Finally, the still molten lava at depth rises beneath the filled caldera, bulging its surface, presumably setting the stage for the next eruption. That bulge is the resurgent dome.

Several rhyolite eruptions during historic time, modest by the monstrous standards of resurgent calderas, have caused damaging environmental effects that lasted as much as several years. Ash lingering in the upper atmosphere blocked sunlight, causing abnormally cool and wet weather in large regions. One of those eruptions caused the year without a summer, 1815, that brought famine to New England. A major eruption of a resurgent caldera would almost certainly cause extremely severe environmental effects that would probably last several years. A Yellowstone eruption the size of the one that produced the Huckleberry Ridge ash about 1.8 million years ago would spread ash over large parts of Wyoming, Colorado, Nebraska, Kansas, Oklahoma, and Texas. It is well that those monsters typically erupt just a few times, at intervals of hundreds of thousands of years.

Will It Blow?

So far, the Yellowstone volcano has staged three major eruptions at intervals of approximately 600,000 years, most recently about 600,000 years ago. The last major eruption spread rhyolite ash across the High Plains south all the way to Mexico—tens of feet of ash in large areas.

The Yellowstone volcano now shows many of the classic symptoms of a volcano getting ready to erupt: Two large resurgent domes now bulge the caldera floor, slowly swelling like enormous tumors, apparently because fresh magma is rising beneath them. Frequent swarms of small earthquakes tell of magma moving at depth. Earthquake waves that arrive at seismographs after having passed beneath Yellowstone Park have lost most of their shear waves, the kind that make the ground shimmy and twist during an earthquake. Shear waves cannot pass through liquids. Their disappearance beneath the park is good evidence of a large mass of molten rock within a few miles of the surface.

Even though an eruption seems likely, no one can now predict whether it will come in the near future or thousands of years hence. Neither can anyone predict its size. And it is quite possible that the

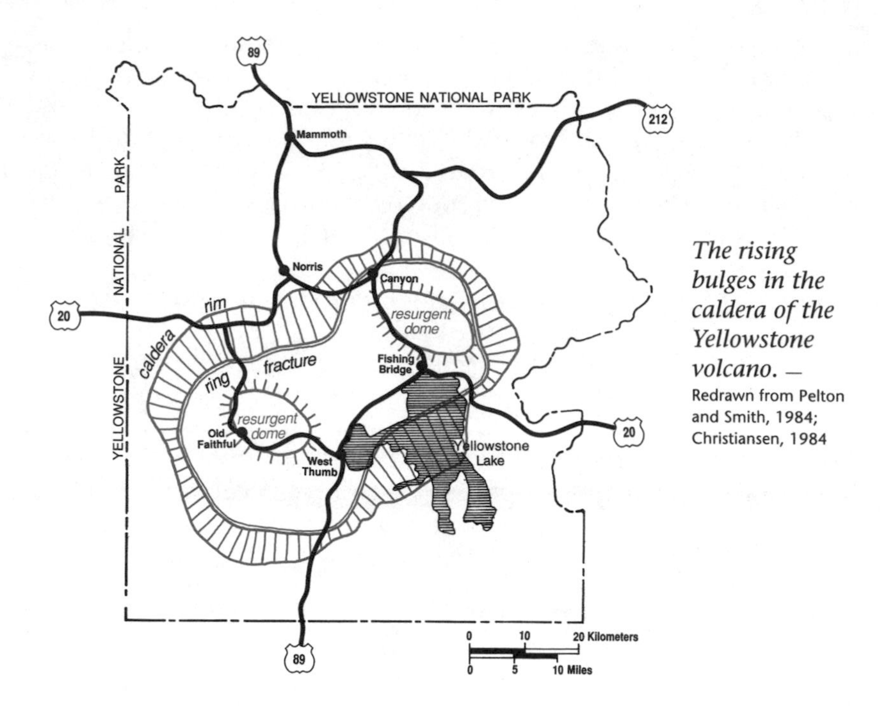

The rising bulges in the caldera of the Yellowstone volcano. — Redrawn from Pelton and Smith, 1984; Christiansen, 1984

magma now rising within the volcano may quietly crystallize into granite without erupting.

Without eyewitness accounts, it is hard to say much about how a major eruption of a resurgent caldera might start. Geologists generally assume that the Yellowstone volcano, like most volcanoes, would probably give ample warning over a period of weeks or months: Numerous swarms of small earthquakes would come in rapid succession, the two resurgent domes would rise more rapidly, and the thermal areas would become more active. Of course, it is quite possible that the Yellowstone volcano we know today will never erupt again. The next eruption may start a new resurgent caldera volcano a few miles to the northeast, taking the Snake River Plain another step on its long northeasterly trek.

Riding High on the Mantle Plume: The Yellowstone Plateau

The famous reports of bitterly cold weather from West Yellowstone tell you more about the elevation of the Yellowstone Plateau than about the climate of Montana. It seems fairly clear that the Yellowstone Plateau stands high because heat rising from the mantle plume has expanded the continental crust. It will gradually subside and the area to the northeast will rise as the continent continues to move across the plume. The area to the southwest is subsiding now as the moving continent carries it beyond the plume. The effect of carrying the continental crust across the mantle plume is a bit like that of dragging a blanket across the top of your extended thumb.

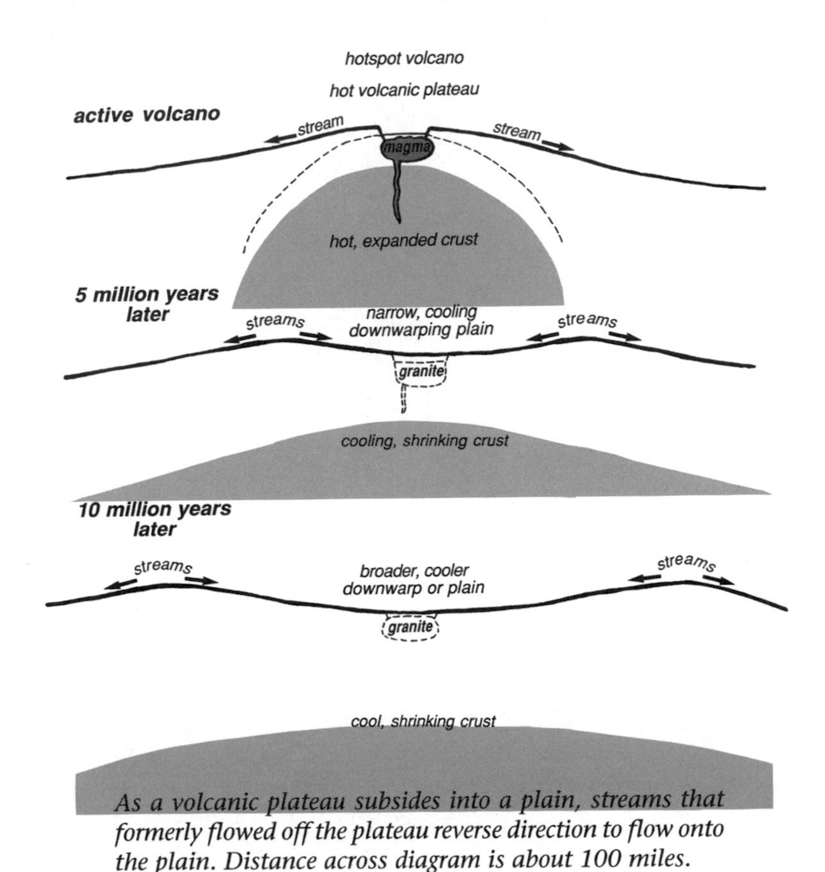

As a volcanic plateau subsides into a plain, streams that formerly flowed off the plateau reverse direction to flow onto the plain. Distance across diagram is about 100 miles.

Rivers now drain in all directions from the high Yellowstone Plateau. It seems reasonable to assume that each of the series of resurgent calderas ancestral to the Yellowstone volcano stood on its own high plateau, which sank as the continent moved on. Streams drained off each in its time. Can we still see some of that in the modern landscape?

Imagine the streams draining away from the volcanic plateau while it stands high above the rising plume of hot rock in the mantle. After four or five million years, the area has moved southwest beyond the mantle plume, the continental crust cools, and the plateau subsides. Streams that were flowing away from the plateau to the northwest and southeast lose their gradients and may eventually reverse their flow. The modern streams generally drain an area about 50 miles wide toward the Snake River Plain. That may reflect the width of the area that rose and fell as the continental crust heated as it passed over the plume and then cooled as it passed beyond it.

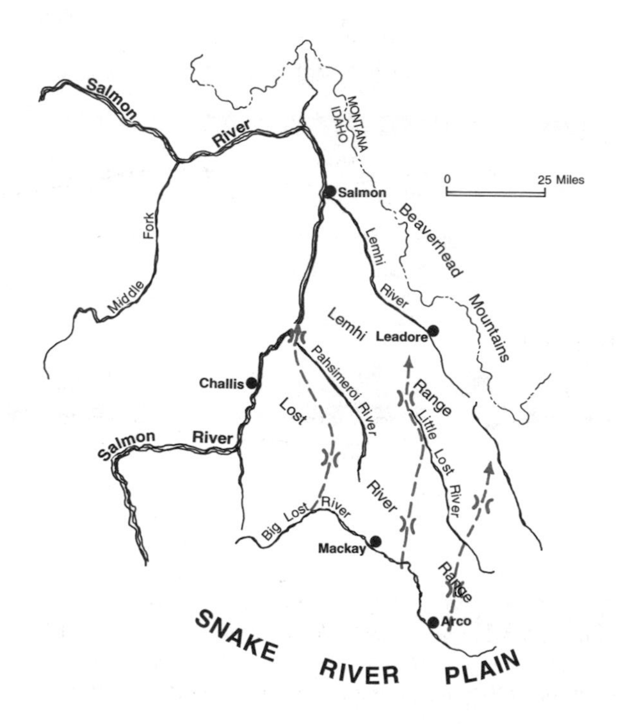

Old streams that flowed northeast cross the Lost River and Lemhi Ranges where they are preserved in deep passes through the ranges. The present Salmon River also flows northeast before turning abruptly west through a deep, narrow canyon.

Chapter 39

FOLDS AND FAULTS IN THE COLUMBIA PLATEAU

15.5 Million Years Ago to the Present

When the big flood basalt flows were freshly erupted, most of the surface of the province must have been nearly as flat and level as a platter of milk. The northern end of the province, the Columbia Plateau, probably sloped ever so gently down to the north and west, away from the big feeder fissures in its southeastern part. Now we see quite a varied landscape, which owes much of its origin to faulting and folding, amazingly little to erosion.

A digital view of Washington and much of Oregon. The smoother area in the eastern part of the region is the Columbia Plateau. —G. P. Thelin and R. J. Pike, U.S. Geological Survey

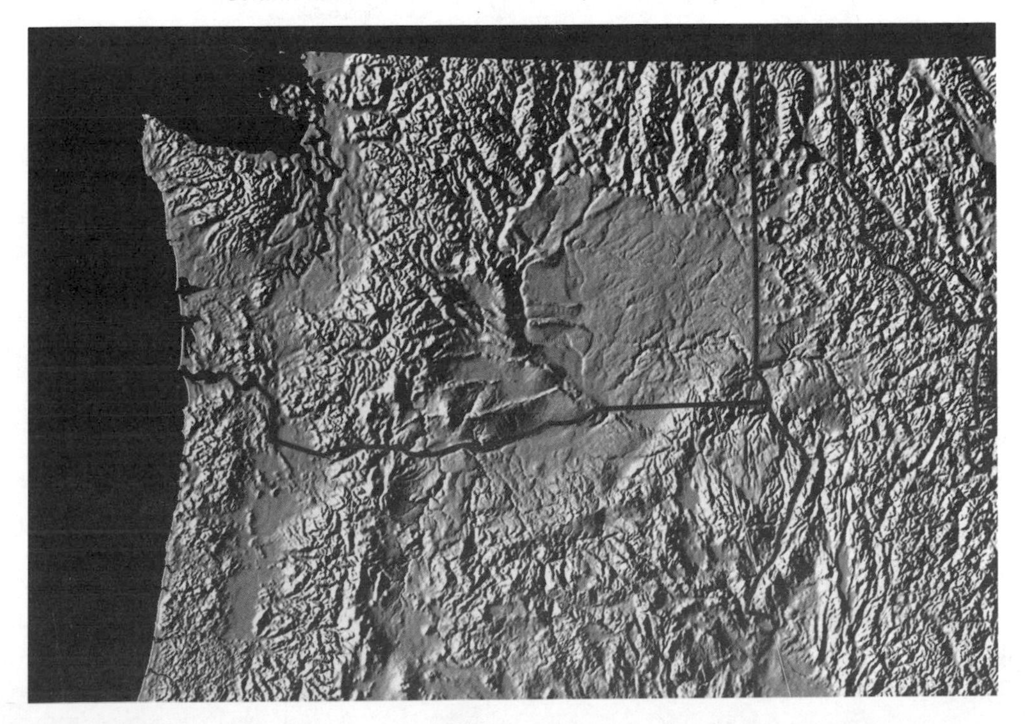

The lava flows in the western part of the Columbia Plateau, the part generally north of the Brothers fault zone, are now rumpling into an impressive series of anticlinal arches and synclinal troughs that trend generally east. The arches rise above thrust faults, most conspicuously near a line called the Olympic-Wallowa lineament. They stand high as narrow ridges that overlook and separate the broad synclinal valleys, where people live and farm.

The Brothers Fault Zone and the Olympic-Wallowa Lineament

The Brothers fault zone is a dense swarm of faults that trends northwest from the southeastern corner of Oregon to the High Cascades. The faults slip horizontally, the south side moving to the

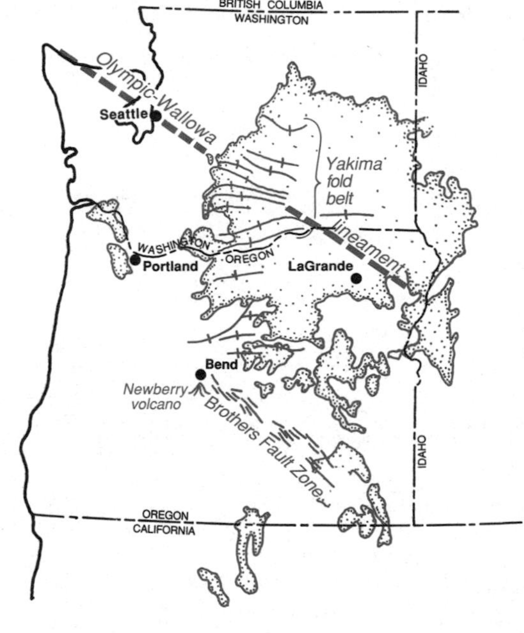

The Brothers fault zone and the Olympic-Wallowa lineament angle northwest across the Columbia Plateau.
—Modified from Reidel and Campbell, 1989

northwest relative to the north side. The Brothers fault zone is also a trend of small volcanoes, which become generally younger to the northwest, ending in Newberry Volcano.

The very look of the landscape changes across the Brothers fault zone. South of it, the Basin and Range is stretching along faults that break the flood basalt flows into tilting blocks that trend generally north. Their rising edges make mountains, and their sinking edges broad valleys. North of it, some force is shoving large slabs of the

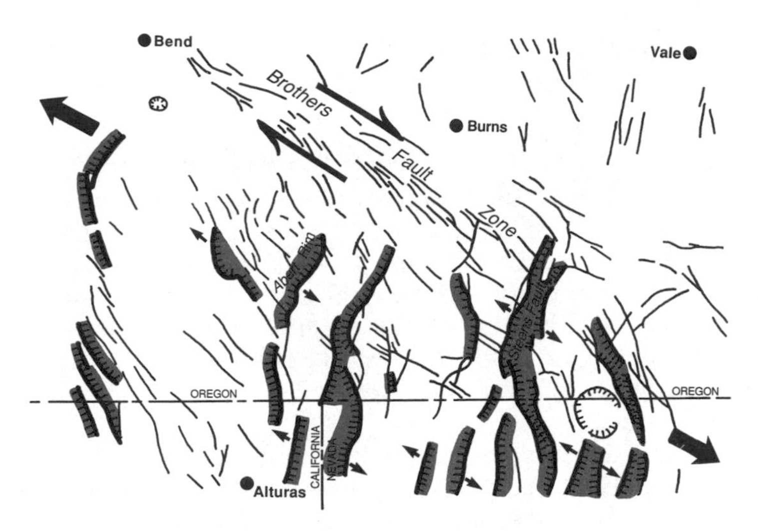

The Brothers fault zone is the northern limit of crustal stretching. —Modified from McKee and others, 1983

Columbia Plateau north along thrust faults, crumpling it from south to north into great folds that trend west to east. So it seems that the earth's crust is stretching from east to west south of the Brothers fault zone, compressing in the opposite direction north of it.

Landscapes both north and south of the Brothers fault zone owe their origins almost entirely to crustal movements, much less to erosion. Mountains south of the boundary are rising along fault scarps that streams have hardly begun to dissect. The dropping basins contain salty desert lakes on a floor of basin fill sediments. The long mountain ridges north of the boundary are folds sharply arched above thrust faults in the flood basalts and almost uneroded. The broad valleys between those ridges also contain a fill of sediments.

The Olympic-Wallowa lineament is a topographic line that cuts across the Columbia Plateau from the North Cascades southeast to the general vicinity of the Wallowa Mountains. It shows up nicely in satellite images. It is north of the Brothers fault zone and parallel to it. And it is another of those nagging enigmas that neither sort themselves out nor quietly vanish.

Some geologists think the Olympic-Wallowa lineament is a fault. That straight line across the map certainly does suggest a fault, but the geologic map shows no breakage of the rocks. If it is a fault, it has not moved much since the flood basalts erupted, perhaps just enough to explain why some of the folds southwest of the lineament trend to the east, then turn to a southeast trend within it.

Other geologists have suggested that the lineament is some sort of rift that stretched the crust, but no good evidence of rifting exists.

Thrust Faults, Anticlines, and Synclines

The long ridges in the western Columbia Plateau north of the Brothers fault zone are anticlinal arches. Layers of gravel deposited as recently as two or three million years ago tilt as much as 20 degrees on the flanks of some of the arches. Layers of flood sediment deposited as recently as 15,000 years ago are slightly tilted. That tilting of such young sediments leaves little doubt that the folding is continuing. We suspect that it began before the last flood basalt eruptions ceased.

All the folded arches are steepest on their north sides, where thrust faults clearly break many of them. It seems that a compressional force directed from the south is breaking the plateau basalts along thrust faults and shoving them north in great slabs. The anticlines arch as the rocks above each thrust fault rumple into a big fold. Geologists call such folds hanging wall anticlines.

The faulting and folding diminish eastward, which suggests that their cause lies in the west. Many geologists attribute it to the general northward drag that the sinking remnant of the Farallon plate exerts on the western edge of the continent. That seems reasonable. And you would expect that drag to diminish eastward as the plate sinking beneath the Pacific Northwest descends to greater depths. But why should that drag manifest itself so differently on opposite sides of the Brothers fault zone?

Idealized section through a typical thrust fault and anticline in the western Columbia Plateau. Rocks stretch in the crest of an anticlinal arch.

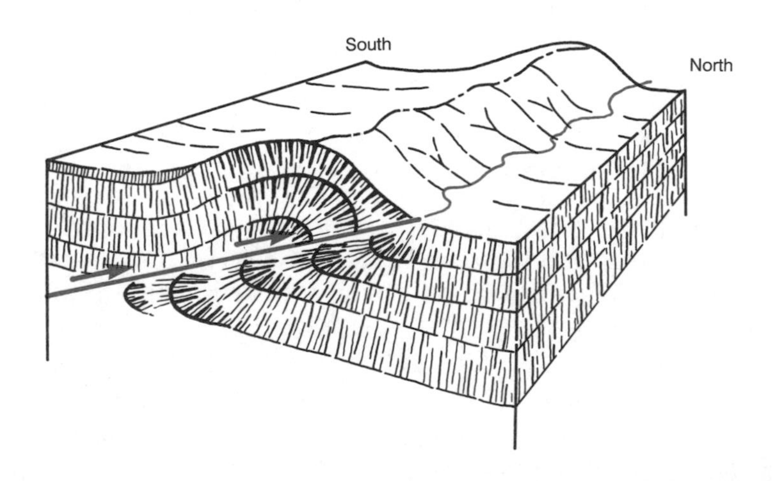

Folds, Fractures, and Weathering

At first thought, it seems reasonable to expect anticlinal arches to make ridges. In fact, erosion normally opens them into valleys. Such folds generally stand as ridges only if they are extremely young, as they are in the western Columbia Plateau.

Bend a telephone book into an arch. The pages tend to pull away from each other at the crest of the fold because it is in tension, tending to stretch. That same tension cracks the rocks along the crest of an anticline. Water penetrating the open fractures attacks and decomposes the rocks, making them vulnerable to erosion. That explains why fold crests tend to become valleys. Basalt flows in the crests of the anticlinal ridges in the western Columbia Plateau are full of open fractures, but water needs more time to weather the rocks. Wait a few tens of millions of years, then come back to see how things are going. You will probably find narrow valleys where you now see ridges, broad uplands where you now see valleys.

The ridge in the distance is a fold in the surface of the Columbia River Plateau basalts, just south of Kennewick, Washington.

Chapter 40

THE OLYMPIC MOUNTAINS

15 Million Years Ago

Mountains form in all sorts of ways. Most have long, complicated, and more or less obscure histories all of their own that make them unique. Like most mountain ranges, the Olympic Mountains are easy to understand in some ways, a terrible problem in others.

Mountains from the Ocean Floor

First, the easy part. All the rocks in the Olympic Mountains are clearly oceanic: big slices of the bedrock oceanic crust and crumpled masses of the muddy sediments that were laid down on it. The oceanic crust is mostly basalt, including pillow basalt, darkly greenish rocks in blobby forms. Reaction with water converted some of the original minerals to chlorite, which gives the normally black basalt its greenish cast. Most of the sedimentary rocks are muddy sandstones in various shades of dark gray. It is a somber assemblage.

The sedimentary rocks in the Olympic Mountains have a generally abused look. They were roughly handled when the sinking slab jammed them into a trench. Those scrambled rocks lie structurally beneath the Crescent terrane, the same slab of oceanic crust that continues south through the Willapa Hills and into the northern part of the Oregon Coast Range. The covering slab of oceanic crust wraps steeply around the southern, eastern, and northern sides of the Olympic Mountains, but lies flat farther south.

It seems clear that the oceanic lithosphere sinking through the trench off the west coast stuffed the tortured rocks of the Olympic Mountains under the Crescent terrane, the slab of oceanic crust in the coastal hills. The Olympic Mountains are the only place where you can actually see the rocks that understuff the entire length of the coastal hills.

Tight anticline in the Olympic Mountains. Massive sandstones crack; shaley interlayers thinly cleave into slates. —R. W. Tabor, U.S. Geological Survey

Eventually, of course, oceanic lithosphere will stop sinking through the trench off the west coast. Then all those scrambled but light sediments that have been stuffed into the trench will float up to become a new range of mountains along the coast. They will push aside the rest of the coastal slab of oceanic crust, tilting it steeply down to the east. Imagine that slow rotation of the land we now know, the west walls of the valleys becoming gradually steeper until they finally fail in great landslides. The final effect will be like what you now see in the California Coast Range, where a trench filling lies immediately west of a slab of oceanic crust tilted steeply down to the east. The Olympic Mountains are the first installment of a range that will eventually extend south all the way to the Klamath block in southern Oregon.

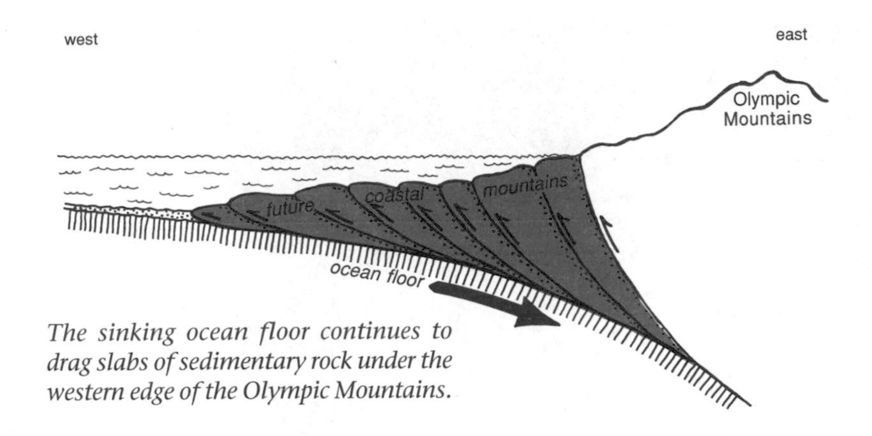

The sinking ocean floor continues to drag slabs of sedimentary rock under the western edge of the Olympic Mountains.

Now, the hard part. What are the Olympic Mountains doing standing high while the modern trench is still active and the rest of the rocks stuffed into it are still far below sea level? Why did this one little segment of the future coastal mountains rise so prematurely? Maybe the trench tried to swallow something it could not quite manage, such as an old volcano. Even oceanic trenches have their limits. Maybe it just stuffed too much sedimentary rock, perhaps from the delta of a major river system, under this stretch of the coastal slab of oceanic crust and then regurgitated the Olympic Mountains.

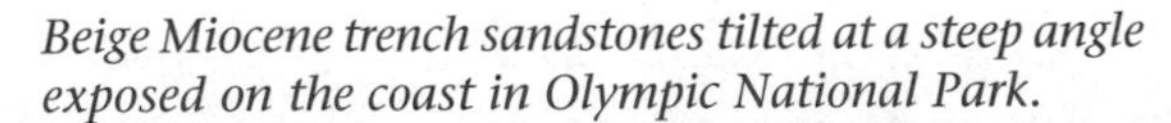

Beige Miocene trench sandstones tilted at a steep angle exposed on the coast in Olympic National Park.

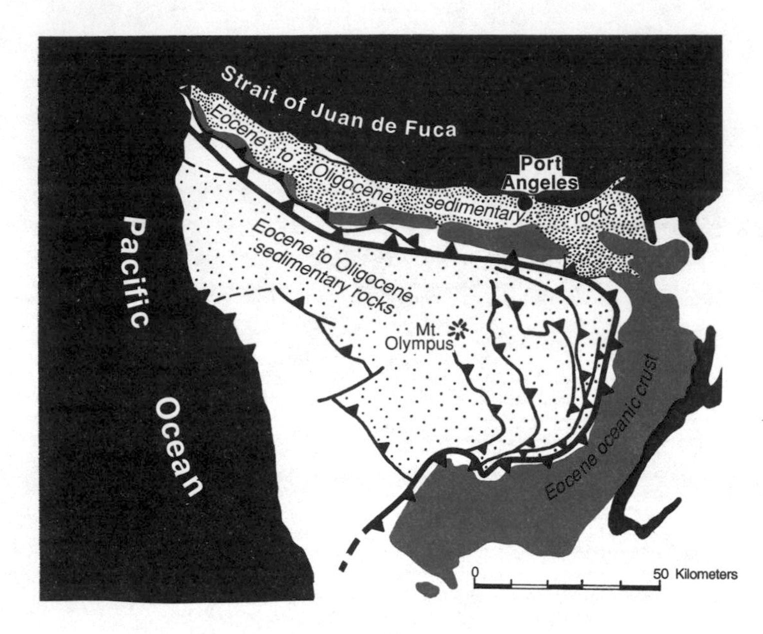

The coastal slab of oceanic crust, the Crescent terrane, wraps around the sedimentary rocks stuffed under it (stippled) *as though it were a sheet of wax wrapped around a piece of cheese. The wavy lines with teeth are thrust faults.*
—Modified from Heller and others, 1992

Whatever the cause, the rising Olympic Mountains rudely pushed aside the coastal slab of oceanic crust. The geologic map shows the coastal slab turning up on edge and wrapping tightly around the crushed oceanic rocks in the core of the Olympic Mountains. That turned-up edge precisely defines all but the western boundary of the Olympic Mountains. Some geologists maintain that the oceanic crust there is the thickest known anywhere.

However you explain them, the Olympic Mountains certainly rose very recently—for mountains. Some of the sedimentary rocks stuffed under their western edge contain fossils of animals that lived in the Pacific Ocean during Miocene time, probably as recently as 15 or 20 million years ago. Since then, those rocks were stuffed under the coastal slab, then raised in the Olympic Mountains. Although the fossils do not tell when the mountains rose, they leave no doubt that they have been standing high for less than 15 million years. These are among the youngest high mountains in the country.

Some people contend that you can tell the relative ages of mountains by looking at them—the craggier the younger. Forget it. Mountains owe their cragginess or smoothness to the processes of erosion that shaped them, not to their age. The Olympic Mountains owe their extreme cragginess and ragged skyline entirely to their glaciers, not to their extreme youth. The Olympic Mountains are far more heavily glaciated than the Rocky Mountains, and they maintain much larger modern glaciers at much lower elevations in a

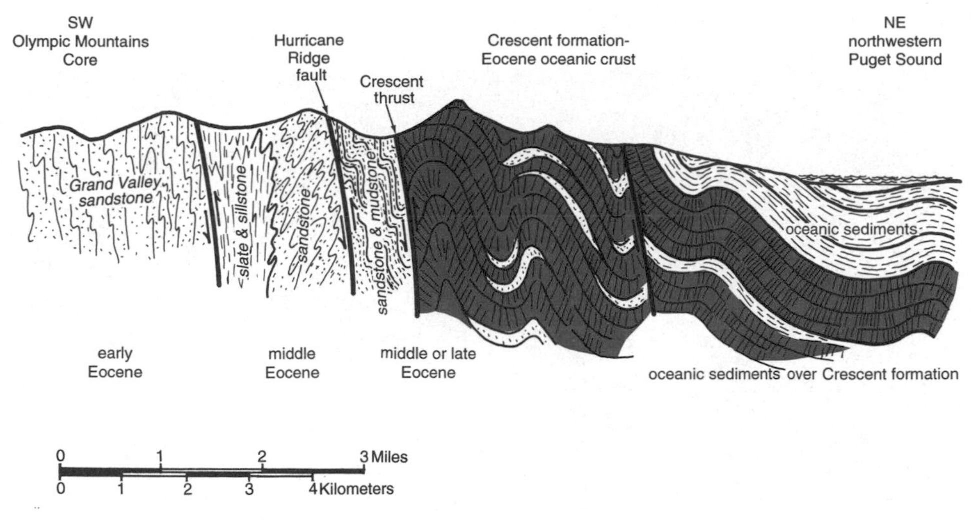

Section through the Olympic Mountains, from the Olympic core to northern Puget Sound. —Modified from Tabor, 1987

much warmer climate. They owe their ice to the wet wind that blows off the ocean, dumping much of its moisture on the first high mountains it meets.

Small Showings of Oil

The mangled rocks in the core of the Olympic Mountains are most accessibly exposed along the coast, where you can see them in the sea cliffs and stacks. Hikers who venture into the barren heights of the range see basically similar rocks. In both areas, watch especially for masses of sheared and broken rocks, hopeless jumbles that geologists call melanges.

Oil actually seeps out of the ground in a number of places along the western side of the Olympic Peninsula, so it must exist in the rocks at depth. Wildcat drilling early in this century brought a minor but exciting boom to the Hoh River Valley, where several wells produced very small amounts of oil. But the boom soon turned into a bust as it became clear that none of them could produce anything remotely resembling commercial quantities. We do not know of any oil fields that produce from rocks as badly mangled as those in the Olympic Mountains. In fact, it is hard to imagine how such rocks could hold much oil without letting it all leak out along faults. But the possibility may exist. Oil seeps powerfully suggest that oil is down there somewhere.

Glaciers filling Puget Sound scoured these deep grooves in bedrock.

Chapter 41

THE SHEARED WESTERN FRINGE

15 Million Years Ago to the Present

Ever since the modern trench was established off the west coast some 50 million years or so ago, North America has been moving generally west, and the rocks sinking into the trench have been moving northeast. They collide at an oblique angle, which causes a shearing motion between the continent and the oceanic crust. You can see that shearing expressed in the landscape of the western part of our region.

The Coastal Lowlands

The Willamette Valley and southern Puget Sound Lowland separate the rugged coastal hills and high Olympic Mountains from the darkly forested ridges of the Western Cascades. Most of the Puget Sound Lowland lies west of the North Cascades. It continues north into the Georgia Depression of southwestern British Columbia. These depressions stand between the volcanic chain of the High Cascades and the offshore trench.

Geologic maps of the Willamette Valley and southern Puget Sound Lowland show a series of rather obscure basins with more or less parallel sides. Pliocene gravel fills those in the Willamette Valley. Glacial debris covers nearly everything in the Puget Sound Lowland and Georgia Depression. Those aligned lowlands are probably a continuous trend, a series of basins that opened as blocks of the coastal slab of oceanic crust dropped along faults. They break the flood basalt flows that cover the Willamette Valley as far south as Salem, so they must have dropped since the basalt erupted between 17 and 15 million years ago. Occasional shallow earthquakes suggest that they are still dropping. The magnetic evidence may help explain what is happening, and why.

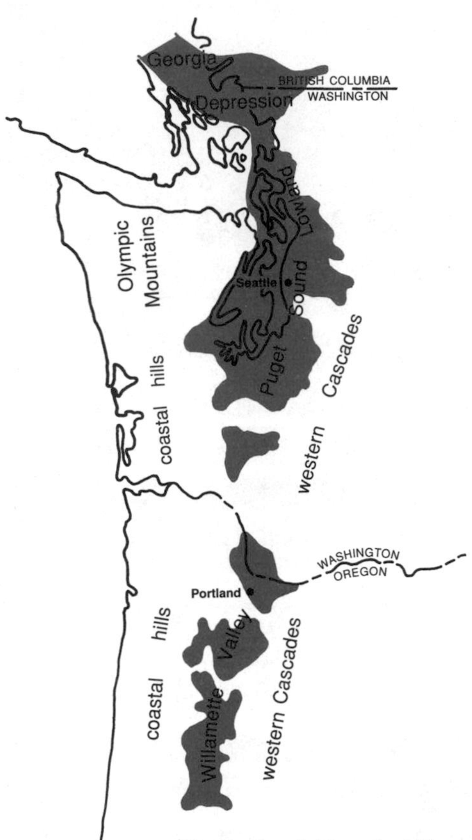

The Willamette Valley, Puget Sound Lowland, and Georgia Depression are probably a series of dropped fault blocks between the coastal hills and the Western Cascades.

Rotated Rock Magnets

The rocks in the Oregon Coast Range and Willapa Hills generally look as though nothing much has happened to them. One road cut after another reveals layers of muddy sandstone lying almost as flat as the day they settled out of clouds of muddy water pouring across the ocean floor. Others are only gently tilted. But appearances deceive. Their magnetism suggests that the rocks are more deformed than you would guess from an occasional glimpse through the brush.

Crystals of the strongly magnetic mineral magnetite align themselves with the earth's magnetic field as though they were miniature compass needles. They make most rocks weak permanent magnets polarized in the direction of the earth's magnetic field. Rocks retain their original magnetic polarization as a permanent memory of the earth's magnetic field, where and when they formed. That is, they record the direction of the magnetic north pole and the magnetic latitude, which crudely approximate the geographic north pole and the latitude. If a moving lithospheric plate carries the rock to a new place on the earth's surface, its magnetic polarization will tell something about that movement.

People who study rock magnetism find that the rocks of the coastal hills all have their magnetic poles pointing well east of the present north pole, as though they were all rotated clockwise. Many point as much as 45 degrees east, some as much as 75 degrees. In general, the amount of rotation increases as you go east or north. Hundreds of those magnetic measurements leave no doubt. Their pattern is much too consistent to be the result of some mistake, and the rotation is far too great to ignore. The magnetic fields of the rocks really do point east of north.

Most geologists imagine the rocks rotating in small blocks caught in the shearing motion between the North American plate moving southwest and the Pacific plate moving north. To understand how that might work, imagine covering part of a table with dominoes, then gently shoving those along the left side away from you. As you move the dominoes, they spread to cover a larger area and gaps open between them. In principle, the gaps could correspond to the basins in the Willamette Valley, the Puget Sound Lowland, and the Georgia Depression.

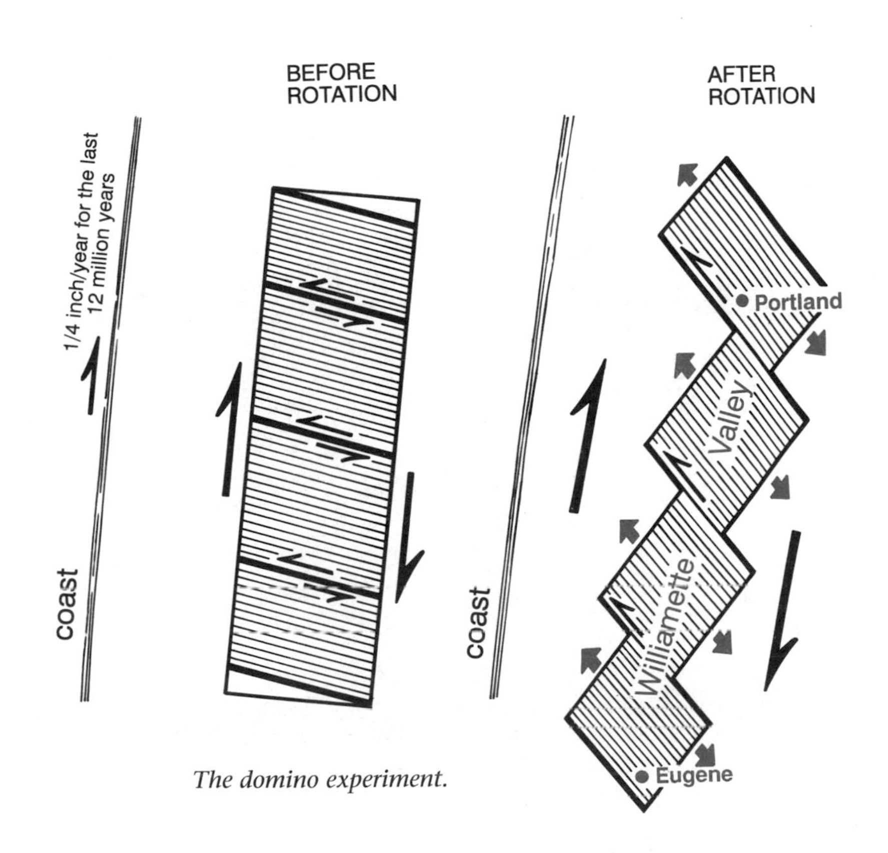

The domino experiment.

Occasional earthquakes suggest that such movements may be happening in the coastal hills and the lowlands east of them, but the geologic maps do not show a clear pattern of faults. They may be misleading. All the soil and plants that cover those rocks make it hard to recognize faults. Nevertheless, geologists still work their painstaking way through the poison oak, refining their maps, finding more faults. Someday, those improved geologic maps will tell whether the domino model is really as valid as we now think it is.

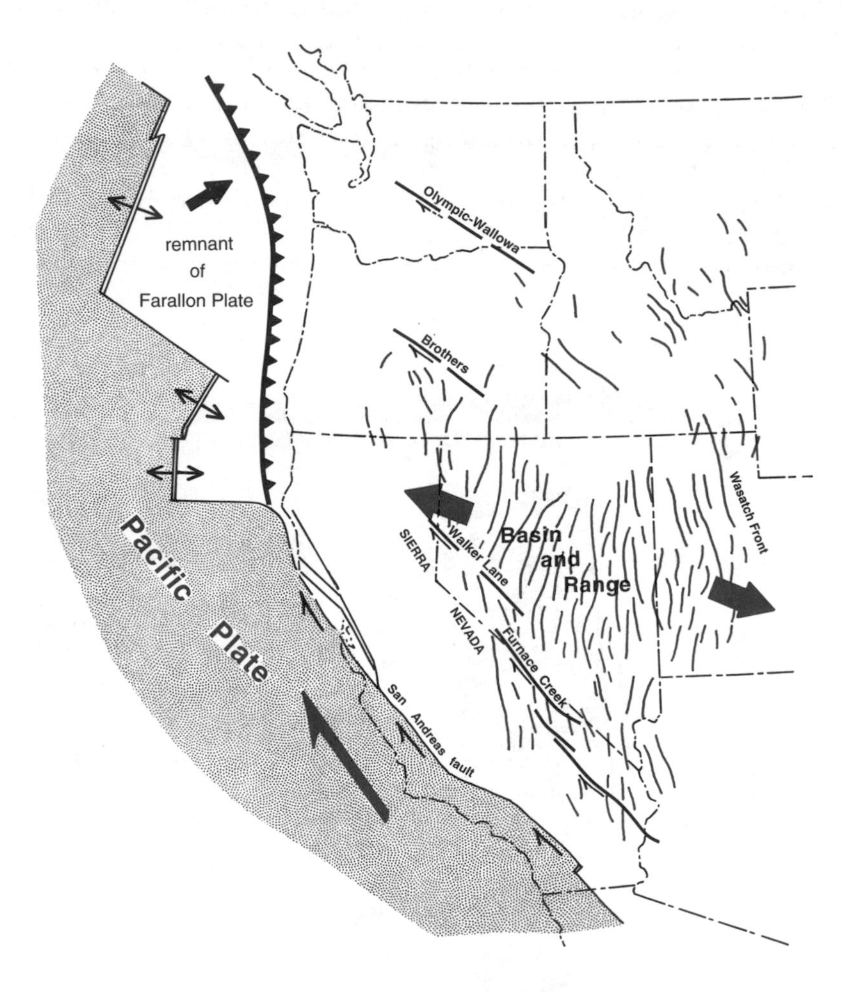

Northward motion of the Pacific plate may be dragging along the western part of the continent, shearing it along faults that move horizontally.

Chapter 42

THE SECOND LONG DRY SPELL

15 to 2 Million Years Ago

The wet and tropically warm climate that accompanied eruption of the flood basalts ended with those eruptions. The climate again became dry, probably even drier than before. The dwindling plant cover again exposed the soil to erosion, letting the occasional rains wash it off the slopes and into the streams, which no longer had enough flow to carry it to the coast. Sediment accumulated on land as new deposits of gravel, sand, and silt were laid down here and there from the Pacific coast to the High Plains.

Fossil bones and teeth of vertebrate animals buried in the younger gravels leave no doubt of their late Miocene and Pliocene age. Many of the larger bones are the remains of ancestral horses, camels, and rhinoceroses, all of which fit comfortably into a desert scene. None of the fossils are younger than about two million years, when the Pliocene period ended as the first of the great ice ages began. Evidently, the climate again became wetter then and new streams began to flow, carrying sediment to the ocean.

The most impressive deposits of desert sediments are on the High Plains.

High Plains Gravel on an Enormous Desert Plain

Westbound travelers first sense a western feel in the landscape as they watch the land rise into the broad reaches of the High Plains and the trees thin to galleries of cottonwoods along the streams. This flat floor under the arching sky is the easternmost of our distinctively western landscapes. Its story begins during the long dry spell of latest Miocene and Pliocene time and hinges on the way raindrops splash desert soil.

The skimpy plant cover of dry regions spreads such a ragged umbrella of leaves over the ground that much of the soil lies open to the sky, naked to the splashing impacts of raindrops. Anyone who parks a car on an unpaved driveway during a hard rain sees how falling raindrops splatter the soil. And everyone has seen the hard crust that develops on untilled gardens and fields during the summer. It forms as raindrops make a slurry of mud on the surface and then hammer it tight. Rain that falls on that hard surface tends to run off, rather than soaking into the ground. That is why desert hills shed rain like a roof, causing sudden flash floods in streams that are dry between rains.

Rain-splattered particles of soil land in the film of water running off the hard crust and ride it down the slope. Water draining off the base of such a rapidly eroding slope typically carries an overwhelming load of sediment. Deposits of mud and sand soon choke the channel, forcing the water to overflow and carve a new channel, which soon chokes with more sediment. The constantly shifting channels that carry water away from the base of the slope leave deposits of sediment that coalesce into a broad alluvial surface, a desert plain.

A miniature desert plain at the base of a rapidly eroding slope in the Makoshika badlands near Glendive, Montana.

We read these days of fractals, of things that recur in many scales from very small to very large but look the same regardless of their size. Desert plains are good examples: They form around nearly any bit of bare ground eroding in the rain, and they look the same regardless of their size. They form in any climate, provided only that raindrops find places to fall on barren ground. You can see miniature desert plains spreading from the bases of eroding road cuts. Much larger desert plains spread for several miles from the flanks of many western mountain ranges. The desert plain that spread east from the Rocky Mountains during late Miocene and Pliocene times reached into the central parts of the plains states and prairie provinces, from Saskatchewan to Texas. The landscape must have resembled the modern scene in the plains deserts of central and western Australia.

Imagine the scene when that ancient desert plain was forming: no entrenched stream valleys because the streams constantly filled their channels with sediment. Sheets of muddy water poured across the surface in sudden flash floods, eroding the high places and filling the low places. Those floods left the surface covered with deposits of sediment: the Flaxville gravel of Montana, the Oglalla gravel of states to the south.

Rain and muddy surface runoff graded that enormous desert plain into a smooth surface that sloped gently down to the east from an elevation of four or five thousand feet along the front of the Rocky Mountains to less than 2,000 feet at its eastern margin. Except near the mountains, the slope was too gentle for the eye to detect. Large remnants of that nearly flat desert plain survive in the modern High Plains.

Pliocene Gravel West of the Plains

Meanwhile, more gravel, the Six Mile Creek formation, was filling the broad valleys within the Rocky Mountains. As the climate dried and the streams stopped flowing, sediment washing down from the mountains began to accumulate in the valleys. The new

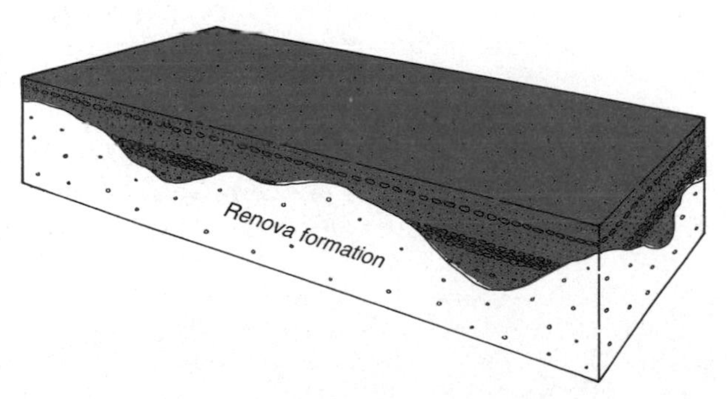

New deposits of gravel buried an erosional landscape developed on the Renova formation in the mountain valleys of western Montana.

Thin layers of fine sediments deposited in a desert playa lake during Pliocene time. —J. D. Love, U.S. Geological Survey

deposits first buried the erosional landscape that had developed on the older valley fill deposits when the climate was wet. Then, sheets of sediment spread across the valley floors, converting them into broad surfaces that swept smoothly from the mountains on one side down to a shallow playa lake in the valley floor, then up to the mountains on the other side.

None of the pebbles in those younger gravel deposits came from distant sources, as you would expect if a connected network of streams had flowed through the region. The mountain valleys of the northern Rocky Mountains were once again closed basins of internal drainage like those you see today in Nevada.

And more gravel was accumulating farther west. It filled the old valley of the ancestral Columbia River and spread across large reaches of the Columbia Plateau, covering some of the flood basalt flows. Gravel accumulated in the Willamette Valley of Oregon, filling the fault block basins that were opening there. People driving Interstate 5 through the northern Sacramento Valley of California can see the eroded remains of Pliocene alluvial fans banked against the flanks of the mountains along either side of the valley.

All those widespread Pliocene deposits contain layers of sand, silt, and mud, but gravel is the most distinctive material, much of it quite coarse. The large pebbles could have moved only under torrents of water, which suggests a barren landscape where very few plants encouraged water to soak into the ground. That suggests an extremely dry climate. The Pliocene gravels contain very little petrified wood, evidently because few trees grew. They do contain fossils of many kinds of animals, including ancestral camels and horses that probably grazed on open plains, just as their descendants do now.

The Palouse Hills

The Palouse Hills of eastern Washington and nearby Idaho, with their lush crops of wheat, lentils, and peas, are one of the strangest of western landscapes. At first glance, the picturesquely rolling hills seem perfectly normal. But the logical continuity of an erosional landscape is missing; the hills do not continue into ridges, and the valleys do not connect into an integrated network of streams. In fact, most of the valleys are without streams. This bizarre landscape is mostly a creation of the wind. Only a few local details owe their origin to the modern and mostly irrelevant streams.

The Palouse Hills are big dunes of yellowish silt. Watch for road cuts that show complex patterns of tilted layers, the record of shift-

The Palouse Hills northwest of Pullman, Washington.

ing winds. The silt completely fills the valleys and covers most of the hills of the erosional landscape that existed before the wind blew the dunes across it. Occasional exposures of bedrock—basalt in Washington, basalt or granite or Belt rocks in Idaho—are the tops of old hills that the dunes did not quite bury. You can spot them from a distance because they support scraggly forests of pine trees, which refuse to grow on the Palouse silt. Quarries work many

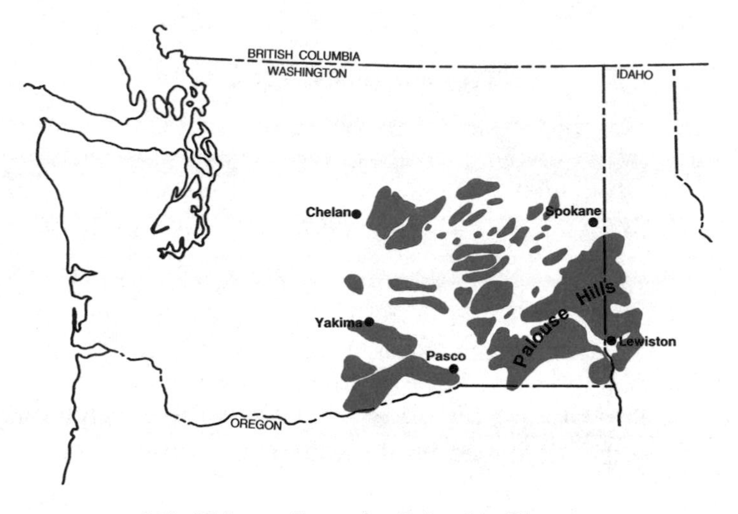

Wind blown silt on the Columbia Plateau.

of those bedrock exposures for road metal and construction aggregate, rare commodities in a landscape of dunes.

Some geologists contend that the silt blew off glacial outwash during the great ice ages. It is true that cold wind blowing off the ice does raise great clouds of dust from freshly deposited glacial outwash. But the steepest slopes of the Palouse Hills face generally northeast, and that is a problem. The steepest slopes of dunes invariably face in the direction they are moving. So the Palouse dunes moved northeast. That puts their most likely source in the area around the Pasco basin.

The Palouse silt becomes coarser toward the southwest and finally grades into sand dunes. That is precisely the pattern you would expect if the dunes that became the Palouse hills moved from southwest to northeast. They appear to be desert dunes that marched northeast out of the southern part of central Washington.

A lack of direct evidence leaves open to conjecture the question of when the wind drove the Palouse dunes. They cover a landscape eroded into flood basalt lava flows, so they arrived long after middle Miocene time. They probably blew in during interglacial periods between ice ages when the climate was very dry.

Palouse silt overlies Columbia Plateau basalt along U.S. 12 northeast of Walla Walla, Washington.

A thirty-foot-deep roadcut in beige loess, north of Dayton, in southeastern Washington.

Chapter 43

THE HIGH CASCADES

12 Million Years Ago to the Present

When we last visited them 17 million years ago, the long chain of the Western Cascades, with its history of 25 or more million years of intense activity, had just snuffed out. The Cascades remained virtually silent while the great flood basalt eruptions continued. Then, sometime around 12 or 13 million years ago, new volcanic activity began to build a new chain of peaks, the modern High Cascades. They rose along a line east of the old Western Cascades. Oceanic lithosphere was evidently again sinking beneath the advancing western edge of North America.

The Eastward Shift of the High Cascades

The High Cascades stand along a moderately straight line about 50 miles east of the old volcanic trend. Their snaggled row of volcanic teeth extends from Garibaldi in southern British Columbia south to Lassen in California. Why did the new line of volcanoes shift east, away from the trench?

Shifts in the trends of volcanic chains quite commonly move the line of activity farther from the trench. New oceanic crust is very hot where it forms at oceanic ridges, and relatively light. It cools and becomes denser as it ages and moves away from the ridge. That change in the temperature and density of the oceanic crust explains why the ocean floor slowly settles to greater depth as it moves farther from the ridge.

As North America moves west, it encroaches on the offshore ridge, the source of the ocean floor sinking through the trench offshore. Every year, the oceanic lithosphere sliding beneath the advancing western edge of the continent becomes a bit younger, a bit hotter, a bit lighter, and a bit more reluctant to dive through the trench. The sinking lithosphere that drives the High Cascades is a bit lighter

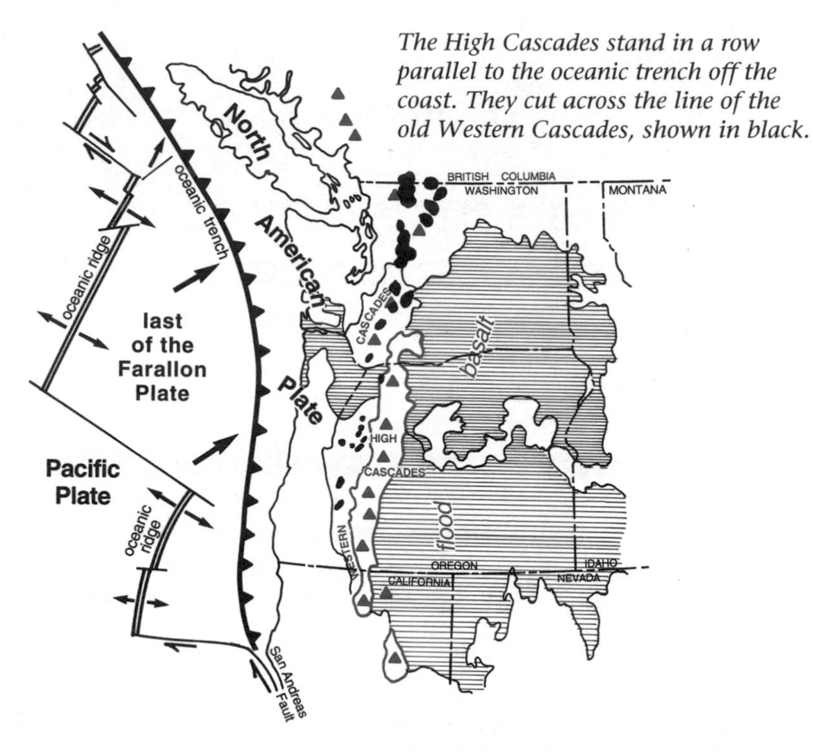

The High Cascades stand in a row parallel to the oceanic trench off the coast. They cut across the line of the old Western Cascades, shown in black.

than the slab that drove the Western Cascades, so it probably sinks at a flatter angle. A volcanic chain erupts above the line where the sinking slab of oceanic lithosphere reaches a depth of 60 or 70 miles. Oceanic lithosphere sinking at a lower angle will move farther inland before it reaches that depth than if it were sinking steeply. That probably explains why the High Cascades lie east of the Western Cascades.

If the sinking slab passes through the trench at a rate of several inches every year, it will reach a depth of about 60 miles in a million or so years. Add another two or three million years for the rising steam to begin generating magma, and you have an interval that approximates that between the end of activity in the old Western Cascades and the beginning of the High Cascades.

A New Crustal Arch

The modern High Cascades, like most active volcanic chains, stand on the crest of a broad crustal arch that rose as magma heated the older rocks below the volcanoes, making them expand. When masses of magma under the volcanic chain crystallize and shrink, the crest of the arch drops along faults as though it were the keystone of a

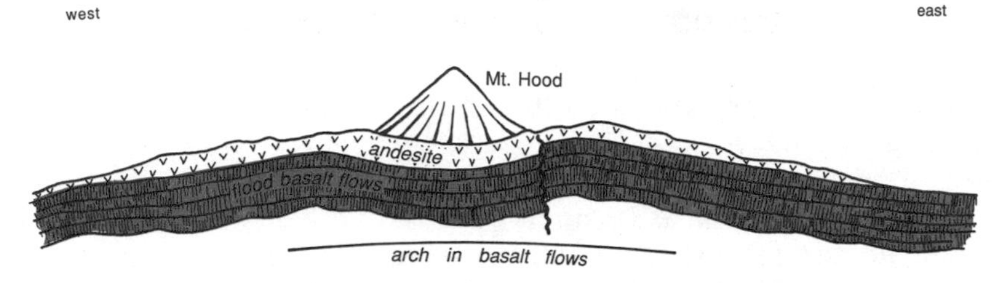

Tilted flood basalt flows make the broad arch beneath the High Cascades easily visible.

sagging arch in an old building. The Western Cascades lie in long fault blocks that evidently dropped as the rocks at depth cooled and shrank after the supply of magma from earlier eruptions stopped.

You can easily see the west side of such a volcanic arch as you drive between The Dalles and Hood River on Interstate 84. Watch the dark and continuous ledges of flood basalt flows, north of the Columbia River, as they rise gently westward. The flood basalt flows were almost level when they erupted, so the arch must have risen since then. Its western flank, west of Hood River, is much harder to see from the highway.

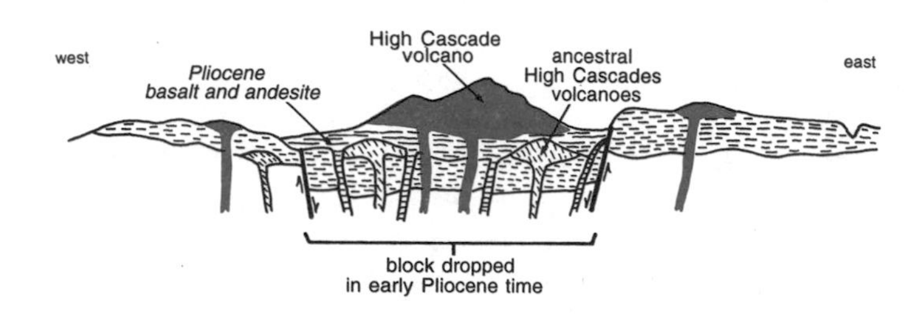

The central block of the Cascades drops as the crustal arch cools. Idealized section through the central Oregon Cascades. —Modified from Taylor, 1981

The Columbia River carved the narrow Columbia Gorge as it eroded its bed fast enough to keep pace with the rising arch beneath the High Cascades. At least the lower part of the Columbia River must have had enough flow during the dry years of late Miocene and Pliocene time to maintain its course through the rising arch.

Breeds of Cascade Volcanoes

Volcanoes come in varieties, each with its own style and habits, which depend upon the viscosities of their lavas and the amount of water dissolved in them, both of which relate to their silica content. Water, when it flashes into steam, is the only instigator of volcanic violence; lavas without it erupt quietly.

Of the common lavas, basalt contains the least silica, the least water, and has the lowest viscosity. Basalt eruptions typically begin as the lava blows off steam, coughing shreds and blobs of lava. The smaller bits drift on the wind as volcanic ash, while the larger pieces settle around the vent to build a cinder cone. Most cinder cones are a few hundred feet high; they have steep sides and a crater in the top. After most of the steam has gone, the dry lava erupts quietly as a flow that pours across the countryside.

The lava flow bursts out through the base of the cinder cone for the same reason that milk pours out of the bottom of a pile of corn flakes. The flow commonly tears away parts of the cinder cone and rafts away big pieces of it. After it produces its lava flow, a cinder cone never erupts again. The next eruption will produce another single-shot volcano.

A long succession of nearby basalt eruptions builds a pile of thin lava flows called a shield volcano, supposedly because it is shaped like a classical Roman shield. In fact, shield volcanoes are shaped about like a vanilla wafer, a somewhat more familiar household object. Shield volcanoes slope gently because basalt lava is quite fluid—the same reason that thin batter makes thin pancakes. The shield volcanoes in the High Cascades attract very little attention because they are so flat. People are far more likely to notice the steeply conical andesite volcanoes.

Andesite is one of those wonderful rock names that is useful mainly because it is so imprecise. Most volcanic rocks consist of such tiny mineral grains that no one can identify them without a microscope. The word andesite covers all those nondescript and gray volcanic rocks that are neither as black as basalt nor as pale as rhyolite. If you find a medium gray volcanic rock in the Cascades, call it andesite.

Very dark andesite looks and erupts almost like basalt because the lava is fluid and contains little water. Increasing the silica content of andesite makes the color of the rock lighter, increases the viscosity of the magma, and increases its ability to absorb water. Intermediate andesite lavas, those that make medium gray rocks, are too viscous to blow off steam. Instead, they swell up like rising bread dough, which is also so viscous that it traps gas. If those lavas contain a lot of water in the form of steam, they swell up so fast

Rubbly surface of basalt lava flow in the High Cascades, east of Medford, Oregon.

when they erupt that they explode. The steam blows the lava out of the volcano in clouds of ash and cinders. If andesite lavas contain little or no steam, they erupt quietly to become thick and blocky lava flows.

Pale gray andesites contain even more silica. They are extremely viscous and able to absorb a lot of water. If they erupt with a heavy charge of steam they explode with tremendous violence, spreading sheets of pale ash across the countryside and sending great clouds of ash towering into the sky. If pale andesite lava contains very little water it erupts as lava domes, mountains of pasty lava that slowly bulge out of the ground like gigantic spring mushrooms.

Beyond pale andesite is rhyolite, the most dangerous lava if it contains much steam. Rhyolite is typically pale gray, yellow, pink, pale green, or white. You can tell it from pale andesite by looking for grains of quartz, glassy grains that generally look dark when you see them in their shadowed settings in the rock.

Pumice blown from Mazama in the climactic rhyolite eruption that produced Crater Lake.

Most of the larger High Cascades volcanoes have erupted basalt and several flavors of andesite. A few have produced some rhyolite. Basalt eruptions generally produced lava flows. Some andesite eruptions produced mostly ash and fragments, others lava flows. Mudflows are common on andesite volcanoes, with or without the help of an eruption.

The varied activity of andesite volcanoes builds a steep and fairly strong edifice—steep because it contains large amounts of angular rubble, which tends to form steep slopes, and strong because lava flows knit the structure together. Their strength enables andesite volcanoes to hold an internal column of lava, which explains why they so commonly erupt from the summit crater. Repeated summit eruptions build towering conical volcanoes. The prominent High Cascades peaks consist mainly of andesite.

Many of the great volcanoes in the High Cascades began their careers erupting basalt that first filled the river valleys, then built a broad shield volcano. Next, they erected a large volcanic cone on that base as they erupted progressively lighter varieties of andesite. And many finished with rhyolite. Some of those volcanoes finally destroyed themselves in a great cataclysm. The famous eruption of 1980 that destroyed the beautiful conical form of St. Helens produced lava that was very close to rhyolite.

Active and Dead Volcanoes

Even though the High Cascades are relatively young for a volcanic chain, their 12 million or so years of activity provides plenty of time for large volcanoes to complete their careers. Some of those big volcanoes are dead; others are active. Which?

Any volcano that has erupted within the last few centuries is almost certainly active. What about those that have not? It is perfectly normal for an active andesite volcano to brood in silence for thousands of years between eruptions. The short span of recorded history in the Cascades does not provide any basis for declaring a volcano extinct. Furthermore, much of that historic record is unreliable. Nineteenth-century newspapers enthusiastically interpreted all sorts of forest fires and curious clouds as eruptions. They even published highly circumstantial accounts of eruptions from Mount Olympus, which is not a volcano.

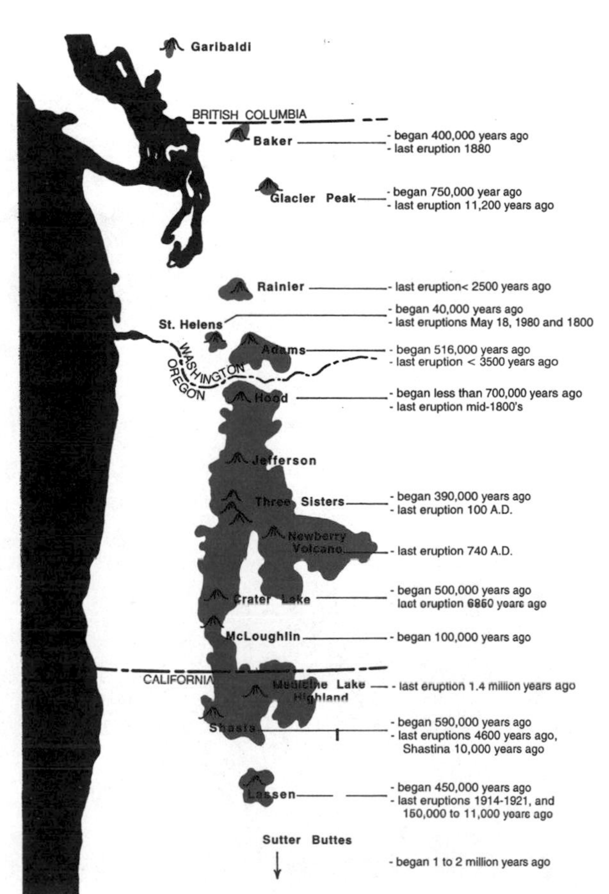

First and last eruptions of Cascade volcanoes.
—Modified from Sarna-Wojcicki and others, 1991

Until the late 1970s, many geologists agreed that the scarcity of eruptions since the middle of the last century was evidence that the Cascade chain was dead. They saw Lassen, which erupted off and on between 1914 and 1921, as a fluke. Then detailed studies of individual volcanoes during the 1970s revealed a much higher level of fairly recent activity than anyone had suspected. The eruption of St. Helens in 1980 convinced everyone that the chain is active, even though some of its volcanoes are surely dead.

You can judge the health of many volcanoes from their appearance. If a volcano has not repaired the scars of ice age glaciation, it has not erupted much since the end of the last ice age—at least 10,000 or so years ago—and is probably dead. Three Fingered Jack is an excellent example of such a glaciated ruin. It is surely dead.

Others are more difficult to judge: Rainier, Hood, and Shasta, for example, bear deeply engraved scars of ice age glaciers, but have produced minor eruptions within the last few centuries. They have not been very active for a long time, but neither are they quite dead. Those giants may well be capable of another great eruption, perhaps one like the extremely violent explosion of rhyolite that converted towering Mazama into Crater Lake just 6,950 years ago. They may be entering the most dangerous periods of their careers.

Rainier as seen from the west. Glaciers have deeply carved its flanks.
—C. L. Driedger, U.S. Geological Survey

Mazama, 6700 B.C.

Crater Lake fills the hole left where Mazama, a giant fully comparable to Shasta, destroyed itself during a calamitous rhyolite eruption of about 4950 B.C. Remnants of that nearly vanished monster make the rim that encloses Crater Lake. If you project their slopes upward at the usual angle of large andesite cones, you reconstruct a volcano that was something like 12,000 feet high, depending upon how much you decide to lop off the top for a crater. Remnants of glaciated valleys scallop the rim of Crater Lake. The farthest moraines of those glaciers lie as much as 18 miles beyond the volcano. Glaciers that long must have descended from very large snowfields high on the mountain.

Mazama started its career sometime around 400,000 years ago, apparently as a cluster of andesite volcanoes. As usual, they grew rapidly at first, then slowed. The volcano probably reached its full size about 50,000 years ago, when the last major series of dark andesite lava flows erupted. About 10,000 years later, pale andesite began to erupt as lava domes. Then nothing erupted for more than 20,000 years, while the volcano quietly brewed a big batch of very pale andesite, almost rhyolite.

Eroded Mazama ash, pale at the base, dark at the top, deposited from the huge ash flow eruption that formed Crater Lake. —Oregon State Highway Department

The end began as that large mass of magma erupted with a full charge of water that flashed to steam that shot a column of ash and larger fragments high into the sky. The blast of steam grew broader and slower as it enlarged the crater until it could no longer support the high eruption column, which then collapsed onto the north side of the volcano. The debris poured down the slopes as a hot ash flow that welded into solid rock when it came to rest.

Meanwhile, Mazama was sinking into the emptying magma chamber beneath. Years ago, geologists assumed that big steam explosions blew the volcano into scattered smithereens. That seemed logical, but the evidence does not support the reasoning—no smithereens. Most of the debris around Mazama is freshly erupted lava, frothy pumice, and ash, not nearly enough broken andesite rubble to account for the missing volcano. The only way to explain the disappearance of Mazama is to conclude that it sank into the emptying magma chamber below, opening a giant collapse caldera at the surface.

Steam and ash flows erupted from the ring fractures around the opening caldera. Some of the flows moved east more than 25 miles,

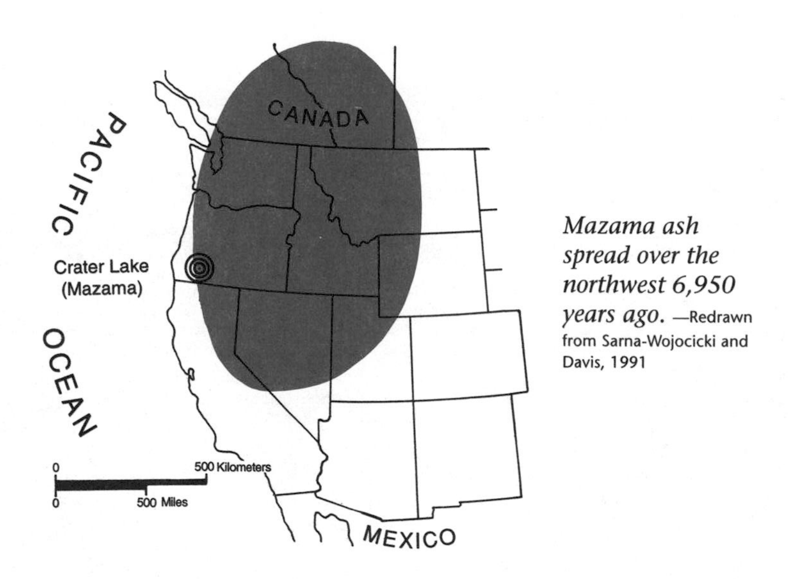

Mazama ash spread over the northwest 6,950 years ago. —Redrawn from Sarna-Wojocicki and Davis, 1991

almost as far in other directions. They left deposits as much as several hundred feet deep, which in places include blocks of pumice the size of small cars. By the time it was all over, Mazama had erupted at least 14 cubic miles of magma.

Cliff exposures of the ash show a color gradation from almost white at the base to very dark gray at the top. Geologists generally interpret that shading as evidence that the body of magma had differentiated before it erupted, that the darker and denser material had settled and the paler and less dense material had risen. Then the eruption emptied the magma chamber from the top down, so the light material is now at the base of the ash deposit and the dark at the top. Some masses of granite that crystallized at depth instead of erupting grade from light rock at the top to darker rock below.

Shortly after the great eruption, two new volcanic cones rose within the steaming caldera. Then the activity stopped, and the caldera slowly filled with water. One of the new cones is entirely submerged; the tip of the other is Wizard Island. A team of geologists who explored the floor of Crater Lake in a research submarine found evidence of warm springs, which must mean that some of the rocks below the lake are still hot. Indeed, the volcano may still be active. Its having collapsed into a caldera 6,950 years ago does not necessarily mean that Mazama is out of business.

The deep sapphire blue of its water vividly expresses the depth and clarity of Crater Lake. That pure water is extremely vulnerable to pollution because Crater Lake has no drainage basin beyond its rim; it depends for its water supply on rain and snow that fall within the embrace of the old volcano. Pollutants can remain in the lake for centuries before its extremely slow replenishment flushes them out.

Wizard Island is the top of a small volcano that rises from the floor of Crater Lake.

Imagine the newspaper headlines if some other Cascade giant were to stage its version of the Mazama eruption. The event would rank as an incredible calamity, a natural holocaust as devastating as any in recorded human history. But the coin has two sides. Some of the best soils for crops and forests in the northern Rocky Mountains are those developed on ash from the big Mazama eruption. The eruption of Mazama, unquestionably a disaster at the time, has been a benevolence ever since.

St. Helens, 1980

All the High Cascades volcanoes erupt andesite. They have a great deal in common, in the same way that all the children in a large family have a lot in common, and they differ in the same way. Small eruptions can be fun to watch, large ones are devastating and tragic. Eruptive products run the gamut from explosive ash that blows east in the wind, to searing ash flows that incinerate everything within a few miles of the eruption, to mudflows that obliterate everything along a valley bottom for tens of miles downstream. St. Helens, on May 18, 1980, produced all three.

For reasons that no one seems able to explain, St. Helens stands well west of the main trend of the High Cascades volcanoes, about 50 miles west of Adams. Radiocarbon dates show that it built most of its edifice in the last 2,000 years, a large part of it since the year 1000. Long before 1980, those dates led nearly every student of the Cascade volcanoes to pick it as the one most likely to go into major eruption.

In March of 1980, St. Helens began to blow off clouds of steam. The steam explosions slowly blasted a new vent in the top of the cone as swarms of small earthquakes rattled the entire edifice. Those classic symptoms of an impending eruption greatly encouraged every geologist in the region. Volcanoes are supposed to erupt; ask any geologist.

The first clouds of steam were white, and their temperature was near that of steam from a teakettle. It seemed quite possible that nothing much was happening, that surface water soaking into the volcano had merely encountered some hot rocks in its interior. Many volcanoes blow off occasional clouds of steam without going on to erupt. But within a few days, the clouds of steam became ominously darker and much hotter. The dark color showed that they were carrying ash, and the rising temperature suggested that fresh magma might be rising into the volcano.

Meanwhile, hastily installed portable seismographs showed that the rattling swarms of small earthquakes were originating miles below the surface. Their deep source suggested that a mass of new

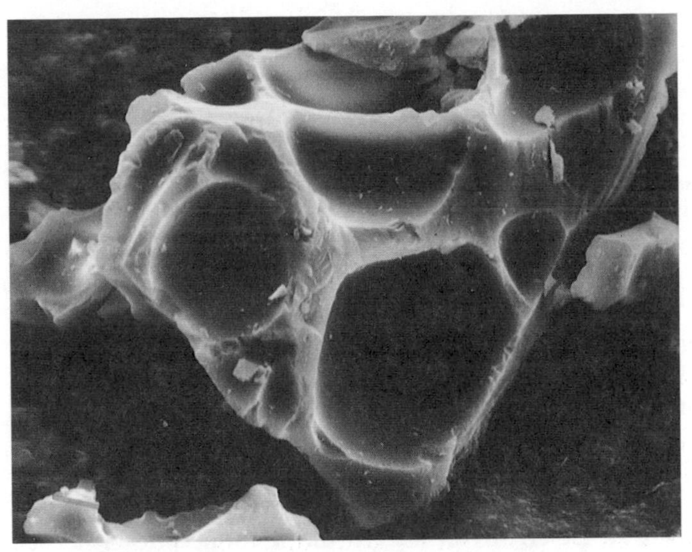

Steam bubbles in a speck of rhyolite ash erupted from St. Helens, magnified about 3,000 times. —J. N. Moore photo

St. Helens blowing off steam in March 1980. —J. N. Moore

magma was indeed rising into the volcano. If the volcano were merely cooking up steam, the earthquakes would originate within the cone.

As April passed, the clouds of steam continued to darken, the source of the earthquakes became steadily shallower, and a bulge began to swell like a tumor on the north flank of the volcano. No one doubted that the growing bulge was rising above an invading mass of viscous rhyolitic magma. The question was whether it contained enough steam to explode or would merely push quietly through the side of the volcano as a rhyolite dome. A quiet dome eruption seemed most likely because snow lingering on the tumor suggested that very little steam was rising from the mass of magma beneath. Concern began to shift to the prospect that the rapidly steepening north flank of the volcano might collapse in an enormous landslide.

Early on a brilliant Sunday morning, May 18, two geologists sightseeing over St. Helens from a light plane looked down to see clouds of steam rising from the bulge on its north flank. The snow was melting, rapidly. Then they saw the north flank of the mountain begin to slip while a cloud of steam black with ash blew out of the fracture at the head of the slide. Right before their eyes, the entire north side of the volcano exploded, as their pilot began a series of frantic evasive maneuvers. At that moment, it became obvious that the weight of the volcano had been exerting enough pressure to keep a tight lid on a steaming mass of magma within. When the slide relieved the pressure, steam exploded the magma.

After a violent lateral blast that blew down much of the nearby forest, most of the eruption cloud of steam and ash fell back on the volcano and poured down its north side as a dense ash flow that filled much of the Toutle River valley. The rest churned into the sky in a towering cloud that drifted almost directly east. Ash fell like snow all day and through the night, blanketing a broad sweep of eastern Washington, northern Idaho, and western Montana. People in those areas woke the next morning to a pale gray landscape drained of all color, almost exactly the shade of dry Portland cement.

Although those were fearful days for many people, the ash caused few lasting problems, except where it fell so heavily that people had to dig out. The ash did kill many flying insects, including bees; birds that feed on those insects were scarce all summer. Few crops were damaged, except where the ash fall was heaviest. Most predictions of damage to lungs and machinery proved groundless. A few days of heavy rain late in the week restored most of the affected area almost to normal.

Spectacular as the 1980 eruption of St. Helens was, it erupted less than one-half cubic mile of magma. Mazama erupted about 18 cubic miles; the last eruption of the Yellowstone volcano erupted something more than 200 cubic miles. By the unreasonable standards of volcanoes, the 1980 eruption of St. Helens was a trifle.

The last act in this eruption of St. Helens was the slow rise of a lava dome in the gaping crater in the north flank of the volcano. The dry magma below the steaming mass that exploded was slowly extruding into the explosion crater, and did so for years. That was the kind of eruption many geologists expected until the volcano exploded.

Newberry Volcano

Newberry Volcano in central Oregon and the Medicine Lake Volcano in California just south of the Oregon border are about as alike as two volcanoes can be. Both stand about 40 miles east of the remarkably linear main trend of the High Cascades. Both are basically enormous basalt shield volcanoes that collapsed to form large caldera basins and then erupted rhyolite. Neither has erupted andesite. Neither fits comfortably into the Cascade picture.

Newberry Volcano is a large basalt shield, the biggest volcano in Oregon by a considerable margin. Nevertheless, its gentle slopes do

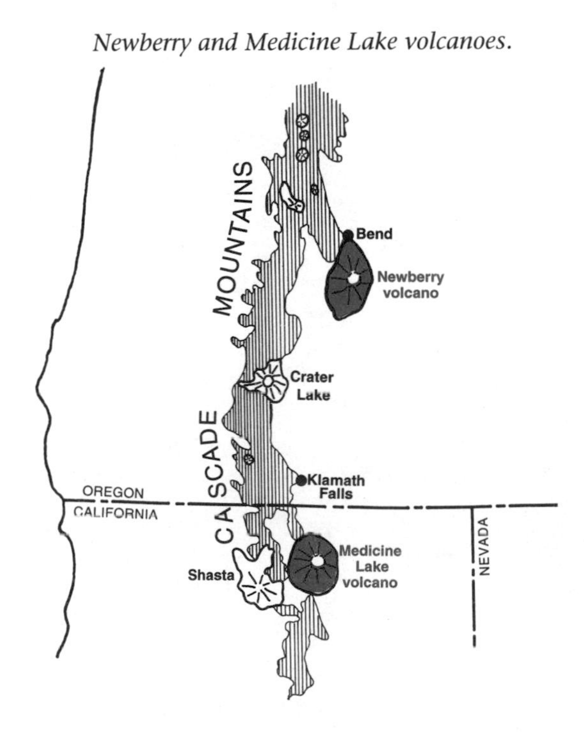

Newberry and Medicine Lake volcanoes.

so little for the skyline profile of the High Cascades that many people drive past on U.S. 97 south of Bend without noticing its enormous bulk.

The top of Newberry Volcano collapsed into an oval caldera basin about five miles long and four miles across. That appears to have happened in several stages during separate eruptions, probably many thousands of years apart. The caldera is now a beautiful and popular recreation area. After its summit collapsed, Newberry Volcano changed its habits and began to erupt rhyolite.

A small and very fresh cone of rhyolite pumice on the floor of Newberry caldera separates Paulina Lake from East Lake. Much more rhyolite lava erupted as obsidian in the crater floor and from the ring faults around the rim of the caldera. Some of it erupted as recently as 1,000 years ago. Obsidian is rhyolite that cooled into shiny black glass instead of crystallizing. It owes its intense blackness to a small amount of iron, which stains the entire glassy mass instead of crystallizing into small grains of black magnetite. The difference is like that between dissolving crystals of food coloring in water to

An obsidian flow in Newberry caldera.

stain it and scattering them through a bowl of sugar to speckle it. Given millions of years, obsidian eventually crystallizes into ordinary pale rhyolite.

If the rhyolite magma that became those obsidian lava flows had picked up some water before it erupted, it might have exploded into sheets of ash. So could some future batch of rhyolite magma. Nothing guarantees that Newberry Volcano will continue to erupt its rhyolite quietly in the form of obsidian flows. If the next batch of rhyolite magma happens to pick up some water, it may erupt explosively.

For more than a century, geology textbooks have insisted that rhyolite magma becomes glassy obsidian if it cools so quickly that no crystals can grow. They must be wrong. The obsidian flows on the slopes of Newberry caldera are hundreds of feet thick. How could such an obese mass of magma possibly cool quickly? It must have cooled slowly, for the same reason that a baked ham cools slowly.

Chemical analyses show that the composition of obsidian differs from that of ordinary rhyolite only in the absence of water. Adding water to magma makes it more fluid. A complete lack of water makes rhyolite magma so extremely viscous that atoms and molecules cannot move within it to assemble themselves into crystals. The same viscosity that makes the obsidian flows so thick also prevents them from crystallizing. They cool into glass, which is a cooled and extremely viscous magma, not an ordinary crystalline rock.

The broken stump of Mt. St. Helens after the big eruption of 1980. Ash from the eruption devastated the forest in the foreground.

▸ Part 9 ◂

PLEISTOCENE TIME

2 Million Years Ago to the Present

Pleistocene time started about two million years ago, when the earth began to experience a series of great ice ages. Great glaciers grew during those ice ages; they melted during interglacial periods an unknown number of times. The glaciers tied up enough water on land to lower sea level several hundred feet during the ice ages. If the experience of the last two million years is any guide, we should expect the glaciers to return.

Nearly every landscape in our region bears, in one way or another, the record of those drastic Pleistocene climatic changes. Continental ice spread from central Canada onto the High Plains of northern Montana. The western regional ice filled the valleys of British Columbia and the northern parts of Washington, Idaho, and western Montana. Alpine glaciers poured down the higher mountain valleys. Most of the rest of our region never felt the scrape of moving ice, but it did experience drastic changes in climate that left their mark on the landscape.

Glaciers fill high valleys in the Rockies along the British Columbia–Alberta border.

Chapter 44

ICE AGES COME AND GO

2 Million Years Ago to the Present

The last ice age ended so recently that its record in the landscape is still almost perfectly fresh and remarkably detailed. So it is frustrating to know so much and understand so little of what happened then. No one knows how many ice ages the earth experienced, to say nothing of when they happened or why. Nor does anyone know what caused the ice ages, or what the weather was like. And no one can tell for sure whether another ice age will happen, or when.

Two Ice Ages

Certainly the great glaciers grew and then melted more than the four times that standard geology textbooks listed for so many years, perhaps as many as 20 times. It is doubtful that anyone will ever know the number. Glaciers quite effectively bulldoze the evidence of their predecessors right off the landscape.

Whatever the actual number of ice ages, the landscapes of our region display clear evidence of only the last two. Here and there you can see scattered deposits of glacial debris left from earlier ice ages, but they add nothing to the scenery. No one has fitted the evidence they contain into a coherent account. Those scattered remains of earlier glacial deposits are easy to ignore, and we intend to do precisely that.

The earlier of the two ice ages that left a clear record in our region was much the greater. Its glaciers were thicker and reached farther south than those of the last ice age. In fact, that is the only reason that their record survives. Had the more recent glaciers been the larger, they would have scraped off the older deposits, leaving evidence of only themselves.

The timing of the earlier ice age is very much in doubt. Its deposits are too old to date by the radiocarbon method with its 50,000-

Ice cover of the last ice age.

year limit, and too young for most of the other methods. A conveniently round number is 100,000 years, but be warned that the number may be wrong by as much as 30,000 years either way. The most recent ice age falls well within the useful range of the radiocarbon dating method. Numerous dates provide good evidence that its glaciers reached their maximum about 15,000 years ago and had melted almost completely before 10,000 years ago.

An Ice Age

What changed our planet's climatic patterns some two million or so years ago? What switch has since toggled the climate between ice ages and interglacial times? Will the earth see more ice ages? When and how will the next one start? Those questions still await convincing answers. Even though ice ages cause tremendous geologic consequences, we suspect that their cause is not geologic. Climatologists seem more likely to create a convincing theory.

Whatever caused them, ice ages must involve a change to a wetter and cooler climate in some combination that makes winter snowfall exceed summer snowmelt so that glaciers can grow. It is easy to imagine an ice age starting as snow begins to linger through the summer in the high mountains and in the Hudson Bay region of central Canada. The scattered patches of snow grow thicker year by year and cover more of the ground. As the snow deepens, it turns into ice. And as the ice thickens, it begins to flow, thus crossing the threshold from a patch of ice to a glacier.

Large areas of Alaska and northwestern Canada were free of ice during the ice ages, as was most of Siberia. Those areas are very cold today and still free of ice. Evidently, cold weather alone does not cause an ice age. Moisture is also needed. Old lake shorelines around undrained desert valleys throughout the northern hemisphere certainly show that the climate was much wetter during ice ages than it is now. A change to a much wetter climate could alone increase snowfall enough to make persistent snowfields grow into glaciers, even without a drop in temperature. An expanding cover of permanent snow and ice would reflect sunlight, thus progressively cooling the climate. Perhaps the colder climate of the ice ages was a result of the glaciers, not their cause.

Once started, an ice age lasted for tens of thousands of years. The growing regional glacier in central Canada slowly spread west into Alberta and southwest onto the plains of northern Montana. Meanwhile, ice accumulated on high mountain peaks throughout our region and filled the lowlands of British Columbia to create another regional ice sheet that pushed south into the northern parts of Washington, Idaho, and Montana. The ice moved on until it reached areas where the summers were warm enough to melt it as fast as it advanced.

We know as little of what ended ice ages as of what started them. Radiocarbon dates leave little doubt that the glaciers of the last ice age reached their maximum size approximately 15,000 years ago. Evidence of a rapid rise in sea level between then and 12,000 years ago shows that the great glaciers melted very rapidly, within a few thousand years. Sea level reached its present stand by about 10,000 years ago, which means that the ice cover was then about the same as it is now. The climate must have changed rather quickly to melt the ice so rapidly.

Imagine the advancing glaciers thinning as the climate changed, then stagnating to melt where they lay. A stagnant icescape existed where the glaciers were melting: hills and valleys of ice. On warm summer days, torrents of muddy meltwater poured off the melting ice, dumping loads of mud, sand, and gravel in the stream beds and in lakes on the ice. As the meltwater poured beyond the melting ice, it dumped its sediment in broad outwash plains. When the last of the lingering ice finally melted, it left a landscape of glacial sediments in which the hills are the deposits that accumulated in low places on the stagnant icescape and the hollows mark the ice hills, where no sediment accumulated. You see that inverted icescape in the heavily glaciated parts of our region.

The Next Ice Age?

Most geologists assume that the ice will return. They think so for no reason more specific than that they know that the earth is in the habit of staging ice ages. Geologists make their predictions by looking around to see what has been happening lately, then forecasting more of the same. In truth, no one really knows whether the future will bring another ice age. Neither does anyone know whether another ice age would start slowly, or as abruptly as the last one ended. Several lesser climatic changes have happened abruptly during historic time.

It is not necessary to assume that the coming of another ice age would bring calamity. All of us are direct lineal descendants of people like ourselves who lived during the last ice age, camping out every day of their lives. To judge from their bones, those ancestors were healthy and robust people. Like any large change, a new ice age would be a blessing for some people, a disaster for others.

Many people in the inland part of our region would certainly consider a large increase in rain and snowfall a blessing. Crops would grow without irrigation, and trees would thrive where sagebrush and prickly pear cactus now parch. The climate would grow quite a bit colder, but most of the crops that now grow in the inland parts of the region can stand colder weather.

Many centuries would pass before ice would finally fill the high mountain valleys and send glaciers creeping into lower elevations. After many thousands of years, the great ice sheet in central Canada would advance into northern Montana. Meanwhile, regional ice would finally fill the valleys of British Columbia and start creeping into northern Washington, Idaho, and Montana.

As the growing glaciers stored water on land, sea level would drop, exposing broad areas of shallow continental shelf off the west coast. At the maximum of the last ice age, sea level dropped more than 300 feet, moving the shoreline some 30 or so miles west of its present location. Our region would gain land from the sea to replace much of that lost to the ice.

Interglacial Episodes

We live during an interglacial episode and see its weather every day, so that kind of climate is easier for us to imagine than that of an ice age. Some evidence suggests that interglacial climates became even warmer than our present climate, and considerably drier. Large deposits of windblown dust were laid down during past interglacial periods, and large tracts of sand dunes that now sleep under a cover of plants probably moved then. Many areas that now support a scanty plant cover may have become barren and lost soil to rain-splash erosion. Many streams that now barely flow during dry summers may have failed completely. Interglacial climates may well have been just as challenging to plants and animals as those of the ice ages.

Glacial Deposits

The edge of the ice stabilized at the balance point between the rate of its advance and the rate of its melting. Large volumes of ice melted at those stable margins, dumping large amounts of debris, which survive as lumpy and bouldery ridges—moraines. Broad outwash plains of debris, dumped from torrents of melt water, spread smoothly below most of the big moraines. Now that the glaciers have melted, their moraines and outwash plains survive almost unchanged as souvenirs of the ice ages. Those additions to the landscape precisely record the map outlines of the vanished glaciers.

Chapter 45

THE MODERN STREAMS BEGIN TO FLOW

2 Million Years Ago

All during the 13 million or so years between most of the flood basalt eruptions and the end of Pliocene time, the climate was very dry. In the inland parts of our region, it was too dry to maintain a connected network of flowing streams capable of carrying eroded sediment to the ocean. Deposits of gravel spread across the lowlands. Their fossils show that gravel accumulated until the very end of Pliocene time, but not into Pleistocene time. Evidently, the climate became wetter when the great ice ages began. The change established a connected network of large streams that flow to the ocean, the modern trunk drainage. That happened differently in different parts of the region.

Imagine the changing scene as the wetter climate spread an umbrella of green leaves across the landscape. As leaves sheltered the soil from rain-splash erosion, the land shed less sediment. And burrowing animals living beneath the plants opened holes in the soil, enabling the land to better absorb water, instead of letting it run off the surface. Streams no longer received heavy loads of sediment, but could depend for their flow upon a steady supply of groundwater, instead of surface runoff during wet weather. Stored groundwater maintained clear streams that ran all year, in wet seasons and dry. Those changes transformed the desert landscape of Pliocene time into the scenery of today.

High Plains Streams

A network of large streams flowed through the northern High Plains during the two million or so wet years that accompanied eruption of the flood basalt flows. They eroded large valleys and carried the sediment to the ocean as the slow processes of erosion carved the soft rocks into gently rolling hills. Then, during the 13

The Bighorn River eroded its deep canyon during the last two million years.
—Montana Department of Commerce

million dry years that followed, gravel first filled the valleys and then spread away from them, covering all but the mountains of the old erosional landscape. By the end of Pliocene time, the northern High Plains were a smooth desert plain with an almost imperceptible slope down to the east.

It is easy to imagine how the newly affluent streams of earliest Pleistocene time simply began to flow generally east, down the gentle slope of the old gravel surface of the High Plains. They began to erode their valleys in the gravel, then cut through it into the older rocks beneath. They still flow in the valleys they began to erode then, sometime around two million years ago. Plateaus of the old

Modern stream valleys in the northern High Plains cut through deposits of gravel laid down as recently as late Pliocene time.

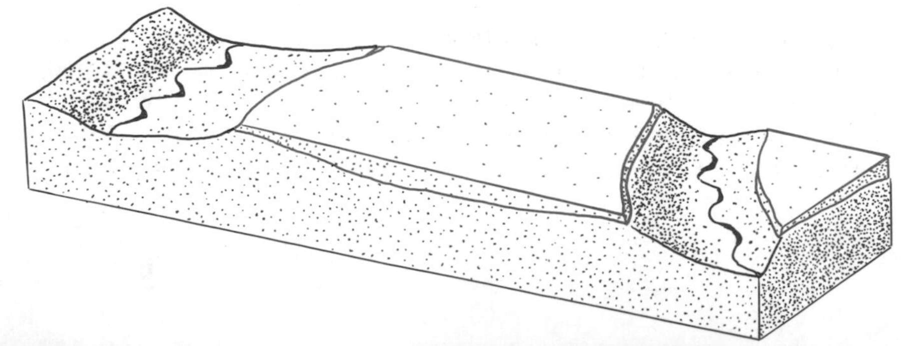

desert plain surface, with its cover of coarse Pliocene gravel, survive perfectly intact between some of the streams. Wheat grows and cattle graze on the surface that Pliocene camels once roamed.

Rocky Mountain Streams

The story is basically the same in the northern Rocky Mountains, except that the mountainous landscape considerably complicated the course of events. Long before Pliocene time ended, the broad mountain valleys of the northern Rocky Mountains had again become undrained desert basins like those that now separate the mountain ranges of southeastern Oregon and Nevada. You can see evidence of that in the pebbles, all of which are pieces of local rocks. Had rivers flowed from one valley to another in those years, they would surely have imported pebbles from distant sources.

We imagine the newly abundant rain of early Pleistocene time filling those undrained desert basins as though they were so many giant bathtubs. Each soon held a lake. The lakes rose until water spilled over the lowest point on the drainage divide into another lake, thence into another, thus establishing a connected network of streams flowing from lake to lake.

Water flowing through those spillway streams finally eroded them deeply enough to drain the lakes. The canyons that drained the lakes still slice through the mountains, carrying the modern streams

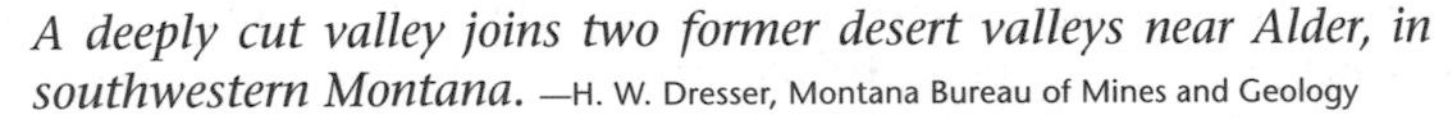

A deeply cut valley joins two former desert valleys near Alder, in southwestern Montana. —H. W. Dresser, Montana Bureau of Mines and Geology

that flow from one broad valley to the next. They owe their existence to the location of the lowest points on the drainage divides between valleys when Pleistocene time started. The continental divide separates basins that happened to overflow to the east from those that spilled to the west.

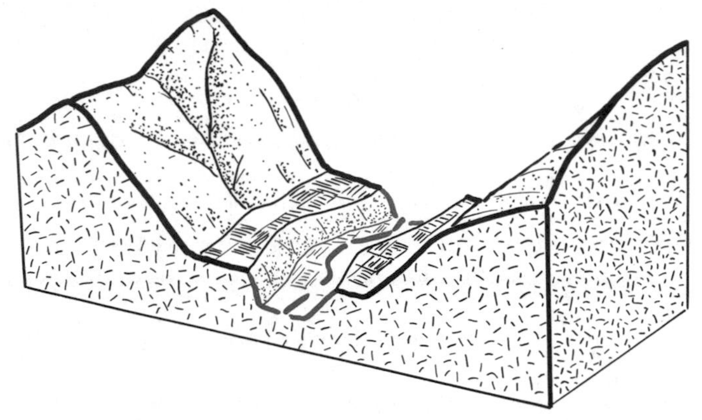

Modern stream eroding its valley in the floor of a much larger valley left from an earlier wet period.

Some of the modern streams in the northern Rocky Mountains flow through oversized valleys. We think those were first eroded during the wet years of middle Miocene time, while the flood basalts erupted. Their streams dried up during the dry period of late Miocene and Pliocene time, but the valleys remained; in many cases the modern streams followed them. If the modern stream is smaller than its vanished ancestor, you see its valley nestling within a larger valley.

Fault movements during the dry years of late Miocene and Pliocene times raised some of the older valleys above the reach of the modern streams. Those old valleys still survive, still dry. One of the most obvious is Elk Park, which Interstate 15 follows for about 20 miles through the mountains north of Butte, Montana. It is a perfectly good stream valley, large enough to hold the Missouri River and complete in every detail—except for the stream. About 20 miles north of Helena, Montana, Interstate 15 passes through the Hilger Valley, another abandoned stream valley large enough to hold the Missouri River, perhaps another segment of the same old valley.

The smaller streams that drain from the mountains to the valleys in the northern Rocky Mountains have a much longer history than the large streams that connect the big valleys. We think that streams

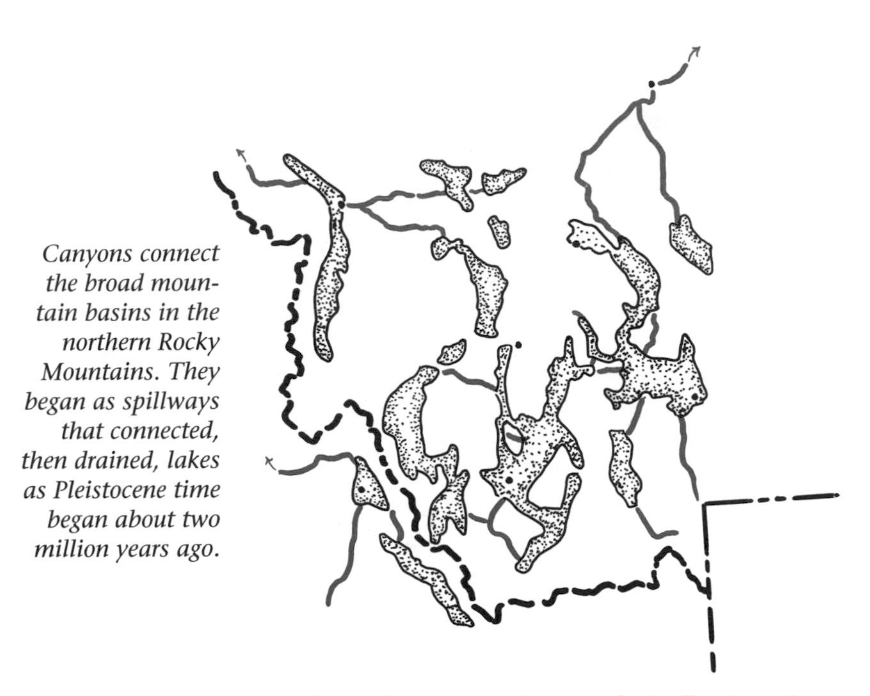

Canyons connect the broad mountain basins in the northern Rocky Mountains. They began as spillways that connected, then drained, lakes as Pleistocene time began about two million years ago.

began to erode the smaller valleys sometime early in Tertiary time. As evidence of their antiquity, consider the streams that drain west from the mountains of central Idaho to the Columbia Plateau; the big basalt flows flooded into every one of them. Most of those streams still flow on basalt in the western part of Idaho, so they still have not eroded back down to their level of about 16 million years ago. Those valleys clearly existed in essentially their present form before the flood basalts erupted. We know of no evidence that tells how long before, but suspect that it was a very long time, tens of millions of years. Perhaps they date back to the wet years of Eocene time.

The Snake River

Most of the Snake River Plain grew during the long dry period of late Miocene and Pliocene times, long after the two million wet years of middle Miocene time. Thick deposits of sediments, preserved between the layers of volcanic rocks, show that no streams carried sediments off the Snake River Plain then. When Pleistocene time finally brought more rain and snow, water flowing west from the Yellowstone Plateau started the Snake River along an entirely new course across the uneroded volcanic surface.

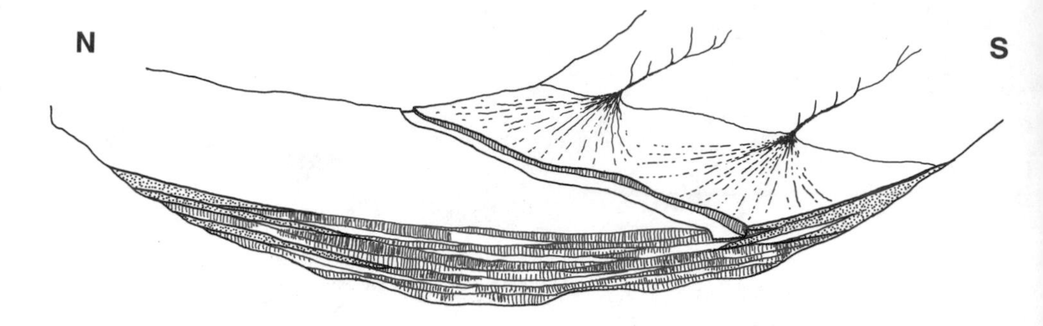

Section across the Snake River Plain showing how basalt flows and deposits of sediment along its northern margin and sediments along its southern margin fixed the position of the Snake River.

The bedrock volcanic surface of the eastern part of the Snake River Plain is slightly convex, highest along the center and sloping gently down to the north and south. Streams flowing south from the mountains of central Idaho lose their water to the porous basalt along the northern margin of the Snake River Plain. They dump their load of sediment where they stop flowing, giving the northern part of the Snake River Plain a gentle southward slope. And basalt lava flows add to the accumulation. Streams flowing north from the mountains of southern Idaho left smaller deposits of sediment along the southern edge of the Snake River Plain, giving it a northward slope.

When water finally began flowing west along the Snake River Plain, it followed the lowest available course. That was where the broad surface sloping south from the northern margin of the plain met the much narrower surface sloping north from its southern margin. That explains why a long stretch of the Snake River almost, but not quite, follows the southern margin of the Snake River Plain.

The situation becomes more complicated north of the Snake River Plain where the river enters Hells Canyon, the deepest in North America: The rim at Hat Point, on the Oregon side, is about 5,600 feet above the river. The upper part of the canyon slices through flood basalts, so it must have been eroded since they erupted. The lower part cuts deeply into the contorted rocks of the Seven Devils complex. What happened to make the river erode this chasm?

Sediments in the general area between Twin Falls, Idaho, and Vale, Oregon, show that a lake existed there during late Pliocene or perhaps early Pleistocene time. Geologists call it Lake Idaho. A major topographic break at an elevation of 3,800 feet may mark the

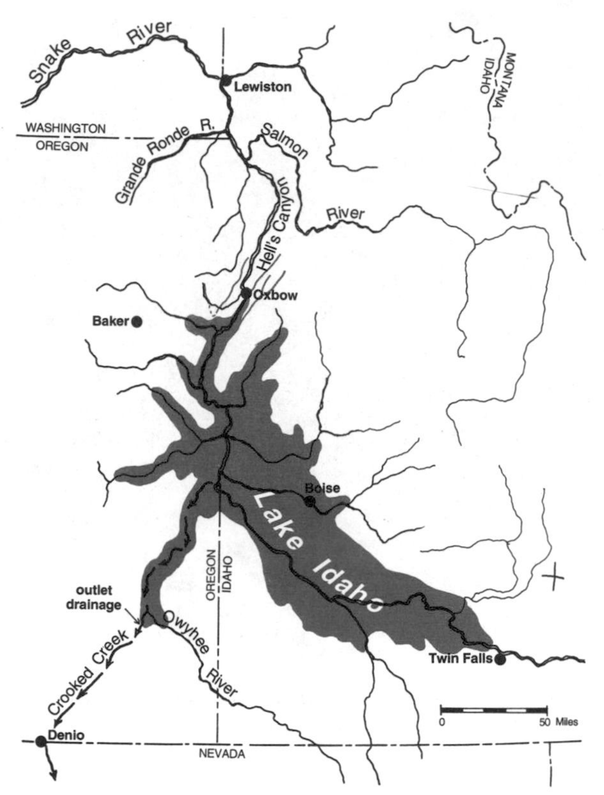

Approximate outline of Lake Idaho in Late Pliocene time. —Adapted from Jenks and Bonnichsen, 1989

shoreline. Fossil fish and snails that resemble those in California lead some geologists to argue that Lake Idaho had an outlet stream that flowed southwest. It is easy to imagine the fish and snails acquiring their distribution during the wet years of middle Miocene time, when plenty of water was available to maintain a stream flowing from Idaho to California.

Other geologists argue that Lake Idaho overflowed north along the route that became Hells Canyon. That would probably have happened when the climate became wetter, at the beginning of Pleistocene time. Snail fossils like those in the Columbia and lower Snake River drainages appear in the latest Lake Idaho sediments, suggesting that a connection was established then.

Once established in the route that became Hells Canyon, the Snake River eroded its channel fast enough to keep pace with the

Thin-bedded sand and silt deposited in Lake Idaho between Mountain Home and Twin Falls. —H. E. Malde, U.S. Geological Survey

Seven Devils fault block rising across its path. Removal of so much rock from the canyon took a heavy load off the continental crust, causing it to float higher on the dense rocks of the earth's mantle. The river had to erode its bed fast enough to keep pace with that rise. As the earth's crust floated higher, it raised the canyon rim along with the canyon itself. That explains why the flood basalts along large parts of the canyon rim now stand above the general level of those a few miles away. It also explains why streams near the canyon tend to flow away from it instead of into it.

The Columbia River and Its Tributaries

Long before the flood basalt eruptions began, streams drained from the surrounding mountains into a broad basin that was to become the Columbia Plateau of eastern Washington; in fact, it is still a basin, even though most people call it a plateau. It is not clear whether the area was still below sea level then, a broad and shallow bay east of the Cascades, or above sea level, accumulating sediments like those in the John Day formation of central Oregon.

Hells Canyon, looking upstream. —F. W. Cater, U.S. Geological Survey

The earliest flood basalt flows, the Imnaha basalt, probably poured into that basin; the Grande Ronde basalt flows certainly did and covered any Imnaha basalt flows that preceded them. Those flows built a broad basalt surface that sloped very gently down to the north and to the west, away from the fissures of the Joseph dike swarm, which erupted the lava in southeastern Washington. They almost certainly pushed the main drainage to the northern and western parts of the Columbia Plateau, against the Okanogan Highlands to the north and the North Cascades to the west. The Columbia River still traces that path from east of Grand Coulee Dam to Wenatchee, entrenched between the older rocks and the front of the basalt flows. Farther east, the Spokane River traces the northeastern edge of the plateau for the same reason.

As central and eastern Washington subsided under the enormous weight of the great flood basalts, the Columbia River turned below Wenatchee to drain into the sinking area. And streams draining west from Idaho also followed the sinking lowland west across Washington. Each succeeding flow poured into the low areas the streams were following, filling their valleys. Then the stream would shift to

Stream drainage on the northern part of the Columbia Plateau during eruption of the flood basalts. Numbers 1 through 5 refer to changes in the river systems through time. —Modified from Fecht and others, 1987

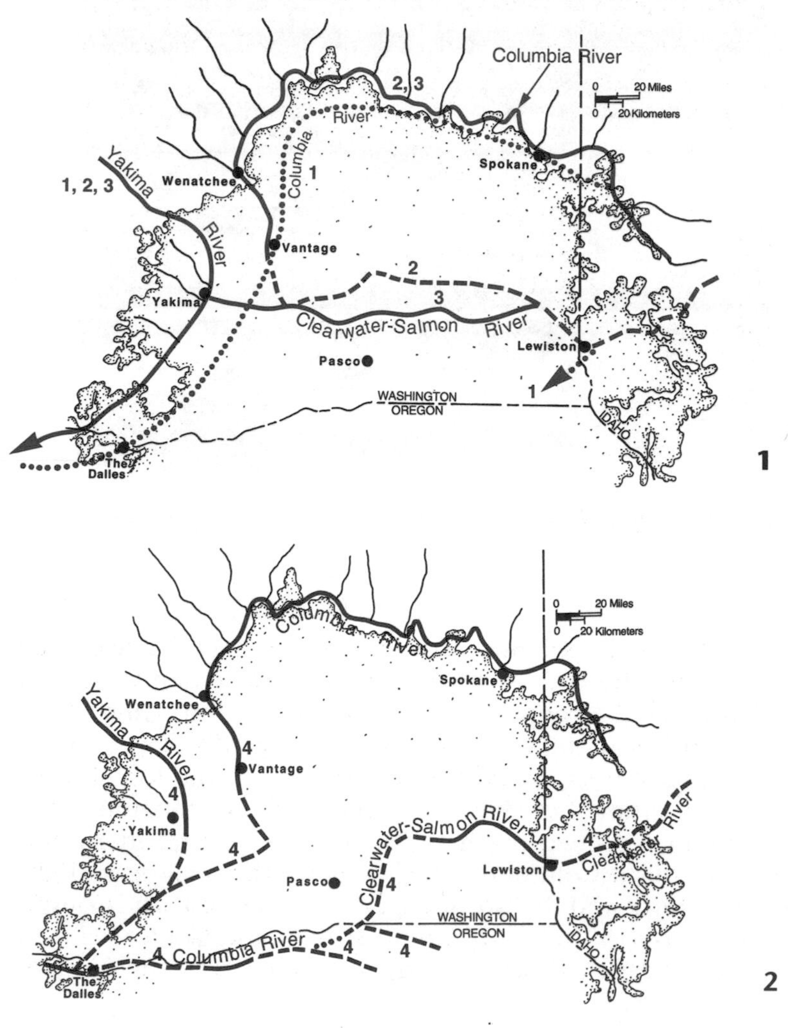

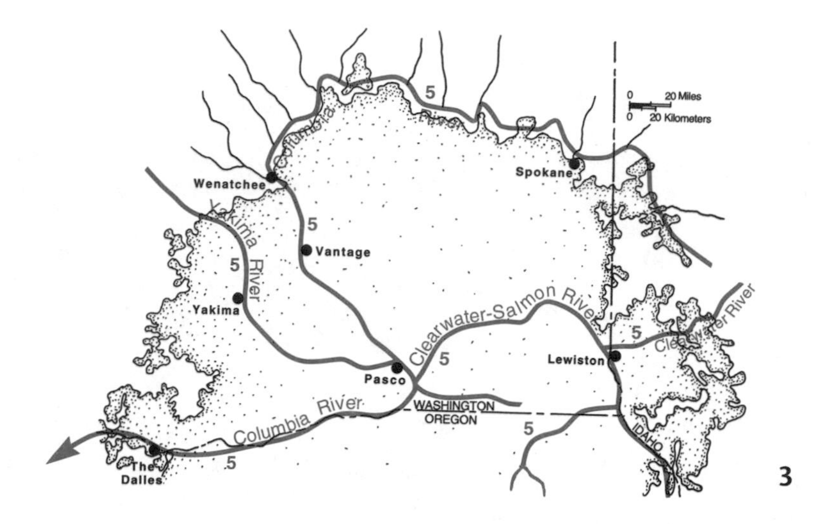

the next available low area and erode a new canyon. Careful mapping of individual flows of the late Wanapum and Saddle Mountain basalts has revealed many of those filled canyons.

The Columbia and Yakima Rivers and other streams in the western part of the Columbia Plateau now flow through canyons they eroded across the anticlinal ridges in the western part of the Columbia Plateau. Those ridges began to rise soon after the flood basalts erupted, or perhaps while they were still erupting, and they are still rising. Meanwhile, the Columbia River also eroded its deep gorge through the rocks as they rose to become the arch beneath the High Cascades.

Water cannot flow up one side of a ridge and down the other, so those streams must have been flowing in their present courses before the ridges began to rise. They began to flow in the wet climate that prevailed while the flood basalts erupted. If the climate had become dry enough during late Miocene and Pliocene times to stop them, they could not have maintained their courses through the rising folds. The Cascades caught enough moisture to maintain flowing streams through the dry years of late Miocene and Pliocene times. In fact, both the Columbia and Yakima Rivers were diverted to the southeast from courses that took them south from Vantage through the areas of Priest Rapids Dam and Satus Pass to the present lower Columbia River. Deposits of gravel record their former courses.

Canyon of the Snake River, downstream from Twin Falls, Idaho.

Chapter 46

THE MANY LANDSCAPES OF THE HIGH PLAINS

2 Million Years Ago to the Present

People who suppose that the northern High Plains are monotonous and boring are sadly mistaken. The landscape is a richly varied tapestry that differs greatly from place to place, a patchwork of distinctive scenes, each with its own character, its own story. Some of those landscapes inherit their form directly from the Pliocene desert plain; others owe their origin to erosion of different kinds of bedrock during Pleistocene time. We will consider just a few of those scenes.

The Relic Plain

Picture the scene as the millions of desert years ended at the beginning of Pleistocene time: The northern High Plains lay under their veneer of gravel, sloping ever so gently down to the east. The mountains of central Montana rose above that surface, as did the Black Hills of South Dakota. Elsewhere, so far as we can tell, it was a nearly featureless plain. The climate became wetter eastward then, as it does now, so we imagine that plain blending into the more rugged erosional landscape of the Midwest somewhere in the eastern part of the Dakotas.

If you should decide to erect a monument to yourself and you want it to last for millions of years, be sure to build it of gravel. Nothing endures quite like a pile of gravel, provided it is above the reach of waves or streams. Gravel absorbs surface water so eagerly that it sheds no runoff, permitting no surface streams to form. Gravel lasts until the pebbles finally weather into clay, which seals the pore spaces between the pebbles, thus permitting surface streams to flow.

Its veneer of gravel saved large parts of the old Pliocene desert plain from erosion during the wetter years of Pleistocene time. Where the gravel veneer was thick, the original desert plain surface survives nearly intact, almost exactly as it was when the climate changed some two million years ago. You see those remnants of the old desert alluvial plain high on the drainage divides between the valleys of the modern streams. Some are merely hills, their flat tops capped with gravel. Others are broad plateaus that stretch their patterned fields of wheat to the far horizons. Geologists call those high and gravelly flats, large or small, the High Plains surface.

You can see the gravel in the fields, but road cuts provide a better view. Watch for it where the edge of the High Plains surface meets a stream valley, where the road just starts to wind down from the wheat fields into a valley.

When the High Plains surface was an active desert plain, its upper margin ramped onto the front of the Rocky Mountains. Erosion has since removed that highest part of the High Plains surface everywhere along the entire long front of the Rocky Mountains except west of Cheyenne, Wyoming, where Interstate 80 crosses the continental divide without leaving the High Plains surface.

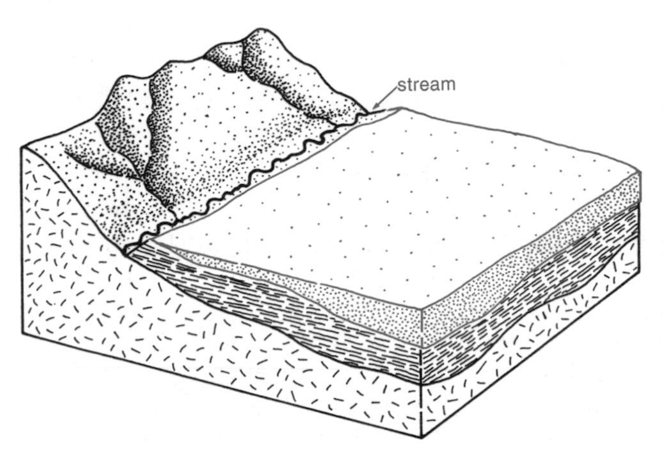

A blanket of sand and gravel spreads east from the Rocky Mountain front in the dry climate of Pliocene time. Year-round streams of Pleistocene time eroded valleys in the old surface.

Wind Blowouts and Buffalo Wallows

The wind picks at the ground, digging a shallow hole here, another there. Tens of thousands of those shallow depressions, blowouts, dot the High Plains, probably an average of at least one per square mile. They are most visible during rainy weather, when many become shallow ponds. They range in size from a few that are miles across and tens of feet deep down to a great many that are mere dimples. Typical blowouts are a few hundred feet across and a few feet deep. Some have a low rim of sand around their eastern margins, a residue left as the wind blew the finer fractions of the soil all the way into the Midwest.

Many people who live in the High Plains insist that the blowouts are old buffalo wallows. They probably were. A wandering buffalo would have thought a blowout a logical place to have a nice dusty wallow. The ensuing commotion would certainly loosen the soil and cast it into the breeze for the long trip east.

The Eroded Plains

Do not assume that every more or less flat landscape east of the mountains is a remnant of the old High Plains surface lying beneath a protective veneer of ancient gravel. Many are not.

The veneer of gravel lay deep on some parts of the old desert plain, thin on other parts, and not at all on still others. When the climate changed, the areas that lacked a thick coat of protective gravel began to erode in the usual manner of regions wet enough to maintain a connected stream drainage. What happened next depended mostly on the bedrock.

In large areas, the old desert plains surface was eroded across soft rocks: shales, mudstones, and weakly cemented sandstones. They make few bedrock outcrops and lack the strength to support steep slopes. They offer precious little resistance to erosion.

Wherever those weak formations were exposed without a cover of gravel, they eroded into plains almost as flat and level as the old High Plains surface, but at a lower elevation. These younger plains stretch and roll gently away into the distance, a new and slightly

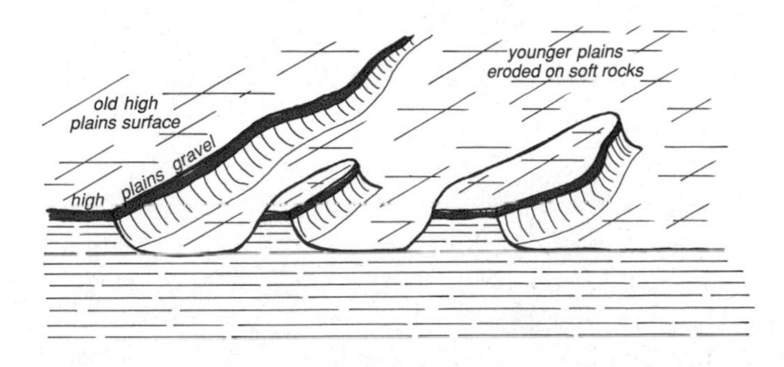

The old High Plains surface and the younger plains beneath it eroded in areas where a gravel cap is absent and the rocks are soft.

lower floor under the same broad western sky. Occasional bodies of more resistant rocks, typically sandstones, rise above them as isolated buttes, or groups of hills.

In other areas, the bedrock beneath the old High Plains surface was hard sandstone, solid enough to make bedrock outcrops and strong enough to stand on steep slopes. Some of the Cretaceous sandstones are both thick and hard, especially those deposited along old beaches during late Cretaceous time. And the upper part of the Fort Union formation contains thick sections of the massive Tongue River sandstone. You see it especially well-exposed in the high bluffs of yellowish rock along the Musselshell River in central Montana.

Where those resistant sandstones were exposed without a protective cap of gravel, they have since eroded into tracts of ruggedly picturesque hills dotted with bouldery outcrops of sandstone. Most hold enough water to support a thin cover of scraggly pine trees.

Coal Landscapes

It is no coincidence that the coal fields of the northern High Plains are in picturesque regions of rugged, colorful, and forested hills. High Plains coal burns in the ground, and in so doing creates its own charming landscapes.

An outcrop of red clinker near Decker, southern Montana.

As a stream erodes through a coal seam, the groundwater drains out of it, leaving the coal dry and vulnerable to a lightning strike or prairie fire. Once ignited, coal seams may smolder for centuries. The fire burns until it reaches the end of the coal, wet rock, or the far side of the hill. Heat from the burning coal bakes the sandstone and shale above it into a hard rock generally called clinker, although some people call it scoria. Clinker has all the hardness, color, durability, and mechanical strength of kiln-fired brick. It resists erosion exceedingly well.

Resistant and colorful clinker horizons protect the ridge crests from erosion, thus maintaining the upper level of the landscape. The slopes below them erode into rugged hills, their steep slopes boldly painted with horizontal stripes of red, yellow, or black clinker. Like sandstone, clinker generally holds enough water to support a forest of gnarled and straggly trees, mostly pines and junipers.

All the sedimentary rocks in large parts of the northern High Plains are soft enough to dig with a shovel. Clinker is the only hard and resistant rock, the only convenient source of road metal and construction aggregate. Watch for the gravel pits in the clinker beds, especially on the ridge crests, and for the bright red and yellow roads they surface.

Gravels of the High Plains surface in central Montana.

Chapter 47

MOUNTAIN GLACIERS

2 Million Years Ago, and Continuing

Moist air expands and cools as it rises against high mountains, bedecking them with clouds and dropping far more rain or snow than falls on the surrounding countryside. Mountains snatched even more moisture out of the clouds during the ice ages, enough to maintain big glaciers. Most of the glaciers are gone now, but you can see their effects in jagged peaks, deep valleys, and craggy skyline profiles.

Most of the high mountains supported alpine glaciers during the ice ages.

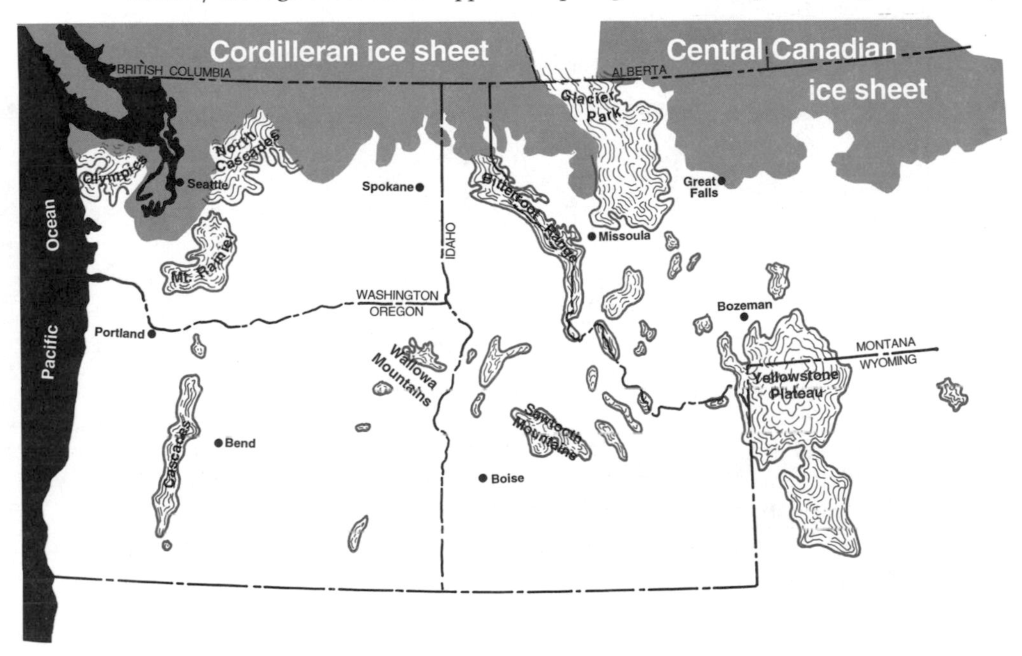

Glaciated Mountains

Even if the ice is gone, it is easy to recognize heavily glaciated mountains. High on the peaks where the glaciers eroded bedrock, they carved a jaggedly serrated skyline of gnarled peaks and deeply gouged valleys. Cirques make deep hollows at the heads of those valleys. Little lakes sparkle in bedrock basins in the floors of many cirques and spangle the floors of the gouged valleys beneath them.

Glaciers are much too stiff to take sharp curves in stride, so they tend to straighten their valleys, leaving them much straighter than any that have held only streams. That is most apparent in the long views up those straight valleys, across lakes that reflect distant peaks—postcard scenery.

Glacial ice contains embedded sediment of all sizes, so it scrapes across bedrock surfaces like a big sheet of cheap sandpaper. The fine particles polish the bedrock, while the larger pieces cut scratches and deep grooves that exactly record the direction of ice movement. Many bedrock surfaces in glaciated valleys are more or less polished and covered with sets of parallel scratches.

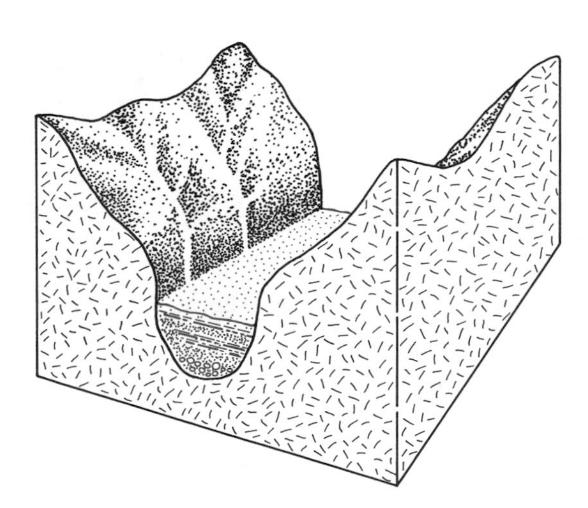

The steep valley walls and flat floor of a valley that once held a glacier.

Most glaciers gouge their valleys to a rounded form quite unlike that of typical stream valleys. The bedrock valley walls tend to be fairly smooth where blocks of rock embedded in the ice scraped them and much rougher above the ice levels where freezing and thawing tore the rock apart.

Craggy peaks rise above deep cirques around Mount Stuart in the North Cascades of Washington. This is the typical look of a heavily glaciated mountain landscape. —A. Post, U.S. Geological Survey

The long view through the glacially straightened valley of Swiftcurrent Creek in Glacier National Park, Montana. —M. R. Campbell, U.S Geological Survey

The Chelan Lake glacier polished and scratched this bedrock surface just south of Chelan, Washington.

Glacial till beside Wallowa Lake, north-eastern Oregon.

Melting ice leaves a distinctive kind of sediment called till, a disorderly dump of boulders, cobbles, pebbles, sand, and clay, all scrambled together without much show of internal layering. Anything made of till is called a moraine. Glaciers drop especially large quantities of till around their edges to make lumpy morainal ridges that precisely record the former position and outline of the glacier. Most mountain glaciers leave a morainal ridge shaped like an irregular hairpin that outlines the lower part of the glacier.

Glacial meltwater pouring off the ice sweeps large volumes of sediment beyond the moraines, then dumps it in deposits of glacial outwash. Those sediments look like any other stream deposits; geologists recognize their glacial origin mainly through their relationships to moraines. Many outwash deposits are smooth, broad alluvial fans that slope gently away from the moraine. Others fill valley floors below the moraines.

Melting glaciers generally leave a deep fill of till and outwash in the lower part of the valley, giving it a broad and nearly flat floor enclosed between steep bedrock valley walls. Tributary streams tend to come down those valley walls in waterfalls or steep cascades. The valley floors generally contain lakes, some of which are quite large.

Large lakes in the lower parts of many glaciated valleys lie behind natural dams made of either glacial till or outwash. Moraines make ridges across the valley floor that look as though they were

Glacial erratics near Tower Junction, Yellowstone Park. —J. P. Iddings, U.S. Geological Survey

Wallowa Lake nestles within the embrace of a moraine dumped around the lower margin of a large valley glacier. —P. L. Weiss, U.S. Geological Survey

designed to become natural earth-fill dams, but few of them actually impound lakes. Large lakes more commonly fill basins in outwash, which generally are places where the last remnants of the glacier lingered. The ice prevented accumulation of outwash sediments, and so left an empty basin when it finally melted. Other large glacial lakes lie behind big alluvial fans, most commonly deposits of outwash swept in from a tributary valley. Alluvial fans are hard to erode because water flows between the pebbles rather than washing them away.

Chapter 48

THE GLACIATED NORTHERN PLAINS

15 Thousand Years Ago

The great North American continental ice sheet spread in all directions from a center in the Hudson Bay region of central Canada. At its farthest reach, it covered most of the High Plains of Montana north of the Missouri River. That area was among the last parts of our region to feel the ice, and one of the first freed from it when the ice age ended. The ice thinned as it spread to its outermost edge and moved slowly, lying gently on the land. That most distant ice eroded hardly at all; it left its mark mainly in subtle features much harder to recognize than those in places where the ice was heavier and moved faster.

Bedrock under the High Plains is mainly soft shales, sandstones, and limestones laid down in horizontal layers during the last 100 million or so years. Scattered across the broad surface north of the Missouri River are boulders of pale granite and gneiss dropped from the melting ice. They are basement rocks eroded from the Canadian Shield more than 500 miles to the northeast.

Glaciers and Lakes

The plains of central Montana south of the Missouri River slope gently down to the north, and the surface of the thinning ice north of the river sloped gently down to the south. Those opposing slopes trapped glacial meltwater in a series of large lakes along the ice margin. They drained through a stream that flowed along the edge of the ice, setting the present course that the Missouri River follows in its canyon across central and much of eastern Montana.

Too bad that we all came along too late to visit central Montana during the maximum of the ice age, suitably equipped of course with a video camera and an aluminum canoe. It would have been

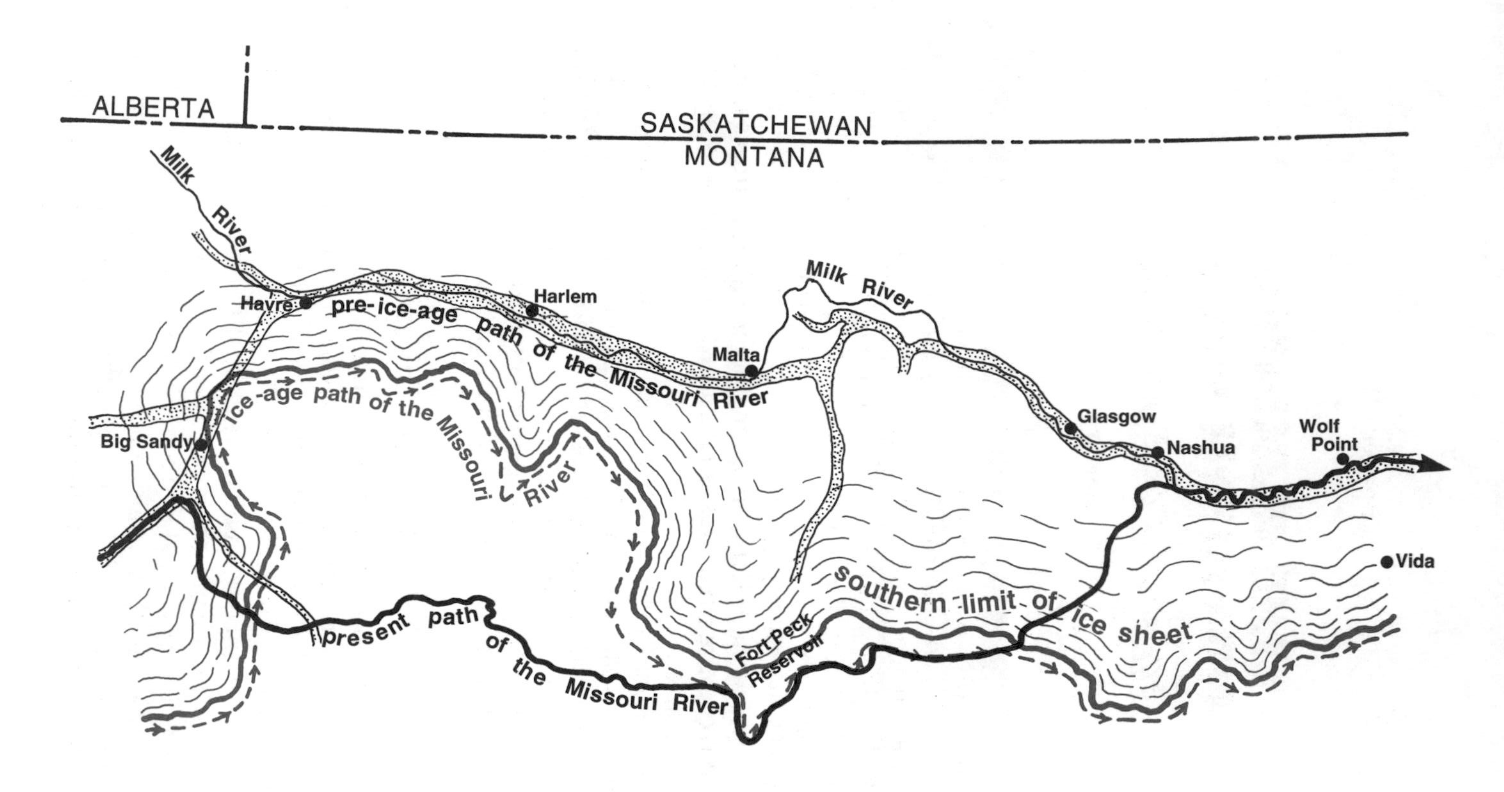

Before the ice came, the Missouri River flowed through the valley that the Milk River now occupies.

good fun to drift in the flow of meltwater from Glacial Lake Cut Bank in the northwest along the general course of the future Missouri River, through a long succession of big lakes, all the way to North Dakota. Some of those lakes were more than 100 miles across in places, and as much as several hundred feet deep.

Imagine the vast ice fields, the southern edge of the continental ice sheet, glittering along the northern shores of those lakes. The climate was wet during the ice ages, so it is reasonable to imagine the southern shores of those lakes cloaked in dark forests. Mastodons came out of the forests to drink and bellow at each other. Oversized bison roamed the open meadows; beavers the size of grizzly bears gnawed the trees.

Icebergs cracked off the floating edge of the ice sheet along the northern side of the lake, then drifted across the lake and down the current. As they melted and where they finally ran aground, they left rocks and clusters of rocks, many of which had come all the way from northern Manitoba. You see those basement rocks littering the fields now, great rounded blocks of pink gneiss and granite, or gray schist, lying as much as 1,000 miles from home. Where several of those stray rocks cluster, they give you valid scientific license to imagine a great chunk of drifting ice running aground and slowly melting in the warm summer sunshine.

The southern shorelines of the lakes along the ice margin survive intact in every detail, except for the water. All you need do to map one is to find its exact elevation in one place, then watch a map of the vanished lake emerge as you trace that elevation on an ordinary topographic map. The lakes lapped against ice along their northern edges, so no old shorelines survive in that part of the modern landscape. In some places, the deposits of sediment in the old lake floors make a land surface as flat as the proverbial billiard table.

Displaced and Misplaced Streams

Before the ice came, the Missouri River flowed north from the vicinity of Helena, Montana, almost to the Canadian border, then turned east. Ice age glaciers covered that long northern leg of the original river. Meanwhile, meltwater flowing through the ice marginal lakes and the channels that connected them eroded a valley that nearly traced the southern edge of the ice. By the time the ice melted, that meltwater channel was so deeply entrenched that the Missouri River continued to follow it, instead of returning to its old valley.

The Missouri River still flows through that former meltwater channel in a long canyon that reaches from the area about 50 miles east of Great Falls to the Fort Peck Reservoir, a distance of about 100 miles. Its steep walls and narrow floor look most unlike the broad original valley above and below. Virtually no flood plain exists in the floor of that narrow canyon, and no road follows the river through it.

The coming and melting of the ice left the part of the old valley that trends to the east just south of the Canadian border almost unscathed. The little Milk River flows through it now, wandering almost lost in the broad floor of a valley that a much larger Missouri River eroded long ago.

The Missouri River eroded this valley northwest of Fort Benton, Montana, after the ice pushed it out of its original valley.

Chapter 49

THE WESTERN REGIONAL ICE

15 Thousand Years Ago

Had you flown over British Columbia during an ice age, you would have seen the tops of occasional high mountains rising above the sea of moving ice that geologists call the Cordilleran ice sheet. It formed as ice accumulated on the mountains, poured down their valleys, and filled the lowlands. That ice pushed far enough south to fill the valleys of northern Washington, Idaho, and western Montana. It filled Puget Sound, the Okanogan Valley, came down the Purcell trench as far south as Coeur d'Alene Lake in Idaho, and down the Rocky Mountain trench to fill the Flathead and Swan Valleys of western Montana.

In many northern parts of our region, you can easily see which mountains the ice covered and which stood above it. Moving ice

The western regional ice sheet filled all the mountain valleys and covered many of the lower mountains.

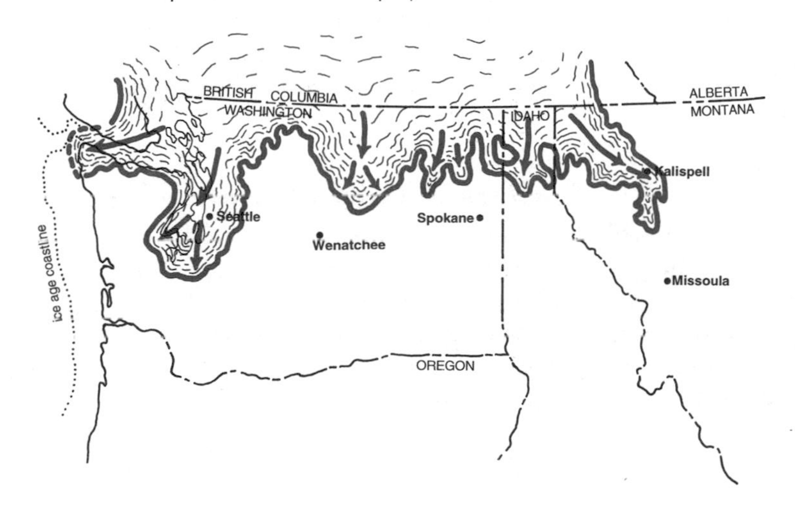

This aerial view of Greenland shows how the western regional ice looked 15,000 years ago.

full of rocks and dirt sandpapers the rocks it crosses, rounds and smooths the mountains it covers. Mountains that stood above the ice are irregular and jagged, their sides scalloped by cirques and deeply gouged valleys, with sharp ridges between them. Valley glaciers pouring down those mountains fed the regional ice that filled the lowlands.

Rocks embedded in moving glacial ice scrape and scour the bedrock walls of the valleys it flows through, steepening the lower valley walls. Parts of the Okanogan Valley of Washington and the Purcell trench of northern Idaho clearly show glacial gouging on a large scale. So do many mountain valleys, such as those that hold Ross Lake in the North Cascades, Lake Chelan north of Wenatchee, and McDonald Creek in Glacier National Park.

Glaciers in Puget Sound

Radiocarbon dates from sediments above and below the glacial deposits show that ice moving south out of British Columbia filled Puget Sound as recently as about 15,000 to 13,000 years ago. At its maximum, it reached a little south of Olympia. You can see high mountains that the rasping ice covered and rounded smooth rising above Interstate 90, U.S. 2, and Washington 20, east of the cities

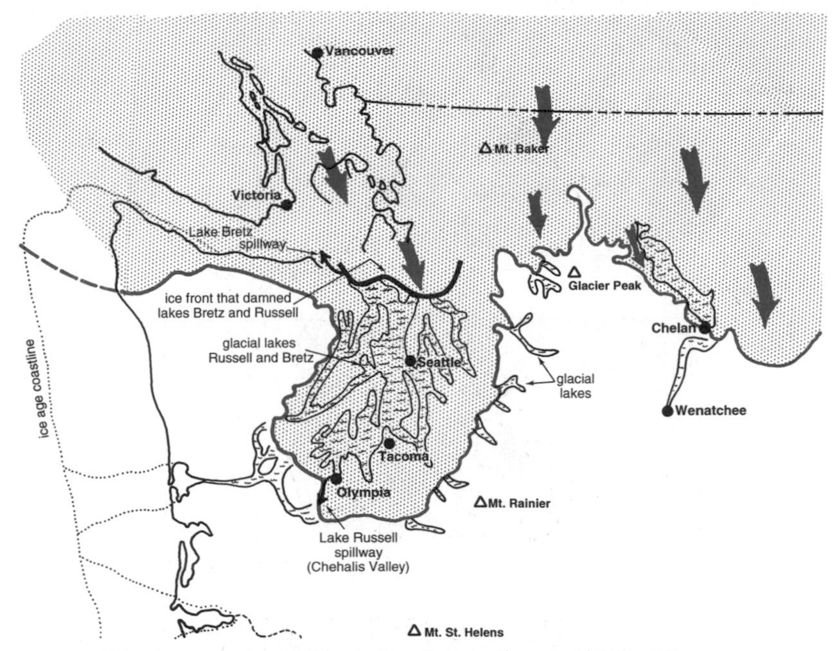

Maximum extent of ice in Puget Sound area and the Olympic Mountains about 15,000 years ago. Glacial lakes formed in the Puget Sound lowland as the ice retreated northward.

along Puget Sound. Ice also filled the Strait of Juan de Fuca north of the Olympic Peninsula.

Streams flowed north into the south end of Puget Sound, much as they do today, then ponded against the southern margin of the ice. When the sound was full of ice, the lake along the ice margin overflowed west to the Pacific Ocean through the lower Chehalis River. Meanwhile, meltwater streams filled broad areas south of the ice with outwash and also drained through the Chehalis River. The greatly diminished Chehalis River now wanders in the oversized valley it eroded when it was flush with all that glacial meltwater. Sea level was some 300 feet lower then; the shoreline was some 30 miles west of its present position, and the Chehalis River was that much longer. The rising sea level at the end of the last ice age flooded its lower course to make Grays Harbor, an estuary.

As the ice began to melt, two curving lakes formed between the ice and the spillover point into the Chehalis River. Lake Russell was along the southwest end of the ice lobe, west of Olympia. Lake Nisqually was along the south end of the lobe, in the area of small lakes about 25 miles south of Tacoma and Puyallup.

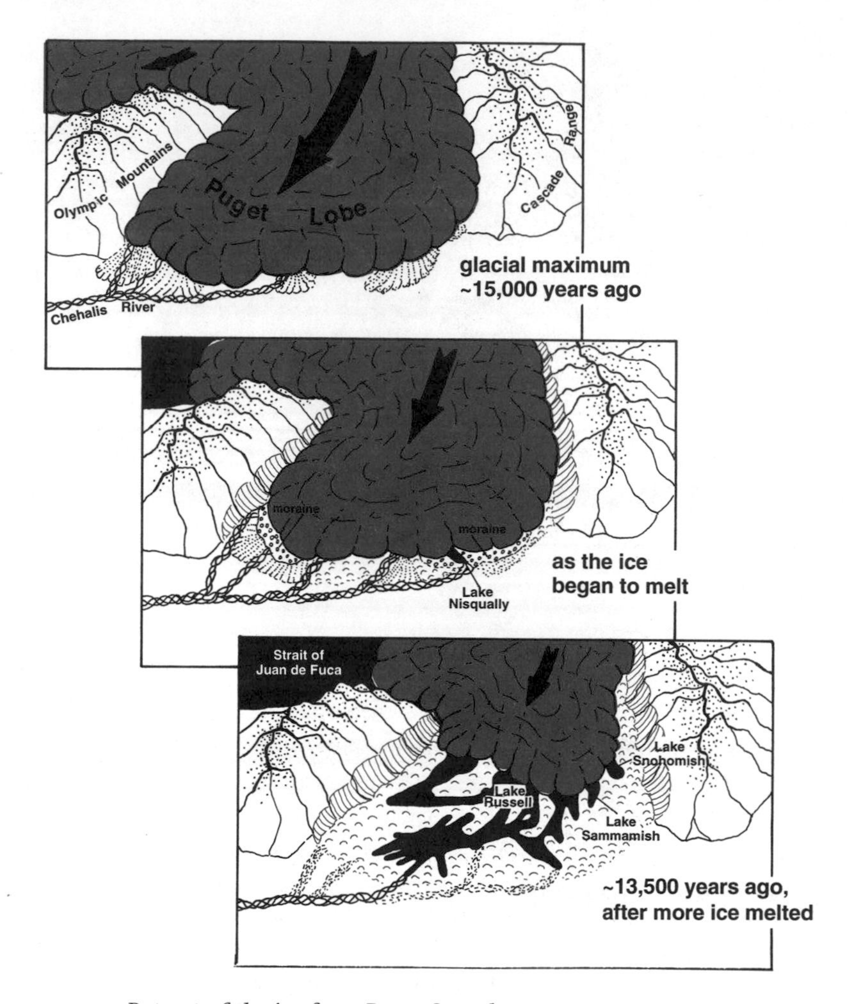

Retreat of the ice from Puget Sound. —Modified from Thorson, 1980

When the ice had melted back to a point about opposite Seattle, about 13,500 years ago, the many fingers of Lake Russell filled the present canals and bays of the south half of Puget Sound and flooded adjacent valleys, including a long arm through lakes Washington, Youngs, and Tapps, and the lower Puyallup River, southeast of Tacoma.

Meanwhile, ice in the Strait of Juan de Fuca still blocked the rivers flowing into Puget Sound. For example, the lower Puyallup River was diverted to the southwest through the site of Olympia into the Chehalis River. The ice in Puget Sound later melted back far enough north to permit drainage northwest into the Strait of Juan de Fuca, which by then was also free of ice. Seawater did not immediately flood back into Puget Sound, because sea level was still well below its present stand.

Hood Canal is not a canal; it is a fjord, a glaciated valley flooded by the ocean. Ice moving down Puget Sound followed the hard ridge of basalt that wraps around the eastern side of the Olympic Mountains, gouging a deep valley in soft sedimentary rocks. Then sea level rose as the glaciers melted at the end of the last ice age, and seawater flooded into the valley, converting it into a fjord.

Ice in the Olympic Mountains

The Olympic Mountains still snatch a tremendous amount of snow from the moist air that blows in off the Pacific Ocean. They still maintain large glaciers that come down to within about 500 feet of sea level. During the ice ages, they maintained much larger glaciers that descended to much lower elevations.

Ozette Lake, near the northwestern corner of the Olympic Peninsula, looks on the highway map like a coastal lagoon impounded behind a beach ridge. In fact, it is a lake impounded behind a glacial moraine, which was well inland from the coast when it was deposited. When sea level rose as the ice age ended, it just reached but did not quite flood the moraine that holds Ozette Lake. It marks the farthest reach of the ice on the northwestern part of the Olympic Peninsula.

Crescent Lake, one of the jewels of the northern Olympic Peninsula, floods another glaciated valley. As the ice melted, it left the valley walls unsupported and they collapsed, dumping the debris in the valley floor to impound the lake. Quinault Lake, on the western side of the Olympic Peninsula, floods yet another glaciated valley. It is probably the legacy of a big patch of stagnant ice that lingered in the valley after the rest had melted. Such remnants prevent glacial sediment from filling the valley floor where they lie, and thus leave an open basin when they finally melt.

Ice in the Okanogan Valley

Canadian ice flowed down the Okanogan Valley about 85 miles into northern Washington. Watch for the glacially rounded mountains, completely overridden by ice, standing high above the Okanogan Valley northwest and northeast of Omak.

The summer sun was as warm during the ice ages as now, and the great glaciers shed large volumes of meltwater. It flowed to the low side of the glacier, which was generally on the warm side of the valley, the sunny east side. There it flowed along the margin of the

The rounded hills of the highlands, steep valley walls, and huge valley glacier in this aerial view of Greenland must resemble the Okanogan Valley as it was while the ice melted.

A terrace of glacial outwash remains at the base of the ice-scoured wall of the Okanogan Valley near Riverside, Washington. Overriding ice rounded the hilltops.

glacier, confined between ice on one side and the bedrock valley wall on the other. Some of those streams eroded valleys into the bedrock, and you see those today as nearly horizontal slots cut along the valley wall. Wonderful examples of those high and dry meltwater stream valleys exist along the east side of the Okanogan Valley, north of Omak.

Meltwater filled the valley floor south of the melting glacier with outwash sand and gravel. Now, the modern Okanogan River has removed a large part of those outwash deposits. The remnants stand now as prominent grassy sand and gravel terraces flanking the valley sides. In most places, U.S. 97 follows the surfaces of those terraces.

Melting of the Okanogan Valley glacier.

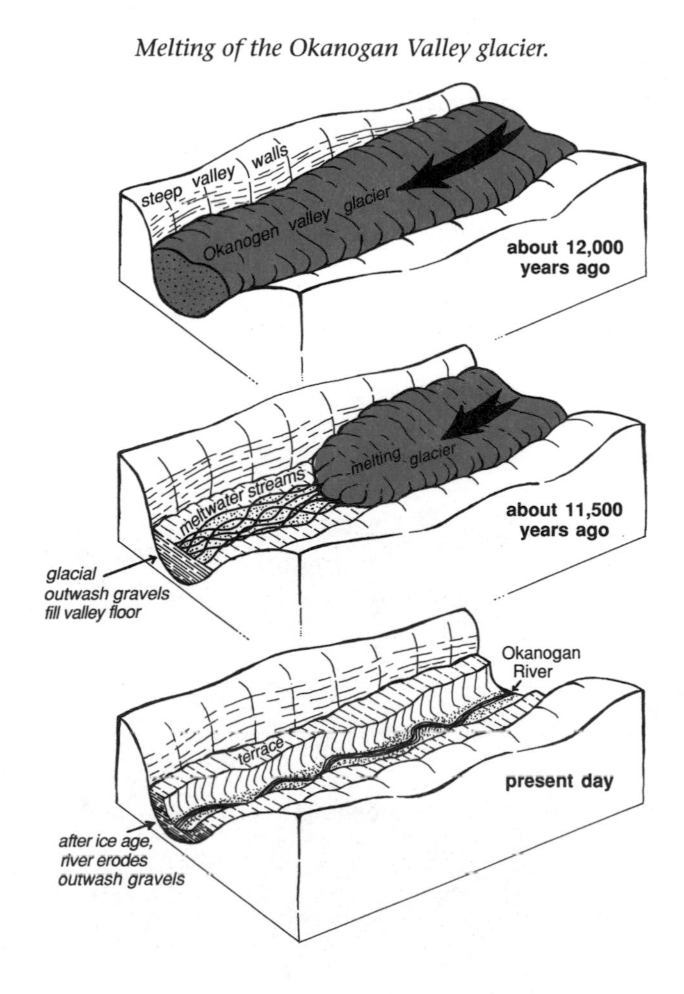

The Withrow moraine rises above the flat outwash plain in the foreground, northeast of Wenatchee, Washington.

The southern end of the Okanogan lobe glacier stuck a broad tongue of ice onto the flood basalt lava flows of the Columbia Plateau. The limit of this tongue is beautifully displayed in the Withrow moraine, a long lumpy ridge of glacial till that trends generally to the east, along the northern edge of the smooth wheat fields on the Columbia Plateau. The best place to see this terminal moraine is at the farm community of Withrow, 30 miles northeast of Wenatchee. Huge boulders of basalt along the outer edge of the moraine were

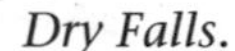

Dry Falls.

probably ripped up by the floods from Glacial Lake Missoula, then dropped where the water ponded at the edges of flood channels.

Glacial Lakes on the Columbia Plateau

The ice that left the Withrow moraine also blocked the Columbia River at the present site of Grand Coulee Dam to form Glacial Lake Columbia. Franklin D. Roosevelt Lake, the reservoir backed up behind the dam, broadly reconstructs a smaller and shallower version of Glacial Lake Columbia, except that the northern arm of the reservoir, along Highway 25, was under the ice and therefore not part of Glacial Lake Columbia.

The outflow from Glacial Lake Columbia eroded Grand Coulee, now an enormous dry valley; it poured across Dry Falls, about 30 miles southwest of Grand Coulee Dam. That defunct cataract was 3.3 miles across and almost 400 feet high; that is 4.7 times wider than Niagara Falls and 2.4 times higher. The ground must have shaken for miles around when the torrent thundered over Dry Falls in one of the great spectacles of the last ice age. Part of the Glacial

Glacial Lake Columbia was dammed at the present site of Grand Coulee Dam. Its outlet stream eroded Grand Coulee.

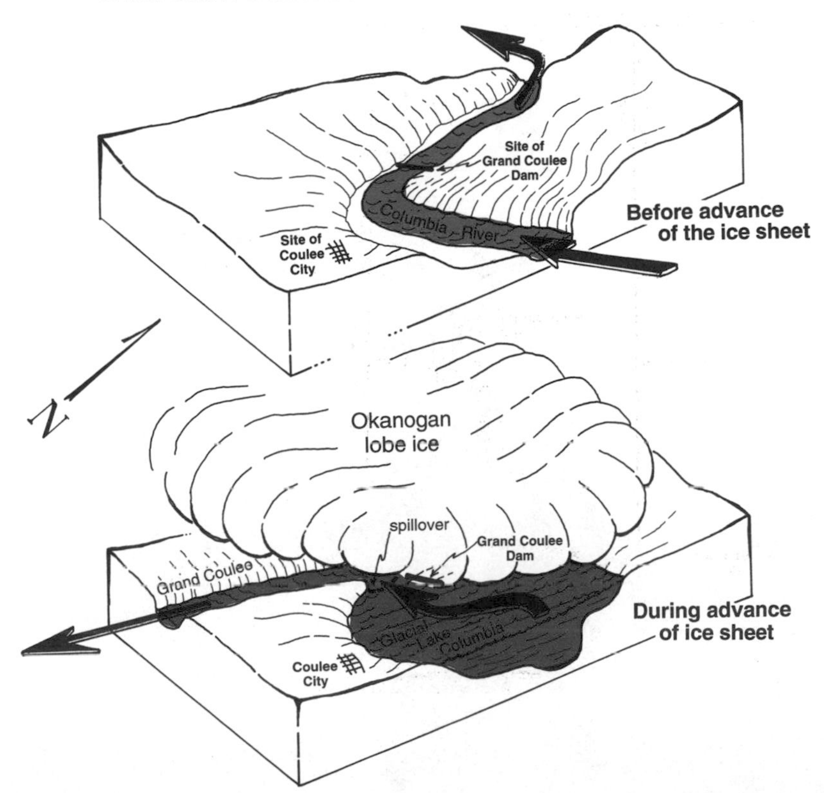

Lake Missoula floods probably entered Glacial Lake Columbia and poured across Dry Falls, too.

When the Okanogan ice lobe was thickest and at its farthest reach, it blocked the outlet and raised Glacial Lake Columbia to an elevation 700 feet higher than its usual level of 1,200 to 1,250 feet. The lake was then about 1,400 feet deep at Grand Coulee. At its thickest, the Okanogan ice lobe stood more than 1,600 feet above the highest level of the lake. That great height made it impossible for the lake to float its ice dam and drain catastrophically.

The deep valley glacier that flowed southeast down the valley that now contains Lake Chelan dammed the Columbia River at the town of Chelan to impound Glacial Lake Brewster. That lake lasted only a relatively short time, while the glacier in the Okanogan Valley was melting, and its end was well north of where it had been when the ice was at its greatest reach.

Ice in the Purcell Valley

During the earlier of the two ice ages recorded in the landscape, ice flowed down the Purcell trench of northern Idaho, filling it to overflowing. Watch for the glacially rounded mountains that stand above both sides of the Purcell trench north of Sandpoint. That earlier ice also filled Rathdrum Prairie all the way to the line of

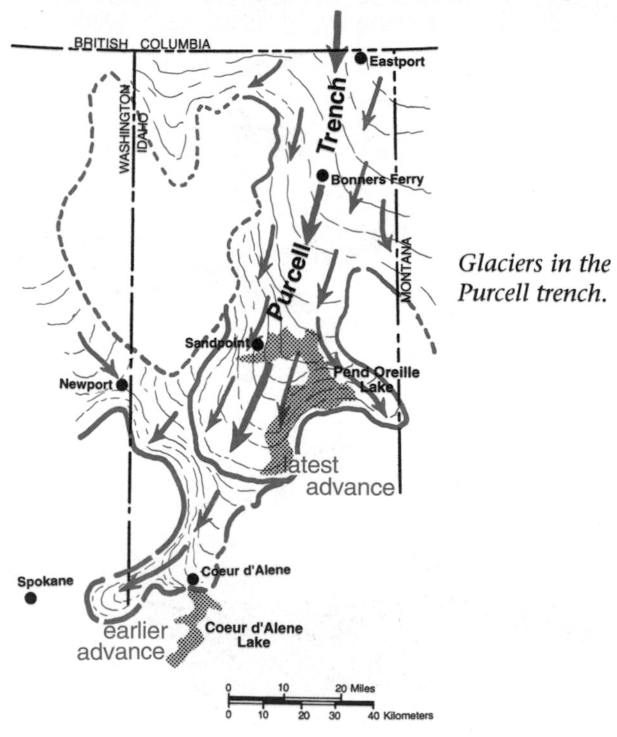

Glaciers in the Purcell trench.

Interstate 90, where deposits of glacial debris blocked the St. Joe River to impound Coeur d'Alene Lake. That is a peculiar glacial lake in that it lies beyond the reach of the ice that created it. The much smaller glacier of the last ice age reached only into the northern part of the Idaho panhandle, where it left the moraine that holds Pend Oreille Lake.

During both of the last two ice ages, glaciers dammed the Clark Fork River at the present site of Pend Oreille Lake to impound Glacial Lake Missoula, which flooded the mountain valleys of western Montana. Glacial Lake Missoula was the largest ice-dammed lake known, anywhere, and the floods that accompanied its sudden drainings were the largest known, anywhere. That story requires its own chapter.

A glacially transported boulder in the Purcell trench southwest of Sandpoint, Idaho.

Glaciation of the Flathead, Mission, and Swan Valleys

Ice filled the Rocky Mountain trench of eastern British Columbia and pushed south into Montana. It was some 6,000 feet deep at the border, but thinned rapidly southward as it approached its farthest reach. That glacier flowed down the Flathead Valley, then split on the wedge of the Mission Range at the northeast end of Flathead Lake. One branch flowed down the Swan Valley east of the Mission Range, the other down the Mission Valley west of the Mission Range. Ice completely buried the northern part of the Mission Range; only the high southern end stood above the sea of groaning ice that filled the broad valleys on either side. Compare the rounded crest of the northern half of the Mission Range, east of Flathead Lake, to the jagged crest of the high southern end of the range, east of St. Ignatius.

Meltwater flowing along the west side of the glacier carved a valley into the bedrock above a long stretch of the west side of Flat-

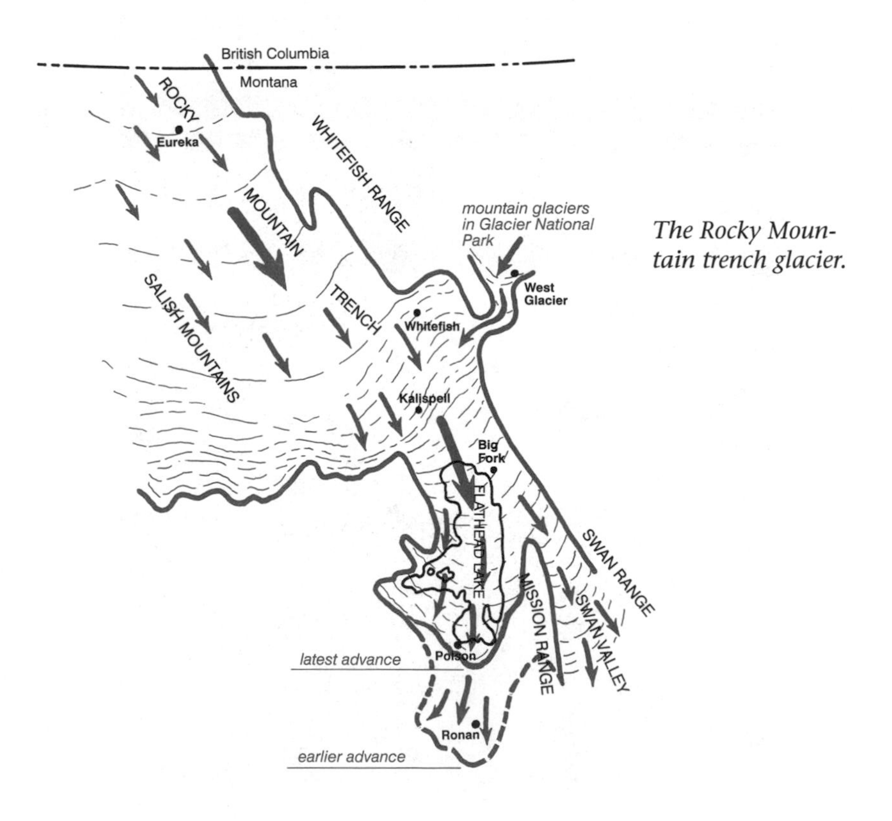

The Rocky Mountain trench glacier.

head Lake between Elmo and Dayton. People driving along Highway 93 pass beneath Chief Cliff, a prominent landmark just north of Elmo. It stands above that abandoned meltwater valley, an old river bluff now without its river.

Some of the best scenery in the northern Rocky Mountains melted with the end of the last ice age. Imagine photographing the view in the Mission Valley of the northern end of Glacial Lake Missoula lapping onto the southern end of the glacier, which rose in white cliffs from the dark water. The end of the glacier must have floated, calving off great icebergs that drifted across the lake. Those icebergs dropped little huddling groups of boulders where they finally ran aground and melted. That litter of stray rocks on the floor of the Mission Valley is all that remains to tell us of an incredible natural scene, now vanished.

Now, Flathead Lake floods the area where the southern end of the glacier once lay. Many people have tried to explain it and Whitefish Lake in the floor of the Flathead Valley as basins that the moving ice gouged into the valley floor. That seems most unlikely. Why should a glacier so near the end of its journey erode deep basins? At that point, glaciers are melting and thinning, hardly moving, far

Chief Cliff.

more likely to deposit sediment than to erode the valley floor. Those lakes and others like them probably mark places where large remnants of ice lingered long after the rest had melted. No sediment could accumulate where those big masses of ice lay, so their eventual melting left open basins in the valley floor. Seeley Lake, Swan Lake, and several others in the Swan Valley probably fill similar basins.

Chapter 50

LAKE BONNEVILLE AND THE GREAT LAKE BONNEVILLE FLOOD

15 Thousand Years Ago

During the ice ages, rain filled the undrained desert valleys in the Basin and Range as though they were bathtubs, creating lakes that dried up with the climate as our interglacial period began. Now, you see the old shorelines low on the mountain slopes around those valleys, perfectly level lines like the ring around the bathtub. They provide compelling evidence that ice ages brought wet climates.

Lake Bonneville filled the Salt Lake Basin. It covered much of northern Utah and lapped into Nevada and Idaho. Think of it as the ice age version of the Great Salt Lake. Its level rose until water overflowing through the lowest point on the drainage divide eroded its spillway, partially draining the lake. That part of its story illus-

Ice age lakes in the Great Basin.

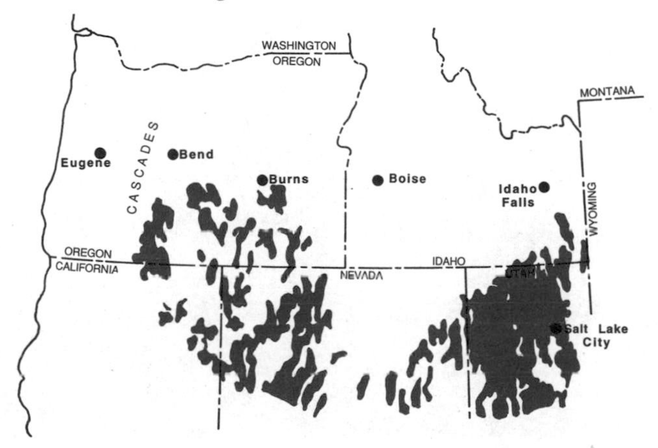

trates the process that started the modern streams flowing through the northern Rocky Mountains about two million years ago, as Pleistocene time began.

But the story of Lake Bonneville includes a special drama of its own. Its partial drainage happened catastrophically, scouring the canyon of the Snake River in one of the greatest floods of known geologic record.

The Great Lake Bonneville Flood

Lake Bonneville spilled through Red Rock Pass, which U.S. 91 crosses south of Pocatello, Idaho. A layer of hard rock in the pass stabilized the lake level for a long time, while the waves carved a fully developed shoreline complete with beaches and sand dunes—everything but the empty beer cans and soda pop bottles. That is the Bonneville shoreline. You can see it all around the Salt Lake Basin, at nearly the same elevation everywhere. It would be perfect for a picnic by the lake, if only the water were still there.

Water flowing through Red Rock Pass slowly eroded the resistant layer of rock and finally cut through it into a much softer formation beneath. The soft rock quickly washed out, abruptly lowering the elevation of the spillway by about 300 feet. It was as though a dam had broken. During the next several months, about 300 feet of water drained off the top of Lake Bonneville through Red Rock Pass,

The Bonneville shoreline east of Tremonton, northern Utah.

down the valleys of Marsh Creek and the Portneuf River, and into the Snake River. That was the Bonneville flood.

It is hard to come along 15,000 years after the event, look at a scoured canyon, and take the measure of the flood that scoured it. Recent estimates place the maximum flow of the Bonneville flood

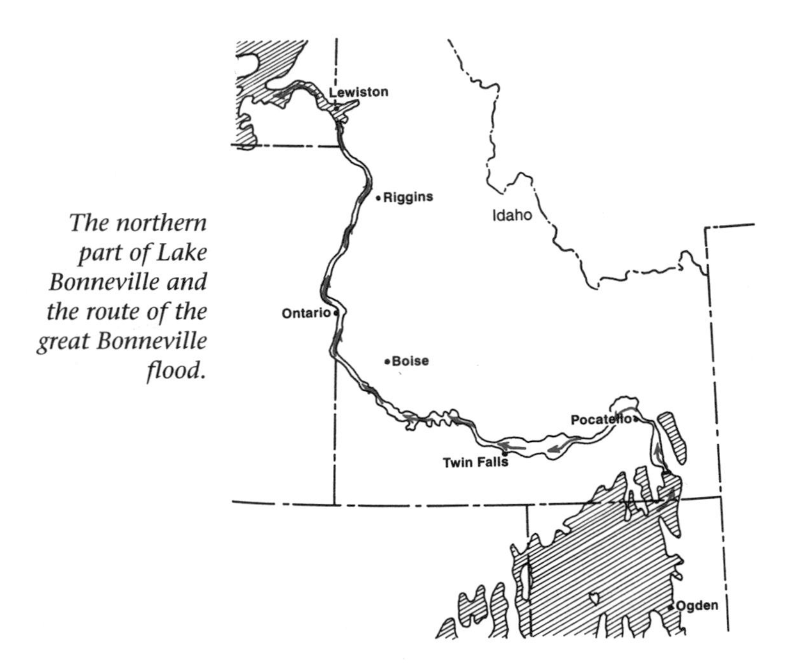

The northern part of Lake Bonneville and the route of the great Bonneville flood.

at approximately 10,800,000 cubic feet per second. That is about 300 times the maximum flow ever measured on the Snake River during historic time, six times the maximum on the Mississippi River, slightly more than the maximum on the Amazon River. But the Lake Missoula floods, which came a few years later, carried as much as 50 times that volume of water.

Had you been there then, you could have stood on a high point near the Snake River Canyon, listening to growling noises rising like muffled thunder from the muddy torrent. That was the sound of rocks rolling and grating along the canyon floor. The Bonneville flood scoured that canyon from the mouth of the Portneuf River near Pocatello all the way to its mouth in the Columbia River. Along most of that long way, water filled the canyon more than brimful, and flooded nearby parts of the Snake River Plain. Water rushed in furious torrents through the narrower parts of the canyon, and ponded in its broader parts.

The largest of the rocks that rolled beneath the flood are rounded boulders of basalt that look like huge black watermelons, the melon gravel. These boulders, most several feet in diameter, now litter the surfaces of giant gravel bars as much as several hundred feet high; they make hills within the canyon. Look for those gravel bars in the broad reaches of the canyon floor immediately downstream from the narrow gorges.

Swirling eddies in the flood scoured deep holes in the canyon floor. The Snake River now pours into some of those in spectacular waterfalls such as those at Twin Falls, Idaho. Other eddies carved deeply recessed alcoves into the canyon walls. The flood left the narrow parts of the canyon floor a ragged waste of jagged basalt, with none of the soft cover of sediments or soil that normally coats a valley floor.

In Hells Canyon, the Bonneville flood flowed hundreds of feet deep, leaving gravel bars along the insides of the bends more than 100 feet above the present river level. Other deposits of sand and gravel partly block the mouths of side valleys, where the water slowed

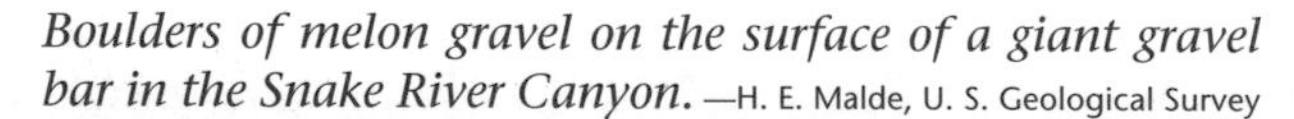

Boulders of melon gravel on the surface of a giant gravel bar in the Snake River Canyon. —H. E. Malde, U. S. Geological Survey

and eddied. Those deposits stand as much as 400 feet above the river.

The deep flood backed into tributary streams, reversing their flow for miles. For example, terraces along the lower part of the Clearwater River above Lewiston, Idaho, contain sediment typical of the Snake River and quite unlike anything the Clearwater River carries. That material must have come down the Snake River in the flood, then backed upstream into the Clearwater River.

When the spillway eroded to the next layer of hard rock, the lake level again stabilized, and the waves cut the Provo shoreline. But the ice age was nearly over then. The climate dried, and the lake level began to drop before the waves could create another shoreline as well developed as the Bonneville shoreline. Lake Bonneville no longer received enough water to maintain overflow through the spillway, and it slowly shriveled through evaporation to its last remnant, the Great Salt Lake.

Look at the mountains almost anywhere around the Salt Lake Basin for the distinctive pair of horizontal shoreline benches. The higher Bonneville shoreline is quite obvious nearly everywhere. The Provo shoreline, about 300 feet below it, is easy to see but not nearly as obvious. If you have an especially good vantage point, you may see dozens of faint shoreline benches below the Provo shoreline. They record stages in the final slow evaporation of Lake Bonneville after the last ice age ended.

If the last ice age had not ended within a few thousand years after the Bonneville flood, and the climate had not dried, the lake would have continued to overflow through Red Rock Pass. Erosion of that spillway would eventually have drained Lake Bonneville and established a new stream draining the Salt Lake Basin. That is probably how the modern trunk stream drainage of the northern Rocky Mountains was established.

Lake Shorelines and the Floating Continental Crust

G. K. Gilbert, one of the great figures in American geology, mapped the shorelines of Lake Bonneville and found the outlet spillway more than a century ago, when most of the region was nearly roadless. In one of the most brilliant achievements of American science, he used the old shorelines to demonstrate that the continental crust floats.

The surface of a lake is perfectly level, so the Bonneville shoreline must have been perfectly level when the waves carved it. Gilbert found that it is no longer level, that the deeper the water had been, the higher the Bonneville shoreline now is. He showed that the weight of the water in the lake had depressed the continental crust: the deeper the water, the greater the depression. Then, when the water finally evaporated, the continental crust rose, carrying the old shoreline highest where the crust had been most depressed.

Gilbert was right, beyond any shadow of a doubt. Ever since then, people have been running level surveys over future large reservoirs before the dam was finished, then running them again a few years after the reservoir was filled. The crust does indeed sink beneath the weight of the water, just as Gilbert concluded from his study of old shorelines.

Chapter 51

GLACIAL LAKE MISSOULA FLOODS

15 Thousand Years Ago

It was the largest lake known to have existed behind an ice dam anywhere in the world, and its sudden drainages caused the largest floods known to have swept any part of the world. The story of Glacial Lake Missoula and the Lake Missoula floods is an exercise in the use of superlatives. It also tells a bit about how science works, or in some cases almost does not work.

Heresy, a Catastrophist

The first geologist known to have visited Missoula, Montana, back in 1878, reported seeing numerous lake shorelines faintly etched on the sides of the mountains around town. He was absolutely right. Sets of lake shorelines are hard to misinterpret. Nothing else is so perfectly horizontal as a lake shoreline, nor is anything so perfectly parallel as two of them, assuming that the lake was too shallow or too temporary to depress the earth's crust under the weight of its water. Glacial Lake Missoula was too temporary.

In 1910, another geologist, J. T. Pardee, described Glacial Lake Missoula in considerable detail. He showed that a large glacier creeping down the Purcell Valley from British Columbia had dammed the Clark Fork River at the present site of Lake Pend Oreille in northern Idaho. He also estimated the size and volume of the lake and the highest elevation of its surface. That, in 1910, seemed to be that.

During the early 1920s, yet another geologist, J Harlen Bretz, was studying the channeled scablands of eastern Washington. Those are dry stream channels that scar the eastern Columbia Plateau, cutting through the cover of windblown Palouse dust and into the dark flood basalt flows beneath. Many of the scablands channels

Grand Coulee, a now dry stream channel eroded by catastrophic flows from ice age lakes. —F. C. Calkins, U.S. Geological Survey

cross local drainage divides, well above the level of the modern drainage. Most geologists had assumed that they are remnants of an ancient stream drainage that existed before the modern streams eroded their valleys.

Finding that the basalt exposed in the scablands channels is almost perfectly fresh, Bretz concluded that they were eroded during the recent geologic past. He proposed that a catastrophic flood, which he called the Spokane flood, had eroded the channeled scablands. How else to explain those young channels eroded across the drainage divides high above the modern streams?

That conclusion aroused a vigorous storm of outraged protest from geologists wedded to the conventional view that the modern landscape is entirely the work of weak forces and slow processes operating over long periods of time. They accused Bretz of catastrophism, widely considered the worst and most offensive of all possible geologic heresies. He grimly pleaded guilty as charged, while vehemently adhering to his views. But he could not explain where his catastrophic flood had come from, and that was an embarrassing problem.

J. T. Pardee then returned to the study of Glacial Lake Missoula and showed beyond any reasonable doubt that it had drained suddenly and was therefore a plausible source of the catastrophic Lake Missoula flood. The best of Pardee's evidence is in Camas Prairie, west of Flathead Lake, where he found gigantic ripples in coarse gravel, tens of feet high and several hundred feet from crest to crest.

Except for their size, they look like the little sand ripples you see in the bed of a stream. Highway 382 passes through those monster ripples south of Markle Pass, just south of Hot Springs, Montana. Similar giant ripples exist on the west bank of the Columbia River in central Washington, north of Vantage, where Interstate 90 crosses the river. In fact, giant ripples exist in quite a number of places along the Clark Fork and Flathead Rivers in western Montana and the Columbia River in Washington.

Pardee studied a narrow stretch of the Flathead River valley to estimate the flow as Glacial Lake Missoula drained. Projecting from experimental results, he estimated that the water must have been moving at a speed of about 45 miles per hour to roll the largest boulders that moved under the flood. Then he found the highest evidence of flood scouring on the canyon walls, measured the cross-sectional area of the flood channel, and multiplied that by the flow speed. The result was a figure of approximately nine cubic miles of water per hour. That is about 450 times the maximum flow ever recorded on the Mississippi River. It is more than the combined flow of all the rivers of the world.

Pardee published his results in 1943. Another 30 years of controversy followed before very many geologists would agree that the evidence really did tell of catastrophic floods. J Harlen Bretz survived his detractors and lived to see most geologists finally convinced that his unpopular flood really did happen.

Glacial Lake Missoula and Its Many Floods

Widely scattered evidence leaves little doubt that earlier glaciers impounded the Clark Fork River long before the last ice age, but no one has assembled it into a coherent story. All we know is what happened during the last ice age.

At its maximum filling, Glacial Lake Missoula rose to an elevation of about 4,350 feet. Its dark waters ran their icy fingers and floated their drifting icebergs deep into all the mountain valleys of the Clark Fork drainage in western Montana. When the lake was at its highest, its water rose to a depth of some 2,000 feet at the ice dam. It then contained approximately 480 cubic miles of water, about as much as Lakes Erie or Ontario. A glacier was all that stood between the lake and the Columbia Plateau.

Ice cubes float in lemonade, and ice dams float in the lakes they impound. Then the dam breaks into great chunks, and the lake drains in a sudden rush. Many anthropologists believe that people crossed from Asia into North America sometime before the last ice age reached its maximum. They give us permission to imagine a fisher-

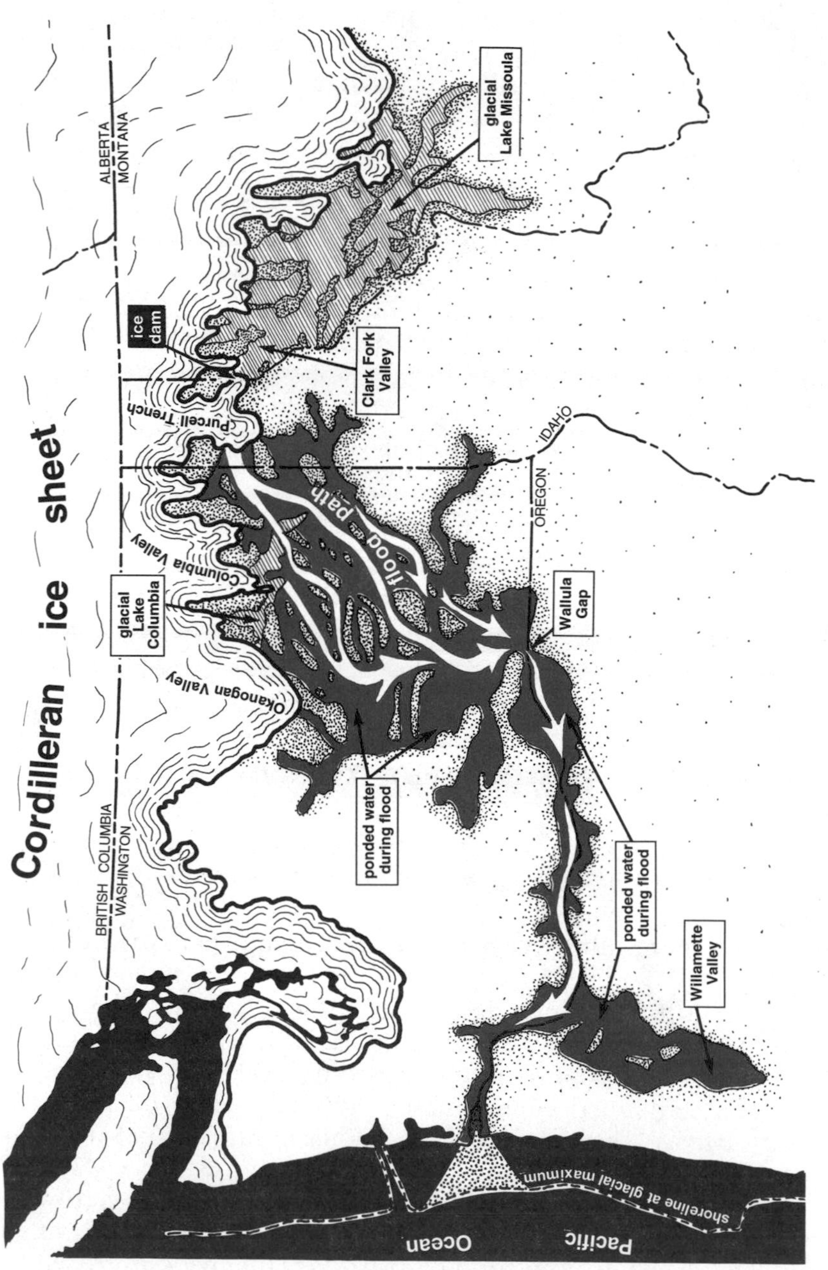

The glacier dam of Glacial Lake Missoula and the path of its floods.

man noticing the icebergs beginning to move on a gentle current. Then the entire lake became a torrent of muddy water, floating ice, and uprooted trees, all in a headlong rush for the Pacific Ocean. We hope our fisherman made it to high ground, but no one in the path of that flood could have.

Those unfortunate people would have seen a wall of water as much as 2,000 feet high with as much as 480 cubic miles of water behind it starting from the site of Pend Oreille Lake, thundering through Rathdrum Prairie, across the site of Spokane, and then southwest across the Columbia Plateau to the Columbia River. Such a flood is as hopelessly out of scale in the subdued landscape of the eastern Columbia Plateau as a washtub of water sloshed across a child's sandbox. Instead of following the streams, those floods filled each valley, then surged west across the drainage divide into the next. Watch along Interstate 90 between Sprague and Spokane for pine trees growing among ragged outcrops of dark basalt, and around irregular channels and little lakes and ponds. That is one of the scablands floodways.

Old dry flood channels of the channeled scablands, near Sprague, eastern Washington.

Each flood filled the Pasco Basin to form a swirling lake of muddy water that soon drained as the water poured through Wallula Gap and down the Columbia River to the Pacific Ocean at 22 to 65 miles an hour. Watch from Interstate 80N or Washington 14 along the Columbia River for the rocky lower valley walls that lost their upholstery of soil to the scouring floods. And you may see widely scattered chunks of utterly distinctive Precambrian mudstones and sandstones, Belt rocks. They could have come only from the Northern Rocky Mountains. Many of them show no sign of rounding, as they would had the flood rolled them to their destinations. Floating icebergs must have carried them along bodily. They lie alone and in small groups that probably mark places where icebergs finally ran aground and melted.

In the narrows of the Columbia Gorge, the angular rocks from the Northern Rocky Mountains lie as much as 1,000 feet above present river level. The flood was that deep. Its rapid erosion of the canyon walls amputated the lower parts of the tributary streams, so they now enter the Columbia River over cascades or waterfalls. Multnomah Falls are the best known of the many you can see from the highway.

An ice jam formed at Portland, where the Columbia River makes a hard right turn. It backed the flood into the Willamette Valley at least as far south as Salem, Oregon. There, too, angular boulders of Precambrian sedimentary rocks, some many feet across, lie scattered where the icebergs that carried them finally ran aground and melted.

After each draining of Glacial Lake Missoula, the Purcell Valley lobe of the regional ice that filled the lowlands of British Columbia again advanced south across the Clark Fork River. That impounded a new version of Glacial Lake Missoula that would drain when its water floated the ice dam. Every sudden drainage released another overwhelming flood, another major catastrophe.

Glacial Lake Missoula Deposits in Western Montana

Glacial lakes typically lay down distinctive deposits composed of alternating thin layers of light and dark sediment. The light layers accumulate during the summers, when melting ice releases large quantities of finely pulverized rock into the lake. After the fall freeze stops the supply of pulverized rock, the remains of the microscopic plants and animals of summer slowly sift to the bottom, depositing a layer of sediment dark with organic matter. Each pair of light and dark layers records one year.

Glacial Lake Missoula shorelines at Missoula, Montana. Those highest on the mountain are the oldest, and they become younger downward.

Some deposits of Glacial Lake Missoula sediment in western Montana contain sequences of glacial lake sediments sandwiched between layers of silt and sand with the internal layering characteristic of river sediments. The lake sediments record times when Glacial Lake Missoula existed; the river sediments were laid down when it was drained. Altogether, they record 36 fillings of Glacial Lake Missoula.

The oldest sequence of lake sediments contains 58 pairs of dark and light layers. Each successively higher sequence contains a record of fewer years than the one below. If each version of the lake lasted fewer years than the one before, each must have risen to a lower elevation. So the highest shoreline must be the oldest, and they become progressively younger as they step down the mountain. Successively lower shorelines must record a thinning glacier forming progressively lower ice dams. It seems clear that Glacial Lake Missoula was filling and draining as the ice was beginning to melt after the maximum of the last ice age.

Flood Deposits

The Lake Missoula floods backed up the rivers that drain into eastern Washington, reversing their flow as they filled their valleys to depths of several hundred feet. Each flood of muddy water left its deposit of sediments in the valleys of those rivers, in many cases far up the valley.

Ice spreading across the Columbia Plateau south of the Okanogan Valley impounded Glacial Lake Columbia, which never rose high enough to float its ice dam. Glacial Lake Columbia reached east to Spokane, confined between mountain ridges and valleys filled with ice on the north and the gentle northward slope of the Columbia Plateau on the south. Lake Missoula flood waters poured into the lake through the valleys of the Spokane and Columbia Rivers and left flood sediments sandwiched between sections of lake sediments with their light and dark annual layers. The deposits in the Sanpoil Valley, a few miles east of Coulee Dam, contain one of the best records of the Glacial Lake Missoula floods. In general, the number of years recorded in the lake sediments laid down between floods decreases upward in the same pattern as that in western Montana.

Glacial lake sediments laid down in Glacial Lake Columbia.

Some geologists interpreted a section in the Sanpoil Valley as a record of no fewer than 89 separate Lake Missoula floods. Stacks of flood deposits elsewhere clearly record at least 41 Lake Missoula floods. The numbers are higher in Washington than in western Montana, probably because some fillings of the lake did not rise high enough to reach the areas of Montana where the lake sediments have been studied.

Counts of the total number of annual layers in the lake sediments between flood deposits in Glacial Lake Columbia vary, but the average comes to about 1,923 layers. Lake sediments above the last flood deposit contain another 200 to 400 annual layers, so Glacial Lake Columbia lasted that long after the last version of Glacial Lake Missoula drained. Radiocarbon dates show that ice in the Purcell trench first blocked the Clark Fork River to form Glacial Lake Missoula 15,500 years ago. Counting down the annual layers in the lake sediments then leads to the conclusion that the last ice dam formed about 13,500 years ago. Further counting down the annual layers shows that the ice coming out of the Okanogan Valley blocked the Columbia River to impound Glacial Lake Columbia until about 13,200 years ago.

All of these clearly recorded floods happened during the last ice age, but piecemeal records exist of much older floods during earlier ice ages. Scraps of glacial lake sediment exist at elevations higher than the highest level of Lake Missoula in the last ice age. The Withrow moraine of the last ice age clearly crosses Moses Coulee, one of the largest glacial outwash channels in the northwestern part of the Columbia Plateau. Dated volcanic ash layers in Pleistocene glacial outwash sediments and Palouse loess record at least six episodes of flooding. They range between 800,000 and 40,000 years ago.

A deep channel cut in the Columbia River Plateau during the Glacial Lake Missoula floods, just below Dry Falls in central Washington.

Chapter 52

THE COAST

10 Thousand Years Ago, and Continuing

People who watch the recording fathometer in a fishing boat often see it trace the profile of a submerged shoreline as the boat passes beyond a water depth of about 300 feet. That was the coast when the last ice age was at its maximum, about 15,000 years ago. Dry land extended that far seaward, and streams flowed across it, eroding their beds to grade them to sea level.

As the last ice age ended, melting glaciers shed great torrents of meltwater into the oceans, rapidly raising sea level to its present stand. People who lived along the coast then must have noticed sea level rising, even without formal measurements or written records. Imagine the old people sitting around the fire, boring their grandchildren with tales of how they had seen the sea rise, drowning the forest along the coast, flooding far into the mouths of the rivers. Breaking waves created most of the detail we see along the modern coast after that rise in sea level.

Estuaries, Bays, and Sand Bars

The rising ocean submerged the lower courses of the coastal rivers, then flooded into their new mouths, converting them into tidewater estuaries, bays. Nearly every large stream becomes an estuary as it approaches the coast. Those are the bays and harbors.

Waves sweeping sediment along the beaches built sandspits across the mouths of most of the estuaries, converting them into nearly enclosed bays now shoal with deep fills of trapped river mud. River water flowing out through the bar maintains an inlet connecting the bay to the open ocean. Inlets should really be called outlets, because it is the outflow of water from the land that maintains them, not the inflow of water from the ocean. The vigorous flow of the Columbia River prevents formation of a bar across the mouth of its estuary.

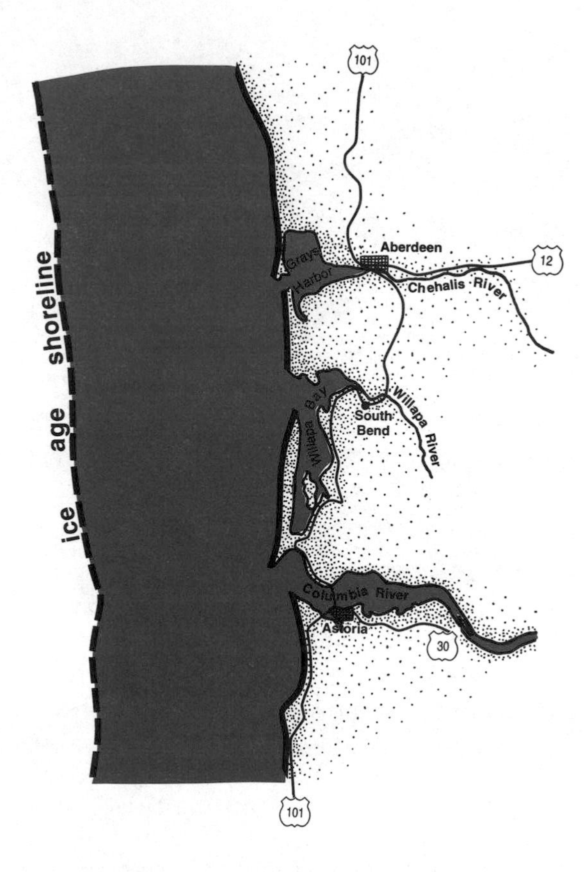

Grays Harbor and Willapa Bay are former river valleys that flooded as sea level rose when the last ice age ended. Wave-built bars now enclose them.

The coastal winds tend to blow out of the southwest in the winter, the northwest in the summer. So the sediment tends to move north in the winter, south in the summer. The winter winds blow harder than those of summer, so the northward drift of sediment dominates along most of the coast. Waves drive sand from south to north across the mouths of bays. The northward drift of sand forces many rivers to turn north as they approach the coast.

Soft Rocks, Sandy Beaches, and Coastal Dunes

Soft sedimentary rocks are too weak to make good raw material for a rocky coast. Waves quickly pound them to sand as they reduce the coast to a smoothly curving shore fringed with sandy beaches that melt landward into broad tracts of shifting dunes.

Three stages in rise of sea level:

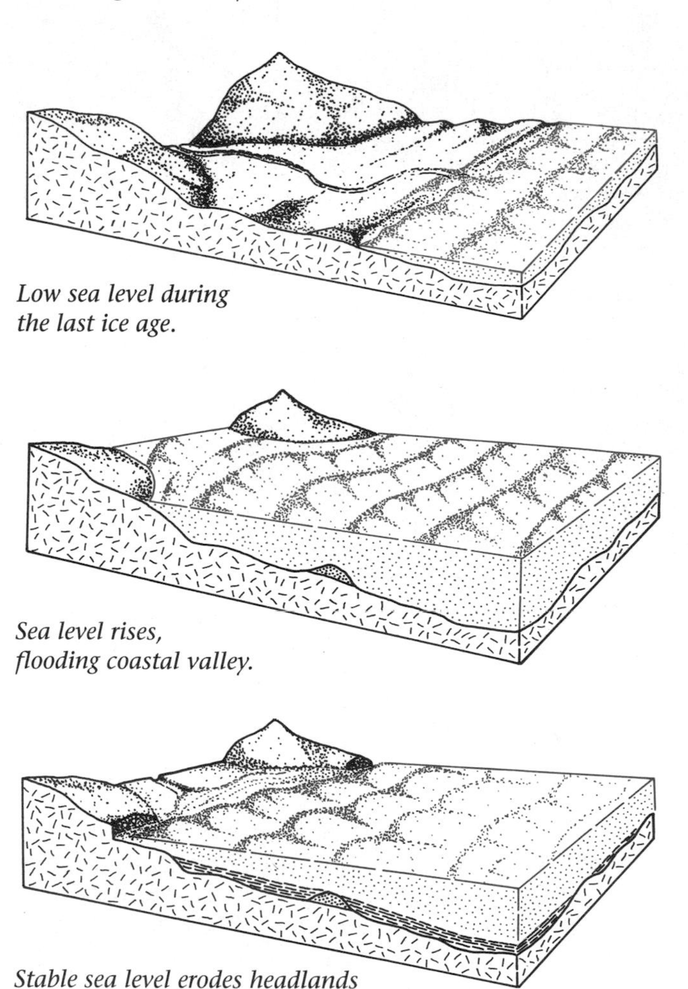

Low sea level during the last ice age.

Sea level rises, flooding coastal valley.

Stable sea level erodes headlands and deposits sediment in bay.

At high tide, breaking waves bring sand ashore and spread it across the upper beach; at low tide, the strong sea breeze whips sand off the drying upper beach and blows it inland, into the dunes. The combination of sandy sedimentary rocks, waves, wind, and tide have given the Oregon coast the largest actively marching dunes in North America. These giant dunes along this rainy coast strongly contradict the notion that sand dunes are creatures of the desert. In fact, dunes form wherever strong winds blow freely across dry sand, regardless of the climate.

The spacing between ripples is the average distance sand grains bounce as the wind drives them along. It changes with the wind.

Along the central Oregon coast, especially around Florence, the encroaching dunes dam streams to impound freshwater coastal lakes. Deep forest typically grows around the eastern sides of those lakes, while the impounding sand advances toward the western end. Trees and brush growing in the wet climate overwhelm the inland parts of the dunes, limiting their landward reach to the zone in which salt spray inhibits plant growth.

The Rockbound Coast

Where the surf beats against hard and resistant rocks, it creates an irregular coastline of great contrast. Gentle waves wash onto sandy beaches nestled in the heads of quiet coves while towering breakers crash into rocky headlands. The waves shape the coast, and the shape of the coast dictates the behavior of the waves.

Stand high on the seaward end of a bold headland to watch the approaching waves wrap themselves around its projecting tip. They do that because waves slow down as they enter shallow water and begin to drag on the bottom. The part of the wave moving across the shallow bottom beyond the projecting tip of the headland lags behind, while the parts in deeper water on either side race ahead. So the approaching wave feels the shape of the coast as it drags on the bottom, and it changes its own shape to conform.

Waves shorten as they wrap the short tip of a headland in their curling embrace, grow higher as they compress their energy into a shorter length. Compare the furious breakers crashing against the projecting tip of a headland with the much calmer surf that the same train of waves lands on a straight length of coast. The headland focuses the energy of wave attack on its tip, making it the point of greatest wave energy. Somewhere down the coast, the same

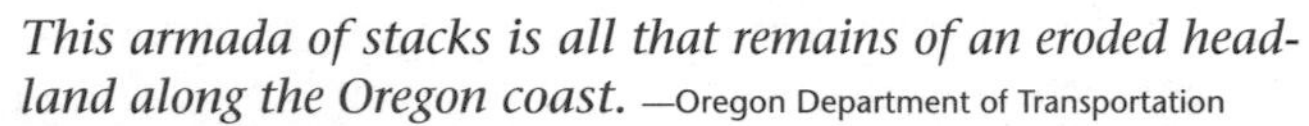
This armada of stacks is all that remains of an eroded headland along the Oregon coast. —Oregon Department of Transportation

wave lengthens as it stretches into the long shoreline of a cove, flattens as it stretches its energy into the longer length.

Breakers erode hard bedrock mostly by pounding pulse after pulse of compressed air into fractures, loosening blocks of rock, then popping them out into the surf. Heavily fractured rocks are most vulnerable to that kind of attack, so the waves dismantle them first, block by block, leaving the less fractured rocks standing. In places, the waves exploit a fracture zone so deeply that they erode a slot, or a sea cave, or perhaps an arch with its roof just above the highest reach of the tide.

As the breakers demolish the more fractured rocks in the headland, they leave the less fractured rocks standing as isolated sea stacks. Eroding headlands eventually dissolve into armadas of stacks standing out to sea. And even those stalwarts finally yield to the untiring surf.

Meanwhile, the waves sweep the debris of the headland along the coast, down the direction of the prevailing wind. The sediment finally comes to rest in a sheltered cove, where it stays. So the waves attack all headlands and fill all coves, eventually converting any coast, no matter how ragged, into a smooth shore that sweeps along

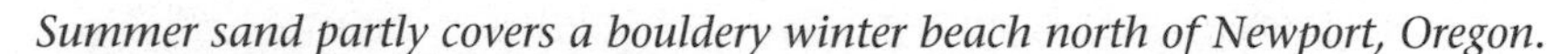

Summer sand partly covers a bouldery winter beach north of Newport, Oregon.

A wave-cut bench exposed at low tide on the coast north of Florence, Oregon.

in a series of uneventful curves. That is the main reason why good harbors are so rare along most of the world's coasts.

Sea Cliffs and Wave-Cut Benches

Waves breaking against the sea cliff hammer a notch in its base as they dislodge one block of rock after another. The deepening notch undermines the cliff above, eventually to the point of making it collapse into the surf. Then the waves wash its debris seaward, while starting a new notch in the new base of the sea cliff. And so it goes as the waves slowly erode the coast back into the land.

The retreating sea cliff leaves behind it a remarkably smooth bedrock surface called a wave-cut bench. You rarely see much of that surface except at dead low tide, especially the spring tides. Then you can walk across it, look for starfish clinging to the sea stacks, and peek into the tide pools. Wave-cut benches rarely become more than a few hundred yards wide because they are so precisely tied to sea level. The slightest rise of the land or drop of sea level sets the process back to its beginning.

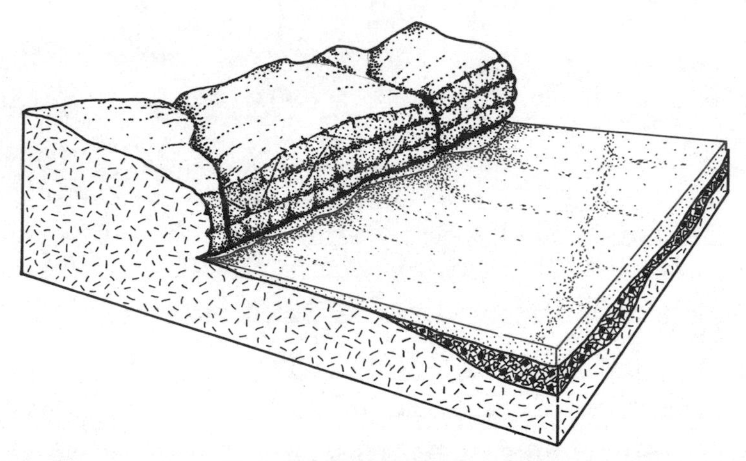

Notched sea cliff and wave-cut bench.

Fluctuating Sea Level, Rising Land, and the Coastal Landscape

Long stretches of U.S. 101 from southern Washington to California follow old wave cut benches now raised above sea level. Watch for the broad and nearly smooth surfaces that slope gently seaward, then end abruptly at the edge of the modern sea cliff. They end landward at much less abrupt scarps where the land rises sharply into hills, old sea cliffs now eroded into softer slopes.

Watch on those old wave-cut benches for occasional, small, rocky hills, former sea stacks. Except for being a bit weathered, they are exactly like the stacks that now stand offshore. Seaweeds grew on those little rocky hills; fish swam among them; sea lions loafed on

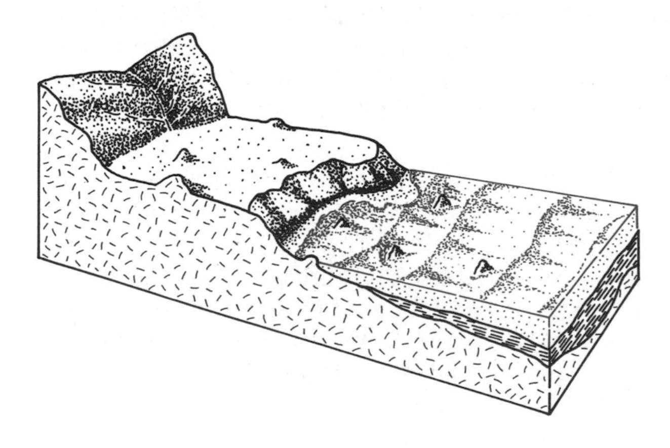

Sketch of an emergent marine terrace like those along much of the west coast.

them; cormorants perched on their high points. Along much of the coast you see old shorelines complete in every detail, except for the water. Long stretches of the coastal highway are on old sea floor.

It is hard to imagine that a gradual rise of the land or drop in sea level could allow those old strips of sea floor to emerge so nearly unscathed from the surf. It is much easier to suppose that they emerged suddenly during the great earthquakes that apparently rock this coast at intervals of a few hundred years.

The Mystery of the Missing Earthquakes

Oceanic trenches are typically associated with fairly frequent earthquakes, some of which are very strong. Plots of their precise sources within the earth define a surface that descends from the trench deep into the mantle, typically at an angle near 45 degrees. In effect, it is a picture of the sinking slab of oceanic lithosphere. The steeply slanting zone of deep earthquakes fades downward, then vanishes at a depth of approximately 300 miles, presumably because that is where the sinking slab becomes too hot and too soft to generate earthquakes.

The historic earthquakes in the Pacific Northwest have been neither frequent nor very strong, nor have they come from deep sources, nor does the plot of their sources define a sinking slab. That seems odd: Why should our west coast have all the other features typical of a collision plate boundary except the earthquakes? Recent research shows that the Northwest is not really as free of earthquakes as everyone had supposed. It seems that they strike less frequently than in most other regions landward of an oceanic trench, but more violently.

Studies of very young coastal sediments in places between Puget Sound and northern California reveal occasional layers of sand that appear to have been laid down beneath huge waves, probably tsunamis that followed very strong earthquakes. In many places, those sand deposits cover the broken stumps of trees that grew in coastal marshes. Radiocarbon dates on wood collected from the stumps show that the sand layers were deposited at intervals of less than 600 years, most recently about 300 years ago. That interval is based on at least 13 sand layers preserved in the 7,600 years since deposition of Mazama ash. Most geologists now interpret the layers as evidence that very strong earthquakes and enormous tidal waves, tsunamis, ravage the coastal northwest about every 500 or 600 years. That is not a schedule, though. The next one could come at any time in the next few hundred years.

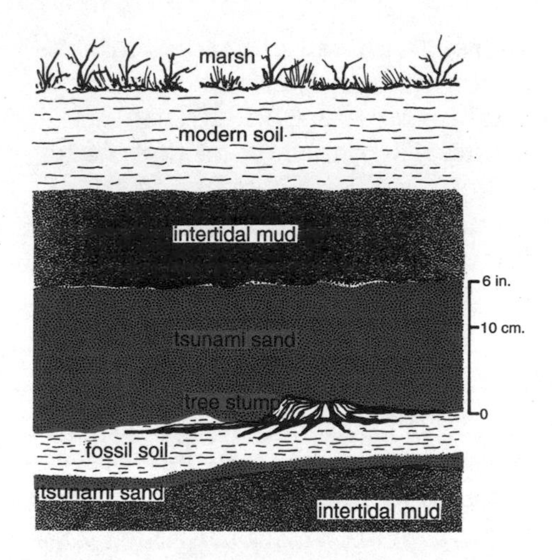

Layers of sand buried in tidal marshes were probably deposited from tsunamis generated by major earthquakes.
—Sketch from photo by John Clague, Geological Survey of Canada.

Precise tidal measurements show that the continental shelf just offshore is now rising at a rate of about two feet per 100 years, while areas about 50 miles inland are sinking. Obviously, the coastal slab of oceanic crust is warping, probably because it is now jammed against the slab of oceanic crust sinking through the offshore trench. The trench is stuck. When it eventually breaks loose, the sinking slab will abruptly slide some tens of feet into the trench, the land now sinking inland will suddenly rise, and the shallow bottom now rising offshore will suddenly sink. All of that to the thundering accompaniment of a great earthquake.

Some of the coastal areas that are now rising will probably drop below sea level. The sudden drop of the shallow ocean floor off the coast will almost certainly cause a great tsunami that will surely

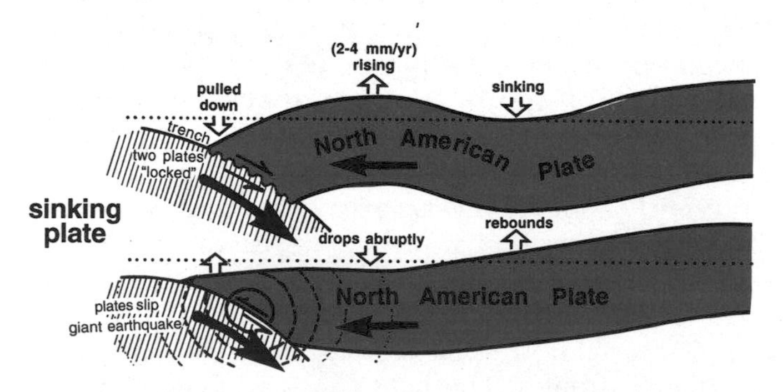

The North American plate buckles when it jams against the sinking ocean floor, then straightens when the two break loose.
—Redrawn from a diagram by Pacific Geoscience Center, Geological Survey of Canada

involve long stretches of the coast. Then we will regret having crowded so much of our coastal communities so close to the shore.

From a layer of sand, it is hard to guess the size of the earthquake that generated the tsunami that laid it down. Most estimates suggest that coastal areas of the Northwest should expect an earthquake of magnitude 8.5 or 9 about every 500 or 600 years. That is fully comparable to the largest ever recorded anywhere. An earthquake of magnitude 8.7 would release about 1,000 times the energy of one of magnitude 6.6, the size of the earthquake that struck Los Angeles in 1994. Violent shaking would continue for at least two minutes, instead of the half minute of shaking that Los Angeles suffered. The next really big earthquake may strike Portland, Seattle, Vancouver, or somewhere in northern California.

West of Puget Sound, the central part of the Olympic Peninsula is sinking and the western edge is rising. When the plates slip, the central part of the Olympic Peninsula will suddenly rise and the coastal areas suddenly sink, in some places below sea level. Giant earthquakes will rock Puget Sound.

A similar scenario exists as far south as northern California, but the details are not as precisely known.

These giant earthquakes seem likely to hit coastal areas 50 or 100 miles west of Seattle or Portland, the distance providing a little, but not much, buffer to the shock. Worse, recent study has uncovered evidence of similar tsunami sand deposits near downtown Seattle. The Seattle fault carries the ground to the south up and over that to the north. Big sand wedges have been generated in various places in the Seattle area, apparently by movement on that fault. How the Seattle fault fits into the plate movements in the Pacific Northwest is not clear.

Sea stacks in the surf below a raised wave-cut bench along the coast of northern California.

Chapter 53

EPILOGUE

The Future

What next for the Great Northwest? What will you find here when you come back 20 million years from now to see how things turned out?

One of the charms of history, human or geologic, is the sudden turn of fate that sends the pattern of events into a new and entirely unforeseen course. Obviously, we cannot possibly know when a great happening, perhaps another great impact, may intervene to set the course of geologic events in a new direction. Those things come as they will, without warning. All we can do is to project what will happen if the present pattern of plate movements persists, and even that seems risky.

The San Andreas Fault

If the San Andreas fault continues to move as it has during the last 15 or so million years, it will detach the part of California west of the fault from the rest of California. The detached slice will then become an island microcontinent moving north with the Pacific plate. It will sail past the northwest coast at a rate of about two inches per year, and eventually arrive at the Aleutian trench to become another piece of the Alaskan mosaic of continental fragments.

Meanwhile, the last remnant of the Farallon plate will have disappeared into the trench off the coast of the Pacific Northwest. As that happens, the San Andreas fault will extend itself north and the Queen Charlotte fault south. The two will meet just as the last remnant of the Farallon plate disappears.

Joining the San Andreas and Queen Charlotte faults will convert the boundary between the North American and Pacific plates into a single fault, a transform plate boundary, that will extend all the way from the East Pacific oceanic ridge off the west coast of Mexico to the Aleutian trench off the south coast of Alaska.

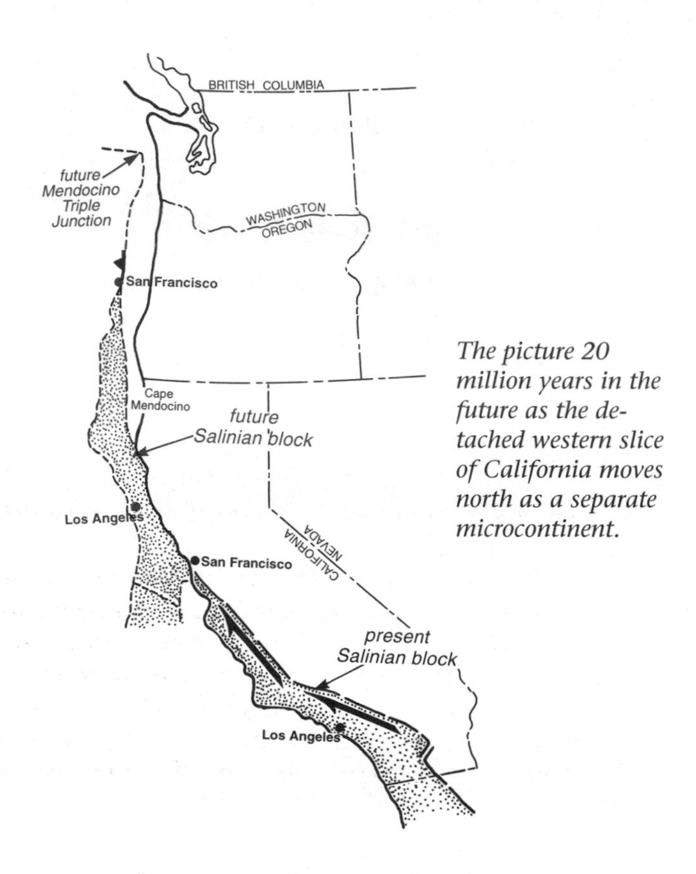

The picture 20 million years in the future as the detached western slice of California moves north as a separate microcontinent.

The piecemeal disappearance of the Farallon plate. —Abstracted from Stock and Molnar, 1988

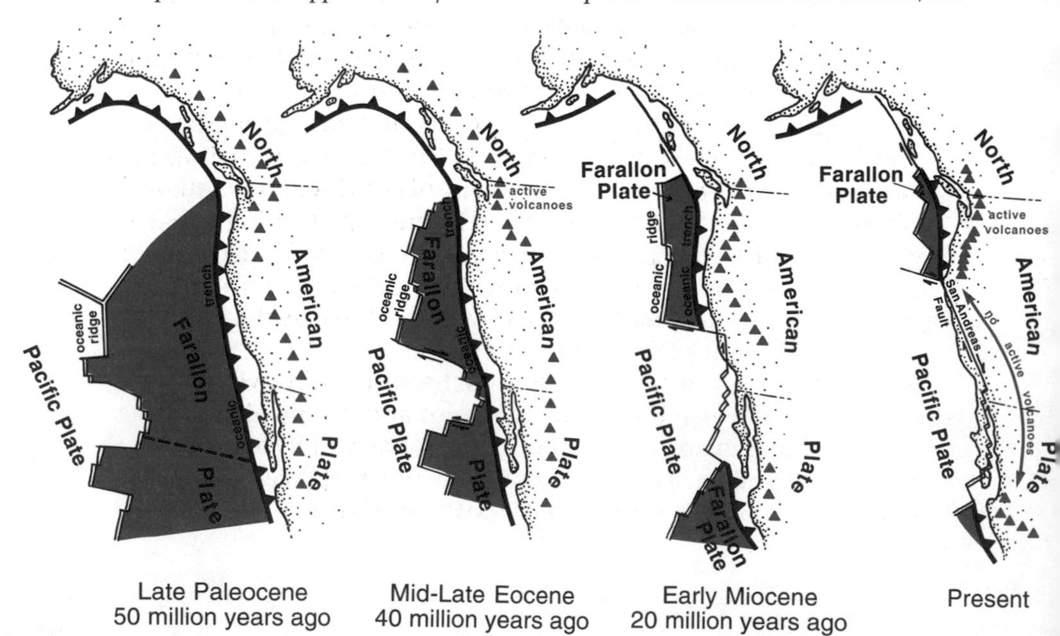

The End of the High Cascades

As the last of the Farallon plate vanishes and a transform fault replaces the trench as a plate boundary, the High Cascades will lose their supply of magma and snuff out, permanently. The trench will disappear from both ends as the San Andreas and Queen Charlotte faults encroach from the south and north, so the High Cascades will die from both ends, with the last active volcanoes surviving in the central part of the chain.

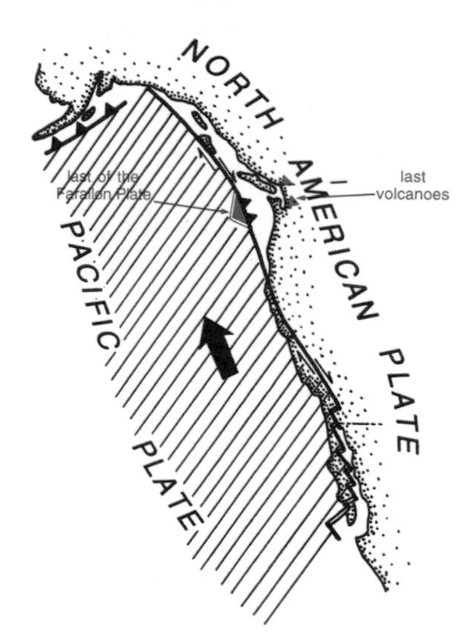

The last of the Farallon plate finally disappears, and with it the mechanism for generation of the magmas for the Cascade volcanoes.

A New Coast Range Rises

As the floor of the Pacific Ocean finally stops sinking through the trench off the northwest coast, nothing will hold down all the sediments in the trench. They will rise, as the Olympic Mountains already have, to become a new range of mountains along the west coast. The rise of that new range will also progress from both south and north as the San Andreas and Queen Charlotte faults encroach upon the trench from the south and north.

Rocks in the new coastal mountains will doubtless be a typical trench filling very much like those in the Olympic Mountains and California Coast Range. They will consist mostly of terribly deformed muddy sandstones that were originally laid down on the floor of the Pacific Ocean. The new range will also contain basalt flows, colorful sections of bedded chert that began as deep ocean deposits of siliceous ooze, and probably masses of intrusive serpentinite.

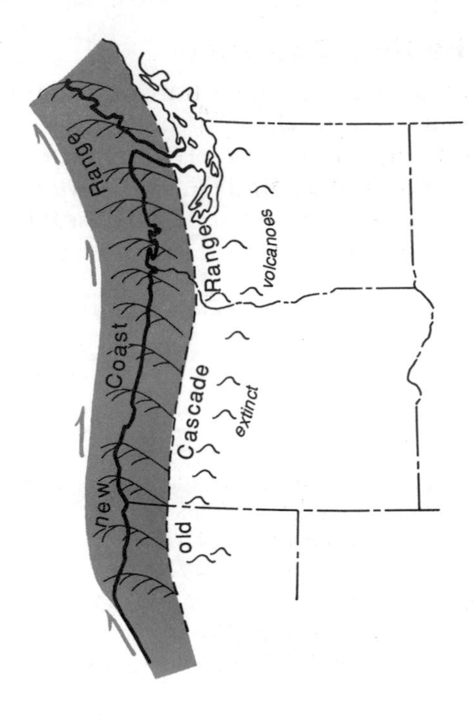

The new coastal range that will rise along the west coast as a fault plate boundary finally replaces the last of the trench.

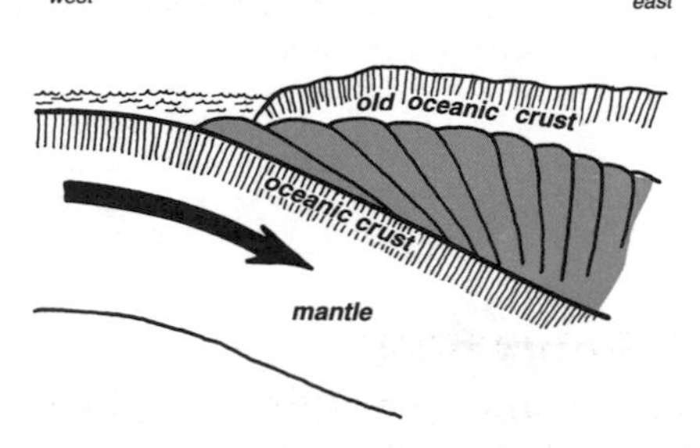

The future coastal mountains are still tucked under the slab of old oceanic crust along the coast.

The Yellowstone Hotspot, Snake River Plain, and Basin and Range

If North America continues to move west, the Yellowstone hotspot will continue to move northeast at the same rate, further elongating the Snake River Plain. If the present rate of movement continues, the Yellowstone hotspot will reach Billings, Montana, in five or six million years. About 17 million years from now, it should reach North Dakota, and the Snake River Plain will extend all that way.

So far, the eastern margin of the Basin and Range has kept pace with the Yellowstone hotspot. It is not certain just how that works, so it is hard to predict where it will be 20 million years in the future. If it continues to keep pace with the Yellowstone hotspot, it will by then have trashed the continent all the way to eastern Colorado.

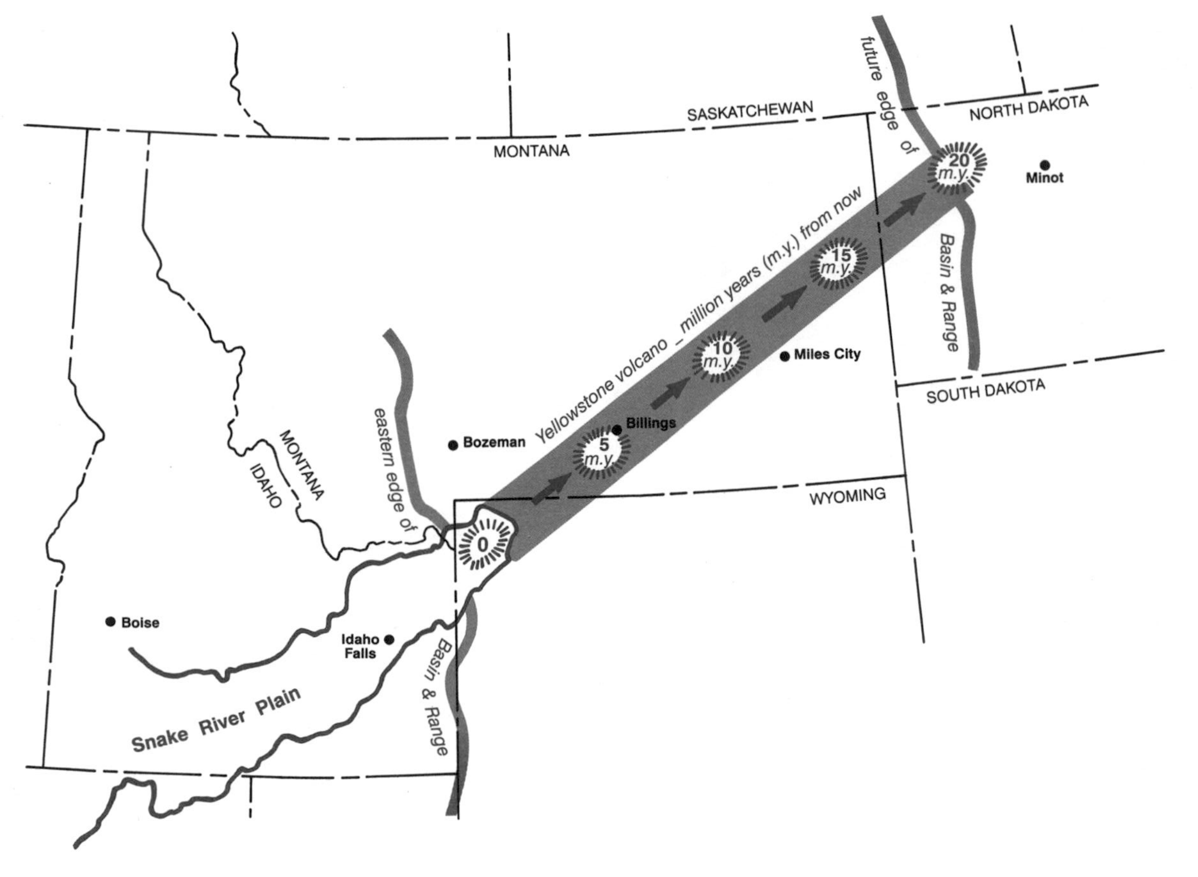

The Snake River Plain of the future.

Glossary

Amphibole. A family of iron and calcium silicate minerals. Hornblende, the common variety, typically crystallizes into glossy black needles.

Andesite. Gray volcanic rock generally intermediate in composition between basalt and rhyolite.

Anticline. A fold that bends layered rocks up into an arch.

Ash. Finely divided volcanic rock, which generally erupts explosively as a volcano vents large volumes of steam.

Ash flow. A dense cloud of volcanic ash and steam that pours across the ground surface. Many ash flows are so hot that the particles of ash weld into solid rock as they stop moving.

Asteroid. A minor member of the solar system large enough to be seen through astronomical telescopes. A very large meteorite.

Asthenosphere. A zone of partially melted rock within the earth's mantle, the lower boundary of the lithosphere.

Augite. A dull black iron and magnesium silicate mineral, the commonest of the pyroxene family. Augite most typically occurs in black igneous and metamorphic rocks.

Basalt. A common volcanic rock composed mostly of the black mineral pyroxene, the greenish white mineral plagioclase, and in some cases small crystals of green olivine. Recognize basalt as hard, black rock in which most of the mineral grains are too small to distinguish without the help of a strong magnifier.

Basement. The complex assortment of granite, gneiss, and schist that forms the bulk of the continental crust.

Basin and Range. A geologic province of western North America in which the earth's crust stretched and broke into a pattern of mountain ranges and valleys that trend generally to the north.

Batholith. A mass of granite exposed over an area of more than fifty square miles. Some cover tens of thousands of square miles.

Belt basin. The large subsiding area within the continent in which the Belt sedimentary formations accumulated.

Boulder batholith. A large mass of granite that invaded western Montana seventy to seventy-five million years ago.

Caldera. A volcanic basin that forms as the earth's surface collapses into the void that forms as large volumes of lava, typically rhyolite, erupt.

Chlorite. A family of green minerals, generally dark green, that are generally similar to the micas. Chlorites most typically form as water reacts with dark rocks such as basalt.

Chloritic breccia. An extremely peculiar pale green rock that occurs above some mylonites. The rock consists of a mass of tiny angular fragments set in a matrix of chlorite.

Cinder cone. A type of basalt volcano. Typical cinder cones are a pile of small pieces of basalt that reaches a height of several hundred feet and then erupts a lava flow.

Clinker. A hard red, yellow, or black rock that forms where sandstone or shale bake above a burning coal seam.

Continental crust. The complex of rocks, mostly granite, gneiss, and schist that makes the bulk of the continent.

Core complex. An assemblage of rocks that typically has a mass of granite near its center, and a mass of older rocks that moved off to one side along a fault or mylonite.

Crust. The upper surface of the lithosphere. Continental crust consists mostly of granite, gneiss, and schist, oceanic crust of basalt.

Diabase. A black igneous rock with the same composition as basalt, but composed of larger mineral grains.

Dike. A sheet of igneous rock emplaced as molten magma filled a fracture.

Diorite. Coarsely granular igneous rock that consists mostly of plagioclase feldspar and pyroxene. Diorite superficially resembles granite but is darker and lacks quartz.

Fault. A fracture in the earth's crust along which the rocks on either side have shifted.

Feldspar. A large family of aluminum silicate minerals, together the most abundant in the continental crust. The major kinds of feldspar are plagioclase and orthoclase.

Flood basalt. An enormous basalt lava flow. Flood basalt flows typically contain more than one hundred cubic miles of basalt and cover many thousands of square miles.

Gabbro. A black igneous rock similar to basalt except in having mineral grains large enough to see without a magnifying glass.

Garnet. A family of silicate minerals that abound in many kinds of igneous and metamorphic rocks. Most garnets are red, but a few are green or white.

Gneiss. A coarsely granular metamorphic rock with a streaky grain that results from the parallel alignment of mineral grains. Most gneisses also show distinct color banding. Pronounce it *nice.*

Graben. A segment of the earth's crust that dropped along faults.

Granite. A coarsely granular igneous rock composed mostly of feldspar, along with lesser amounts of quartz and other minerals such as mica or hornblende. Granites are generally pink or pale gray.

High Cascades. The chain of Cascade volcanoes that have been active during the past ten or twelve million years.

Hornblende. The common variety of amphibole. Hornblende typically forms shiny black crystals shaped like needles. It most commonly occurs in pale igneous or metamorphic rocks.

Hotspot. An isolated volcano that has no apparent relationship to plate boundaries.

Igneous rock. Any rock that crystallized from a molten magma.

Lava. Technically, molten magma becomes lava when it erupts onto the earth's surface.

Limestone. A sedimentary rock composed mostly of calcite, the mineral form of calcium carbonate.

Lithosphere. The relatively cold and rigid outer rind of the earth that extends down to a depth of about sixty to one hundred miles. The lithosphere consists of the outermost part of the mantle, as well as its cover of continental or oceanic crust.

Magma. Molten rock beneath the earth's surface.

Mantle. The part of the earth between the crust on the outside and the core within—the largest part of the planet. The mantle consists mostly of peridotite.

Marble. A metamorphic rock that forms when limestone recrystallizes at high temperature.

Metamorphic rocks. Rocks that recrystallized at high temperature and in many cases under high pressure.

Metamorphism. Recrystallization of a rock to form a new rock composed of new mineral grains. Metamorphism typically occurs at high temperature, and commonly under high pressure.

Mica. A family of silicate minerals that split easily into thin flakes. The most common varieties are the black mica, biotite, and the white mica, muscovite. Micas abound in many kinds of igneous and metamorphic rocks, most commonly in those with a pale color.

Micro-continent. An island composed of continental basement rocks that is too small to qualify as a continent in the ordinary geographic sense.

Moraine. A deposit of glacial till, sediment dumped directly from the ice.

Mudstone. As the name suggests, a sedimentary rock formed from mud.

Mylonite. A type of metamorphic rock that crystallized within a moving fault zone deep enough below the surface that the rocks flow instead of breaking.

North Cascades. A region of northwestern Washington that was added to North America as a series of islands landed against the western edge of the continent.

Obsidian. A glassy form of rhyolite, typically black or dark red in color. Obsidian forms as granitic magma completely lacking in water hardens into a solid rock without crystallizing.

Oceanic ridge. A plate boundary at which two plates pull away from each other, and basalt lava erupts from the opening rift between them to make new oceanic crust.

Oceanic trench. A depression on the ocean floor that develops where it bends down to begin its descent into the mantle at a collision plate boundary.

Olivine. An iron and magnesium silicate mineral that typically forms glassy green crystals. Look for it in black igneous rocks.

Ophiolite. A slice of oceanic crust incorporated into a continent, generally as part of a trench filling.

Orthoclase. A kind of feldspar, a pink or white potassium aluminum silicate mineral. It typically abounds in pale igneous or metamorphic rocks that also contain quartz.

Outwash. Glacial sediments deposited from meltwater streams.

Overthrust fault. A nearly horizontal surface on which a large slab of rocks moved across those beneath. Most overthrust faults place older rocks on top of younger ones.

Palagonite. A nondescript pale yellowish or yellowish brown clay that forms as molten basalt reacts with steam.

Pangaea. A supercontinent that included most of the earth's continental crust in two continents, Gondwana and Laurasia. It existed during Permian and Triassic times, then broke into the modern continents. Pronounce it *pan-JEE-uh.*

Peat. A mucky black deposit of partially decomposed vegetation. If buried, peat turns into coal.

Peridotite. A rock composed primarily of black pyroxene, along with green olivine and lesser amounts of other minerals. Most of the earth's mantle consists of peridotite.

Pillow basalt. Basalt that flowed out into long cylinders as it erupted under water. When seen exposed in section in a roadcut or cliff, the cylinders suggest a pile of pillows.

Placer. A segregation of heavy minerals, typically in stream or beach sands. Pronounce the a as in placid.

Plagioclase. A kind of feldspar that contains sodium and calcium. It is white or greenish, and typically occurs in medium to dark igneous or metamorphic rocks, most abundantly in those that contain little or no quartz. Pronounce it *PLA-jee-o-CLASE.*

Plate. A segment of the lithosphere. The earth's surface is divided into a dozen or more plates of widely various sizes that move about on the earth's surface.

Pumice. A form of rhyolite so full of minute gas bubbles that it is essentially a glass foam.

Pyroxene. A family of silicate minerals that occur mostly in dark igneous and metamorphic rocks. The commonest and most abundant pyroxene is augite, which occurs mainly in black igneous rocks such as basalt, gabbro, or peridotite. It forms stubby black crystals.

Quartz. The mineral form of silica, silicon dioxide. Quartz occurs in many forms, most typically in transparent crystals that look like grains of glass. Although feldspar is volumetrically more abundant, no other mineral occurs in as many kinds of rocks as quartz.

Quartz diorite. A dark igneous rock composed mostly of plagioclase and pyroxene, with a small amount of quartz.

Resurgent caldera. A type of volcano that erupts enormous volumes of rhyolite while developing a large caldera basin, fills the caldera with rhyolite, then erupts again hundreds of thousands of years later.

Rhyolite. A volcanic rock with the composition of granite. Most rhyolites are very pale—white, pink, and yellow are typical colors. Pronounce it *RYE-oh-LITE.*

Sandstone. The name defines it: a sedimentary rock formed from sand.

Schist. Almost any metamorphic rock that contains enough mica to confer a flaky texture, or enough amphibole to make it splintery. Pronounce it *shist.*

Serpentinite. A dark green rock that forms as peridotite reacts with water, typically beneath the oceanic crust.

Shield volcano. A large pile of thin basalt lava flows. Shield volcanoes have gently sloping sides.

Sill. A layer of igneous rock sandwiched between sedimentary layers.

Snake River plain. A broad volcanic province that cuts through southern Idaho, the track of the Yellowstone hotspot.

Stromatolite. An algal growth structure shaped more or less like a dome.

Syncline. A fold that warps layered rocks down into a trough.

Terrane. An assemblage of rocks that share a more or less common origin and history.

Thrust fault. A steeply tilting surface on which a slab of rock moved over and onto the rocks beneath. In most cases the rocks above the fault are older than those beneath.

Till. A disorderly mixture of debris of all sizes deposited directly from glacial ice.

Vein. A deposit of minerals that fills a fracture.

Wave-cut bench. An almost bedrock flat surface that extends seaward from a sea cliff. Commonly exposed only at low tide.

Welded ash. A rock composed of volcanic ash that was still partially molten when it settled to the ground and fused into solid rock. An ash flow deposit.

Western Cascades. The chain of Cascade volcanoes that became extinct about eighteen million years ago.

Reading

People often ask us to suggest something they can read to learn a bit more about a particular subject. Those are tough questions. It is extremely difficult at best and in many cases impossible to assemble good non-technical references on geological subjects. Most of the geologic literature is available only in large university libraries. Nearly all of it is so full of highly technical jargon that it is difficult to read, even for a geologist. Very little of the geologic literature is easily accessible, either physically or intellectually.

The following lists contain a disorderly mixture of different kinds of references. We included a number of primary research papers, most of which contain their own lists of references. People who want to read further on a subject will find them good starting points. We also included a number of symposium volumes that contain numerous research papers on closely related subjects. And we included a few references to old but more or less classic works, which tend to give a good introductory overview of the subject in fairly digestible form. We wish we could have included more books written for the general public, but those are scarce.

GENERAL

Armstrong, R. L. 1975. The geochronometry of Idaho. *Isochron West* 14:1-50.

Atwater, T. 1989. Plate tectonic history of the northeast Pacific and western North America. In *The Geology of North America, Vol. N, Eastern Pacific Ocean and Hawaii,* eds. E. L. Winterer, D. M. Hussong, and R. W. Decker, 21-72. Boulder, Colorado: Geological Society of America.

Bally, A. W., and A. R. Palmer, eds. 1989. *The Geology of North America, Vol. A, An Overview.* Boulder, Colorado: Geological Society of America.

Beus, S. S., ed. 1987. *Centennial Field Guide, Vol. 2, Rocky Mountain Section.* Boulder, Colorado: Geological Society of America.

Bond, J. 1978. *Geologic Map of Idaho.* Moscow: Idaho Bureau of Mines and Geology. 1:500,000.

Burchfiel, B. C., P. W. Lipman, and M. L. Zoback, eds. 1992. *The Cordilleran Orogen, Conterminous United States.* Boulder, Colorado: Geological Society of America.

Daniel, F., and R. B. Berg. 1981. Radiometric dates of rocks in Montana. *Montana Bureau of Mines and Geology Bulletin* 114.

Hill, M. L., ed. 1987. *Centennial Field Guide, Vol. 1, Cordilleran Section*. Boulder, Colorado: Geological Society of America.

King, P. B., and H. M. Beikman. 1974. *Geologic Map of the United States*. U.S. Geological Survey. 1:2,500,000.

Mallory, W. W., editor-in-chief. 1972. *Geologic Atlas of the Rocky Mountain Region*. Denver, Colorado: Rocky Mountain Association of Geologists.

Peterson, J. A. 1986. Paleotectonics and sedimentation in the Rocky Mountain region, United States. *American Association of Petroleum Geologists Memoir* 41.

Stoffel, K. L., N. L. Joseph, S. Z. Waggoner, C. W. Gulick, M. A. Korosec, and B. B. Bunning. 1991. Geologic map of Washington—Northeast Quadrant. *Washington State Department of Natural Resources Map* GM-39. 1:250,000.

Walker, G. W. 1977. Geologic map of Oregon east of the 121st meridian. *U.S. Geological Survey Miscellaneous Investigations Series Map* I-902. 1:500,000.

Walsh, T. J., M. A. Korosec, W. M. Phillips, R. L. Logan, and H. W. Schasse. 1987. Geologic map of Washington—Southwest Quadrant. *Washington State Department of Natural Resources Map* GM-34. 1:250,000.

Wheeler, J. O., A. J. Brookfield, H. Gabrielse, J. W. H. Monger, H. W. Tipper, and G. J. Woodsworth, comp. 1991. Terrane Map of the Canadian Cordillera. *Geological Survey of Canada Map* 1713A. 1:2,000,000.

PART 1: BACK IN THE PRECAMBRIAN

Alt, D., J. W. Sears, and D. W. Hyndman. 1989. Mafic magmatism within intracratonic basins: the impact connection. In *Continental magmatism, abstracts*, General Assembly of the International Association of Volcanology and Chemistry of the Earth's Interior, June 25-July 1, 1989, Sante Fe. *New Mexico Bureau of Mines and Mineral Resources Bulletin* 131:5.

Armstrong, R. L. 1975. Precambrian (1500 m.y. old) rocks of central Idaho—the Salmon River arch and its role in Cordilleran sedimentation and tectonics. *American Journal of Science* 275-A:437-67.

Berg, R. 1979. Talc and chlorite deposits in Montana. *Montana Bureau Mines and Geology Memoir* 45.

Cressman, E. R. 1985. The Prichard Formation of the lower part of the Belt Supergroup (Middle Proterozoic), near Plains, Sanders County, Montana. *U.S. Geological Survey Bulletin* 1553.

Czamanske, G. K., and M. L. Zientek, eds. 1985. The Stillwater Complex, Montana: Geology and Guide. *Montana Bureau of Mines and Geology Special Publication* 92.

Garihan, J. M. 1979. Geology and Structure of the central Ruby Range, Madison County, Montana. *Geological Society of America Bulletin* 90:695-788.

Harrison, J. E. 1972. Precambrian Belt basin of northwestern United States: Its geometry, sedimentation, and copper occurrences. *Geological Society of America Bulletin* 83:1215-90.

Link, P. K., N. Christie-Blick, W. J. Devlin, D. P. Elston, R. J. Horodyski, M. Levy, J. Miller, R. C. Pearson, A. Prave, J. H. Stewart, D. Winston, L. A. Wright, and C. T. Wrucke. 1993. Middle and Late Proterozoic stratified rocks of the western U.S. Cordillera, Colorado Plateau, and Basin and Range province. In *The Geology of North America, Vol. C-2, Precambrian: Conterminous U.S.*, J. C. Reed, Jr., M. E. Bickford, R. S. Houston, P. K. Link, D. W. Rankin, P. K. Sims, and W. R. Van Schmus, 463-595. Boulder, Colorado: Geological Society of America.

McMannis, W. J. 1963. Lahood Formation—a coarse facies of the Belt series in southwestern Montana. *Geological Society of America Bulletin* 74:407-36.

Page, N. J. 1977. Stillwater complex, Montana—Rock succession, metamorphism and structure of the complex and adjacent rocks. *U.S. Geological Survey Professional Paper* 999.

Page, N. J., and M. Zientek. 1985. Geological and structural setting of the Stillwater Complex. In *The Stillwater Complex, Montana: Geology and Guide*, eds. G. K. Czamanske, and M. L. Zientek. *Montana Bureau of Mines and Geology Special Publication* 92:1-8.

Roberts, S. M., ed. 1986. Belt Supergroup: A Guide to Proterozoic Rocks of Western Montana and Adjacent Areas. *Montana Bureau of Mines and Geology Special Publication* 94.

Sears, J. W., and R. A. Price. 1978. The Siberian connection—A case for Precambrian separation of the North America and Siberian cratons. *Geology* 6:267-70.

Vitaliano, C. J., and W. S. Cordua. 1979. Geologic map of southern Tobacco Root Mountains, Madison County, Montana. *Geological Society of America Map and Chart Series* M-C-31.

Winston, D. 1986. Sedimentology of the Ravalli Group, Middle Belt Carbonate, and Missoula Group, Middle Proterozoic Belt Supergroup, Montana, Idaho, and Washington. *Montana Bureau of Mines and Geology Special Paper* 94.

———. 1989. Introduction to the Belt. In *Volcanism and Plutonism of Western North America, Vol. 2,* field trip T334, *Middle Proterozoic Belt Supergroup, Western Montana,* trip leaders D. Winston, R. J. Horodyski, and J. W. Whipple, in the collection *Field Trips for the 28th International Geological Congress,* 1-6. Washington, D.C.: American Geophysical Union.

———. 1989. A sedimentologic and tectonic interpretation of the Belt Supergroup. In *Volcanism and Plutonism of Western North America, Vol. 2,* field trip T334, *Middle Proterozoic Belt Supergroup, Western Montana,* trip leaders D. Winston, R. J. Horodyski, and J. W. Whipple, in the collection *Field Trips for the 28th International Geological Congress,* 47-69. Washington, D.C.: American Geophysical Union.

Wooden, J. L., P. A. Mueller, and D. W. Mogk. 1988. A review of the geochemistry and geochronology of the Archean rocks of the northern part of the Wyoming province. In *Metamorphism and Crustal Evolution of the Western United States, Rubey Volume VII,* ed. W. G. Ernst, 383-411. Englewood Cliffs, New Jersey: Prentice-Hall.

Wooden, J. L., C. J. Vitaliano, S. W. Koehler, and P. C. Ragland. 1978. The late Precambrian mafic dikes of the southern Tobacco Root Mountains, Montana. *Canadian Journal of Earth Sciences* 815:467-79.

PART 2: PALEOZOIC TIME

Burchfiel, B. C., and L. H. Royden. 1991. Antler orogeny: A Mediterranean-type orogeny. *Geology* 19:66-69.

Lapierre, H., F. Albarede, J. Albers, B. Cabans, and C. Coulon. 1985. Early Devonian volcanism in the eastern Klamath Mountains, California—Evidence for an immature island arc. *Canadian Journal of Earth Science* 22:214-26.

Speed, R. C., and N. H. Sleep. 1982. Antler orogeny and foreland basin: A model. *Geological Society of America Bulletin* 93:815-28.

PART 3: MESOZOIC TIME, THE FIRST HALF

Burchfiel, B. C., and G. A. Davis. 1981. Triassic and Jurassic tectonic evolution of the Klamath Mountains-Sierra Nevada geologic terrane. In *The Geotectonic Development of California, Rubey Volume I,* ed. W. G. Ernst, 50-70. Englewood Cliffs, New Jersey: Prentice-Hall.

Jayko, A. S., M. C. Blake, and R. N. Brothers. 1986. Blueschist metamorphism of the eastern Franciscan belt, northern California. *Geological Society of America Memoir* 164:107-23

Paul, G. S. 1988. *Predatory Dinosaurs of the World.* New York: Simon and Schuster.

PART 4: ADDING PIECES TO THE COLLAGE

Blome, C. D., and M. K. Nestell. 1991. Evolution of a Permo-Triassic sedimentary melange, Grindstone terrane, east-central Oregon. *Geological Society of America Bulletin* 103:1280-96.

Brandon, M. T. 1989. Geology of the San Juan-Cascade nappes, northwestern Cascade Range and San Juan Islands. In *Geologic guidebook for Washington and adjacent areas,* editor-in-chief N. L. Joseph. *Washington Division of Geology and Earth Resources Information Circular* 86:137-62.

Brandon, M. T., D. S. Cowan, and J. A. Vance. 1988. The Late Cretaceous San Juan thrust system, San Juan Islands, Washington. *Geological Society of America Special Paper* 221.

Davis, G. A., and G. S. Lister. 1988. Detachment and faulting in continental extension: Perspectives from the southwestern U.S. Cordillera. *Geological Society of America Special Paper* 218:133-59.

Harper, G. D. 1984. Josephine ophiolite, northwestern California. *Geological Society of America Bulletin* 95:1009-26.

———. 1989. Field guide to the Josephine ophiolite and coeval island arc complex, Oregon-California. In *Sedimentation and Tectonics of Western North America, Vol. 2,* field trip T308, *Geologic evolution of the northernmost Coast Ranges and western Klamath Mountains, California,* trip leaders K. R. Aalto, and G. D. Harper, in the collection *Field Trips for the 28th International Geological Congress,* 1-20. Washington, D.C.: American Geophysical Union.

Haugerud, R. A. 1989. Geology of the Metamorphic Core of the North Cascades. In Geologic guidebook for Washington and adjacent areas, editor-in-chief N. L. Joseph, Washi*ngton Division of Geology and Earth Resources Information Circular* 86:119-36.

Hurlow, H. A. and B. K. Nelson. 1993. U-Pb zircon and monazite ages for the Okanogan Range batholith, Washington: Implications for the magmatic and tectonic evolution of the southern Canadian and northern United States Cordillera. *Geological Society of America Bulletin* 105:231-40.

Hyndman, D. W. 1980. Bitterroot dome-Sapphire tectonic block, an example of a plutonic-core gneiss-dome complex with its detached suprastructure. In *Cordilleran Metamorphic Core Complexes,* eds. M. D. Crittenden, Jr., P. J. Cooney, and G. H. Davis. *Geological Society of America Memoir* 153:427-43.

Hyndman, D. W., and S. A. Myers. 1988. The transition from amphibolite facies mylonite to chloritic breccia and the role of the mylonite in formation of Eocene epizonal plutons, Bitterroot dome, Montana. *Geologische Rundschau* 77:201-26.

McGroder, M. F. 1991. Reconciliation of two-sided thrusting, burial metamorphism, and diachronous uplift in the Cascades of Washington and British Columbia. *Geological Society of America Bulletin* 103:189-209.

Miller, R. B., D. L. Whitney, and E. E. Geary. 1993. Tectono-stratigraphic terranes and the metamorphic history of the northeastern part of the crystalline core of the North Cascades: Evidence from the Twisp Valley Schist. *Canadian Journal of Earth Sciences* 30:1306-23.

Vallier, T. L., and H. C. Brooks. 1986. Paleozoic and Mesozoic faunas of the Blue Mountains Province: A review of their geologic implications and comments on papers in this volume. In *Geology of the Blue Mountains Region of Oregon, Idaho, and Washington: Paleontology and Biostratigraphy,* eds. T. L. Vallier, and H. C. Brooks. *U.S. Geological Survey Professional Paper* 1435:1-6.

Whitney, D. L. and M. F. McGroder. 1989. Cretaceous crustal section through the proposed Insular Intermontane suture, North Cascades, Washington. *Geology* 17:555-58

PART 5: THE NORTHERN ROCKY MOUNTAINS

Boyer, S. E., and D. Elliott. 1982. Thrust systems. *American Association of Petroleum Geologists Bulletin* 66:1196-1230.

Brooks, H. C. 1979. Plate tectonics and the geologic history of the Blue Mountains. *Oregon Geology* 41:71-80.

Hamilton, W. B., and W. B. Myers. 1974. The nature of the Boulder batholith of Montana. *Geological Society of America Bulletin* 85:365-78, 1958-60.

Hyndman, D. W. 1983. The Idaho batholith and associated plutons, Idaho and western Montana. In *Circum-Pacific Plutonic Terranes,* ed. J. A. Roddick. *Geological Society of America Memoir* 159:213-40.

Hyndman, D. W., D. Alt, and J. W. Sears. 1988. Post-Archean metamorphism and tectonic evolution of western Montana and northern Idaho. In *Metamorphism and Crustal Evolution of the Western United States, Rubey Volume VII,* ed. W. G. Ernst, 332-61. Englewood Cliffs, New Jersey: Prentice-Hall.

Hyndman, D. W., and D. A. Foster. 1988. The role of tonalites and mafic dikes in the generation of the Idaho batholith. *Journal of Geology* 96:31-46.

Jones, D. L., N. J. Silberling, and J. Hillhouse. 1977. Wrangellia—A displaced terrane in northwestern North America. *Canadian Journal of Earth Sciences* 14:2565-77.

Kilsgaard, T. H., and R. S. Lewis. 1985. Plutonic rocks of Cretaceous age and faults in the Atlanta lobe of the Idaho batholith, Challis quadrangle. In *Symposium on the geology and mineral resources of the Challis 1X2 quadrangle, Idaho,* ed. D. H. MacIntyre. *U.S. Geological Survey Bulletin* 1658:29-42.

Love, J. D., trip leader. 1989. Yellowstone and Grand Teton National Parks and the Middle Rocky Mountains. In *Sedimentation and Tectonics of Western North America, Vol. 3,* field trip T328, in the collection *Field Trips for the 28th International Geological Congress.* Washington, D.C.: American Geophysical Union.

Lund, K., and L. W. Snee. 1988. Metamorphism, structural development, and age of the continent: island arc juncture in west-central Idaho. In *Metamorphism and Crustal Evolution of the Western United States, Rubey Volume VII,* ed. W. G. Ernst, 296-332. Englewood Cliffs, New Jersey: Prentice-Hall.

Molnar, P. and H. Lyon-Caen. 1988. Some simple physical aspects of the support, structure, and evolution of mountain belts. *Geological Society of America Special Paper* 218:179-207.

Mudge, M. R. 1972. Pre-Quaternary rocks in the Sun River Canyon area, northwestern Montana. *U.S. Geological Survey Professional Paper* 663-A.

Mudge, M. R., and R. L. Earhart. 1980. The Lewis thrust fault and related structures in the disturbed belt, northwestern Montana. *U.S. Geological Survey Professional Paper* 1174.

Rhodes, B. P., and D. W. Hyndman. 1988. Regional metamorphism, structure, and tectonics of northeastern Washington and northern Idaho. In *Metamorphism and Crustal Evolution of the Western United States, Rubey Volume VII,* ed. W. G. Ernst, 272-95. Englewood Cliffs, New Jersey: Prentice-Hall.

Rutland, C., H. W. Smedes, R. I. Tilling, and W. R. Greenwood. 1989. Volcanism and plutonism at shallow crustal levels: The Elkhorn Mountains volcanics and the Boulder batholith, southwestern Montana. In *Volcanism and Plutonism of Western North America, Vol. 2, Cordilleran volcanism, plutonism, and magma generation at various crustal levels, Montana and Idaho,* trip leader D. W. Hyndman, C. Rutland, R. F. Hardyman, in the collection *Field Trips for the 28th International Geological Congress* T337:16-31. Washington, D.C.: American Geophysical Union.

Strayer, L. M., D. W. Hyndman, J. W. Sears, and P. E. Myers. 1989. Direction and shear sense during suturing of the Seven Devils-Wallowa terrane against North America in western Idaho. *Geology* 17:1025-28.

Tilling, R. I. 1974. Composition and time relations of plutonic and associated volcanic rocks, Boulder batholith region, Montana. *Geological Society of America Bulletin* 85:1925-30.

Toth, M. I. 1986. Petrology and origin of the Bitterroot lobe of the Idaho batholith. In *Geology of the Blue Mountains Region of Oregon, Idaho, and Washington: The Idaho Batholith and its Border Zone,* eds. T. L. Vallier, and H. D. Brooks. *U.S. Geological Survey Professional Paper* 1436:9-35.

Wilson, D. and A. Cox. 1980. Paleomagnetic evidence for tectonic rotation of Jurassic plutons in Blue Mountains, eastern Oregon. *Journal of Geophysical Research* 85:3681-89.

Wiswall, G., and D. W. Hyndman. 1986. Emplacement of the main plutons of the Idaho batholith. In Geology *of the Blue Mountains Region of Oregon, Idaho, and Washington: The Idaho Batholith and its Border Zone,* eds. T. L. Vallier, and H. D. Brooks. *U.S. Geological Survey Professional Paper* 1436:59-72.

PART 6: THE LATE CRETACEOUS WORLD, AND ITS HORRIBLE END

Alt, D., J. W. Sears, and D. W. Hyndman. 1988. Terrestrial Maria: The origins of large basalt plateaus, hotspot tracks, and spreading ridges. *Journal of Geology* 96:647-62.

Alvarez, L. W., W. Alvarez, F. Asaro, and H. V. Michel. 1980. Extra-terrestrial cause for the Cretaceous-Tertiary extinction. *Science* 208:1095-1108.

Bakker, R. T. 1986. *The Dinosaur Heresies: New Theories Unlocking the Mystery of the Dinosaurs and Their Extinction.* New York: William Morrow and Company.

Bohor, B. F., D. M. Triplehorn, D. Nichols, and H. T. Millard, Jr. 1987. Dinosaurs, spherules, and the "magic" layer: A new K-T boundary site in Wyoming. *Geology* 15:896-99.

Courtillot, V., G. Feroud, H. Maluski, D. Vandamme, M. G. Moreau, and J. Besse. 1988. The Deccan flood basalts and the Cretaceous/Tertiary boundary. *Nature* 333:843.

Horner, J. R., and J. Gorman. 1988. *Digging Dinosaurs.* New York: Workman Publishing.

Rampino, M. R., and R. B. Stothers. 1988. Flood basalt volcanism during the past 250 million years. *Science* 241:663.

Retallack, G. J. 1981. Fossil soils: Indications of ancient terrestrial environments. In *Paleobotany, Paleoecology and Evolution, Vol.1,* ed. K. J. Niklas, 55-102. New York: Praeger Publishers.

Smit, J., and A. J. T. Romine. 1985. A sequence of events across the Cretaceous-Tertiary boundary. *Earth and Planetary Science Letters* 74:155-70.

PART 7: EARLY AND MIDDLE TERTIARY TIME

Armentrout, J. M. 1987. Cenozoic stratigraphy, unconformity-bounded sequences, and tectonic history of southwestern Washington. In *Selected papers on the Geology of Washington,* ed. J. E. Schuster. *Washington Division of Geology and Earth Resources Bulletin* 77:291-320.

Babcock, R. S., and P. Misch. 1988. Evolution of the crystalline core of the North Cascade Range. In *Metamorphism and Crustal Evolution of the Western United States, Rubey Volume VII,* ed. W. G. Ernst, 214-33. Englewood Cliffs, New Jersey: Prentice-Hall.

Bennett, E. H., and C. R. Knowles. 1985. Tertiary plutons and related rocks in central Idaho. In *Symposium on the geology and mineral resources of the Challis 1X2 quadrangle, Idaho,* ed. D. H. MacIntyre. *U.S. Geological Survey Bulletin* 1658:81-95.

Brown, E. H. 1988. Metamorphic and structural history of the northwest Cascades, Washington and British Columbia. In *Metamorphism and Crustal Evolution of the Western United States, Rubey Volume VII,* ed. W. G. Ernst, 196-214. Englewood Cliffs, New Jersey: Prentice-Hall.

Fields, R. W., D. L. Rasmussen, A. R. Tabrum, and R. Nichols. 1985. Cenozoic rocks of the Intermontane basins of western Montana and eastern Idaho: A Summary. In *Cenozoic Paleogeography of west-central United States,* eds. R. M. Flores, and S. S. Kaplan, 9-36. Denver, Colorado: Society of Economic Paleontologists and Mineralogists, Rocky Mountain Section.

Haugerud, R. A., P. van der Heyden, R. W. Tabor, J. S. Stacey, and R. E. Zartman. 1991. Late Cretaceous and early Tertiary plutonism and deformation in the Skagit Gneiss Complex, North Cascade Range, Washington and British Columbia. *Geological Society of America Bulletin* 103:1297-1307.

Hearn, C. B., Jr., F. O. Dudas, D. H. Eggler, D. W. Hyndman, H. E. O'Brien, I. S. McCallum, A. J. Irving, and R. B. Berg, trip leaders. 1989. Montana high-potassium igneous province. In *Volcanism and Plutonism of Western North America, Vol. 2,* field trip T346, in the collection *Field Trips for the 28th International Geological Congress.* Washington, D.C.: American Geophysical Union.

Heller, P. L., R. W. Tabor, and C. A. Suczek. 1987. Paleogeographic evolution of the United states Pacific Northwest during Paleogene time. *Canadian Journal of Earth Sciences* 24:1652-67.

Hyndman, D. W., and D. Alt. 1987. Radial dikes, laccoliths, and gelatin models. *Journal of Geology* 95:763-74.

Hyndman, D. W., D. Alt, and J. W. Sears. 1988. Cited in Part 5.

Lyons, J. B. 1944. Igneous rocks of the northern Big Belt Range, Montana. *Geologic Society of America Bulletin* 55:445-72.

McCarley Holder, G. A., R. W. Holder, and D. H. Carlson. 1990. Middle Eocene dike swarms and their relation to contemporaneous plutonism, volcanism, core-complex mylonitization, and graben subsidence, Okanogan Highlands, Washington. *Geology* 18:1082-85.

Miller, R. B., S. A. Bowring, and W. J. Hoppe. 1989. Paleocene plutonism and its tectonic implications, North Cascades, Washington. *Geology* 17:846-49.

Misch, P. 1988. Tectonic and metamorphic evolution of the North Cascades: an overview. In *Metamorphism and Crustal Evolution of the Western United States, Rubey Volume VII,* ed. W. G. Ernst, 179-96. Englewood Cliffs, New Jersey: Prentice-Hall.

Reynolds, S. J., and G. S. Lister. 1990. Folding of mylonitic zones in Cordilleran metamorphic core complexes: Evidence from near the mylonitic front. *Geology* 18:216-19.

Tabor, R. W., R. A. Haugerud, R. B. Miller, E. H. Brown, and R. S. Babcock, trip leaders. 1989. Accreted terranes of the North Cascades Range, Washington. In *Sedimentation and Tectonics of Western North America, Vol. 2,* field trip T307, in the collection *Field Trips for the 28th International Geological Congress.* Washington D.C.: American Geophysical Union.

Wells, R. E., D. C. Engebretson, and P. D. Snavely, Jr. 1984. Cenozoic plate margins and the volcano-tectonic evolution of western Oregon and Washington. *Tectonics* 3:275-94.

PART 8: LATE TERTIARY TIME

Alt, D., J. W. Sears, and D. W. Hyndman. 1988. Cited in Part 6.

Anderson, J. L., M. H. Benson, R. D. Bentley, K. R. Fecht, P. R. Hooper, A. R. Niem, S. P. Reidel, D. A. Swanson, T. L. Tolan, and T. L. Wright. 1987. Distribution maps of stratigraphic units of the Columbia River Basalt Group. In *Selected papers on the Geology of Washington,* ed. J. E. Schuster. *Washington Division of Geology and Earth Resources Bulletin* 77:183-95.

Bonnichsen, B. 1982. The Bruneau-Jarbidge eruptive center, southwestern Idaho. In *Cenozoic Geology of Idaho,* eds. B. Bonnichsen, and R. M. Breckenridge. *Idaho Bureau of Mines and Geology Bulletin* 26:237-54.

Brandon, M. T., and A. R. Calderwood. 1990. High-pressure metamorphism and uplift of the Olympic subduction complex. *Geology* 18:1252-55.

Budkewitsch, P., and P. Robin. 1994. Modelling the evolution of columnar joints. *Journal of Volcanology and Geothermal Research* 59:219-39.

Carlson, R. W., and W. K. Hart. 1987. Crustal genesis on the Oregon Plateau. *Journal of Geophysical Research* 92:6191-206.

———. 1989. Flood basalt volcanism in the northwestern United States. In Cont*inental Flood Basalts,* ed. J. D. McDougall, 35-61. Dordrecht, Netherlands: Wolters Kluwer Academic Publishers.

Christiansen, R. L. 1984. Yellowstone magmatic evolution: Its bearing on understanding large-volume explosive volcanism. In *Explosive Volcanism: Inception, Evolution, and Hazards,* National Research Council Studies in Geophysics, 84-95. Washington, D.C.: National Academy Press.

Davis, E. E., and R. D. Hyndman. 1989. Accretion and recent deformation of sediments along the northern Cascadia subduction zone. Geological Society *of America Bulletin* 101:1465-80.

Eaton, G. P. 1982. The Basin and Range province: Origin and tectonic significance. In *Annual Review of Earth and Planetary Sciences* 10:409-40, eds. G. W. Wetherill, A. L. Albee, and F. G. Stehli.

Effimov, I., and A. R. Pinezich. 1986. Tertiary structural development of selected basins: Basin and Range Province, northeastern Nevada. *Geological Society of America Special Paper* 208:31-42.

Ekren, E. B., D. H. McIntyre, E. H. Bennett. 1984. High temperature, large-volume, lavalike ash-flow tuffs without calderas in southwestern Idaho. *U.S. Geological Survey Professional Paper* 1272.

Fecht, K. R., S. P. Reidel, and A. M. Tallman. 1987. Paleodrainage of the Columbia River system on the Columbia Plateau of Washington State—A Summary. In *Selected papers on the Geology of Washington,* ed. J. E. Schuster. *Washington Division of Geology and Earth Resources Bulletin* 77:219-48.

Goles, G. G. 1986. Miocene basalts of the Blue Mountain province in Oregon. I: Compositional types and their geological settings. *Journal of Petrology* 27:495-520.

Hamilton, W. B. 1987. Crustal extension in the Basin and Range province, western United States. In *Continental Extensional Tectonics,* eds. M. P. Coward, J. F. Dewey, and P. L. Hancock. *Geological Society of London Special Publication* 28:155-176.

Harris, S. P. 1988. *Fire Mountains of the West: The Cascade and Mono Lake Volcanoes.* Missoula, Montana: Mountain Press Publishing Company.

Hooper, P. R., and D. A. Swanson. 1987. Evolution of the eastern part of the Columbia Plateau. In *Selected papers on the Geology of Washington,* ed. J. E. Schuster. *Washington Division of Geology and Earth Resources Bulletin* 77:197-217.

Iyer, H. M., J. R. Evans, G. Zandt, R. M. Stewart, J. M. Coakley, and J. N. Roloff. 1981. A deep low velocity body under the Yellowstone caldera, Wyoming: Delineation using teleseismic P-wave residuals and interpretation. *Geological Society of America Bulletin* 91:792-98.

Kuntz, M. A., E. C. Spiker, M. Rubin, D. E. Champion, and R. H. Lefebvre. 1986. Radiocarbon studies of latest Pleistocene and Holocene lava flows on the Snake River Plain, Idaho: Data, lessons, interpretations. *Quaternary Research* 25:163-76.

Kuntz, M. A., D. E. Champion, E. C. Spiker, R. H. Lefebvre, and L. A. McBroome. 1982. The great rift and evolution of the Craters of the Moon lava field. In *Cenozoic Geology of Idaho,* eds. B. Bonnichsen, and R. M. Breckenridge. *Idaho Bureau of Mines and Geology Bulletin* 26:423-32.

Leeman, W. P. 1988. Relation between crustal structure and volcanic processes in the Snake River Plain-Yellowstone Plateau (SRP-YP) volcanic province. *Geological Society of America Abstracts with Program* 20, no. 6:4-27.

Luedke, R. G., R. L. Smith, and S. L. Russell-Robinson. 1983. Map showing distribution, composition, and age of Late Cenozoic volcanoes and volcanic rocks of the Cascade Range and vicinity, northwestern United States. *U.S. Geological Survey Miscellaneous Investigations Series Map* I-1507. 1:500,000.

Mabey, D. R. 1982. Geophysics and tectonics of the Snake River Plain. In *Cenozoic Geology of Idaho,* eds. B. Bonnichsen, and R. M. Breckenridge. *Idaho Bureau of Mines and Geology Bulletin* 26:139-54.

Pierce, K. L., and L. A. Morgan. 1992. The track of the Yellowstone hot spot; volcanism, faulting, and uplift. In *Regional geology of eastern Idaho and western Wyoming,* eds. P. K. Link, M. A. Kuntz, and L. B. Platt, *Geological Society of America Memoir* 179:1-53.

Reidel, S. P., and N. P. Campbell. 1989. Structure of the Yakima Fold Belt, central Washington. In *Geologic guidebook for Washington and adjacent areas,* editor-in-chief N. L. Joseph, *Washington Division of Geology and Earth Resources Information Circular* 86:275-303.

Reidel, S. P., and P. R. Hooper, eds. 1989. Volcanism and tectonism in the Columbia River flood basalt province. *Geological Society of America Special Paper* 239.

Reynolds, M. W. 1979. Character and extent of basin range faulting, western Montana and east-central Idaho. In *Basin and Range Symposium,* eds. G. W. Newman, and H. D. Goode, 185-193. Denver, Colorado: Rocky Mountain Association of Geologists.

Rodgers, D. W., W. R. Hackett, and H. T. Ore. 1990. Extension of the Yellowstone plateau, eastern Snake River Plain, and Owyhee plateau. *Geology* 18:1138-41.

Smith, R. B., and L. W. Braile. 1984. Crustal structure and evolution of an explosive silicic volcanic system at Yellowstone National Park. In *Explosive Volcanism: Inception, Evolution, and Hazards*, National Research Council Studies in Geophysics, 96-109. Washington, D.C.: National Academy Press.

Snoke, A. W., and D. M. Miller. 1988. Metamorphic and tectonic history of the northeastern Great Basin. In *Metamorphism and Crustal Evolution of the Western United States, Rubey Volume VII,* ed. W. G. Ernst, 606-49. Englewood Cliffs, New Jersey: Prentice-Hall.

Speed, T., M. W. Elison, and F. R. Heck. 1988. Phanerozoic tectonic evolution of the Great Basin. In *Metamorphism and Crustal Evolution of the Western United States, Rubey Volume VII,* ed. W. G. Ernst, 572-606. Englewood Cliffs, New Jersey: Prentice-Hall.

Taubeneck, W. H. 1970. Dikes of the Columbia River basalt in northeastern Oregon, western Idaho, and southeastern Washington. In *Proceedings of the Second Columbia River Basalt Symposium,* eds. E. H. Gilmour, and D. Stradling, 73-96. Cheney: Eastern Washington State College Press.

Tolan, T. L., S. P. Reidel, M. H. Beeson, J. L. Anderson, K. R. Fecht, D. A. Swanson. 1989. Revisions to the estimates of the areal extent and volume of the Columbia River Basalt Group. In *Volcanism and Tectonism in the Columbia River-Basalt Province,* ed. S. P. Reidel and others. *Geological Society of America Special Paper* 239:1-20.

Wells, R. E., D. C. Engebretson, and P. D. Snavely, Jr. 1984. Cited in Part 7.

Wells, R. E., and P. L. Heller. 1988. The relative contribution of accretion, shear, and extension to Cenozoic rotation in the Pacific Northwest. *Geological Society of America Bulletin* 100:325-338.

PART 9: PLEISTOCENE TIME

Atwater, B. F. 1984, Periodic floods from Glacial Lake Missoula into the Sanpoil area of Glacial Lake Columbia, northeastern Washington. *Geology* 12:464-67.

Atwater, B. F., and D. K. Yamaguchi. 1991. Sudden, probably coseismic submergence of Holocene trees and grass in coastal Washington state. *Geology* 19:706-9.

Baker, V. R., and R. C. Barker. 1985. Cataclysmic late Pleistocene flooding from Glacial Lake Missoula: A review. *Quaternary Science Reviews* 4:1-41.

Breckenridge, R. M., trip leader. 1989. Glacial Lake Missoula and the channelled scabland. In *Glacial Geology and Geomorphology of NOrth America, Vol. 1,* field trip T310, in the collection *Field Trips for the 28th International Geological Congress.* Washington D.C.: American Geophysical Union.

Bretz, J. H. 1969. The Lake Missoula floods and the channelled scabland. *Journal of Geology* 77: 505-43.

Christiansen, R. L. 1984. Cited in Part 8.

Flint, R. F., and W. H. Irwin. 1939. Glacial Geology of Grand Coulee Dam. *Geological Society of America Bulletin* 50:661-80.

Jenks, M. D., and B. Bonnichsen. 1989. Subaqueous basalt eruptions into Pliocene Lake Idaho, Snake River Plain, Idaho. *In Guidebook to the Geology of Northern and Western Idaho and Surrounding Area,* eds. V. E. Chamberlain, R. M. Breckenridge, and B. Bonnichsen, *Idaho Bureau of Mines and Geology Bulletin* 28:17-34.

Jarrett, R. E., and H. E. Malde. 1987. Paleodischarge of the late Pleistocene Bonneville flood, Snake River, Idaho, computed from new evidence. Geological Society of *America Bulletin* 99:127-34.

Sarna-Wojcicki, A. M., K. R. Lajoie, C. E. Meyer, D. P. Adam, and H. J. Rieck. 1991. Tephrochronologic correlation of upper Neogene sediments along the Pacific margin, conterminous United States. In *The Geology of North America, Vol. K-2, Quaternary Nonglacial geology: Conterminous United States,* ed. R. B. Morrison, 117-140. Boulder, Colorado: Geological Society of America.

Smith, G. A. 1993. Missoula flood dynamics and magnitudes inferred from sedimentology of slack-water deposits on the Columbia Plateau, Washington. *Geological Society of America Bulletin* 105:77-100.

Stock, J., and P. Molnar. 1988. Uncertainties and implications of the Late Cretaceous and Tertiary position of North America relative to the Farallon, Kula, and Pacific plates. *Tectonics* 7:1339-84.

Thorson, R. M. 1980. Ice-sheet glaciation of the Puget lowland, Washington, during the Vashon Stade (late Pleistocene). *Quaternary Research* 13:303-21.

Waitt, R. B., Jr., and R. M. Thorson. 1983. The Cordilleran Ice Sheet in Washington, Idaho and Montana. In *Late-Quaternary Environments of the United States, Vol. 1, The Late Pleistocene,* ed. S. C. Porter, 53-70. Minneapolis: University of Minnesota.

Wells, R. E., D. C. Engebretson, and P. D. Snavely, Jr. 1984. Cited in Part 7.

Index

About the Authors

David Alt and Donald W. Hyndman are dedicated to bringing geology to the general public. They founded the popular Roadside Geology series, wrote several of its books, and now help edit others. When they are not working on books, Alt and Hyndman teach geology at the University of Montana in Missoula.